Lecture Notes in Computer Science

Edited by G. Goos, J. Hartmanis and J. van Lee

Springer

Berlin
Heidelberg
New York
Barcelona
Hong Kong
London
Milan
Paris
Singapore
Tokyo

Frits W. Vaandrager
Jan H. van Schuppen (Eds.)

Hybrid Systems: Computation and Control

Second International Workshop, HSCC'99
Berg en Dal, The Netherlands, March 29-31, 1999
Proceedings

Springer

Series Editors

Gerhard Goos, Karlsruhe University, Germany
Juris Hartmanis, Cornell University, NY, USA
Jan van Leeuwen, Utrecht University, The Netherlands

Volume Editors

Frits W. Vaandrager
Computing Science Institute, University of Nijmegen
P.O. Box 9010, 6500 GL Nijmegen, The Netherlands
E-mail: Frits.Vaandrager@cs.kun.nl

Jan H. van Schuppen
CWI, P.O. Box 94079
1090 GB Amsterdam, The Netherlands
E-mail: J.H.van.Schuppen@cwi.nl

Cataloging-in-Publication data applied for

Die Deutsche Bibliothek - CIP-Einheitsaufnahme

Hybrid systems : computation and control ; second international
workshop ; proceedings / HSCC '99, Berg en Dal, The Netherlands,
March 29 - 31, 1999. Frits W. Vaandrager ... (ed.). - Berlin ;
Heidelberg ; New York ; Barcelona ; Hong Kong ; London ; Milan ;
Paris ; Singapore ; Tokyo : Springer, 1999
    (Lecture notes in computer science ; Vol. 1569)
    ISBN 3-540-65734-7

CR Subject Classification (1998): C.1m, C.3, D.2.1, F.3.1, F.1.2, J.2, I.6

ISSN 0302-9743
ISBN 3-540-65734-7 Springer-Verlag Berlin Heidelberg New York

© Springer-Verlag Berlin Heidelberg 1999
Printed in Germany

Typesetting: Camera-ready by author
SPIN: 10703008   06/3142 – 5 4 3 2 1 0   Printed on acid-free paper

Preface

This volume contains the proceedings of the *Second International Workshop on Hybrid Systems: Computation and Control* (HSCC'99) to be held March 29-31, 1999, in the village Berg en Dal near Nijmegen, The Netherlands. The first workshop of this series was held in April 1998 at the University of California at Berkeley. The series follows meetings that were initiated by Anil Nerode at Cornell University. The proceedings of those meetings were published in the Springer-Verlag LNCS Series, Volumes 736, 999, 1066, 1201, and 1273. The proceedings of the first workshop of the new series was published in LNCS 1386.

The focus of the workshop is on modeling, control, synthesis, design, and verification of hybrid systems. A hybrid system is a theoretical model for a computer controlled engineering system, with a dynamics that evolves both in a discrete state set and in a family of continuous state spaces. Research is motivated by, for example, control of electro-mechanical systems (robots), air traffic control, control of automated freeways, and chemical process control. The emerging research area of hybrid systems overlaps both with computer science and with control theory. The interaction between researchers from these fields is expected to be fruitfull for the development of the area of hybrid systems.

The scientific program of the workshop consists of three invited lectures, one tutorial lecture, and 19 contributed lectures. The following researchers have been invited to present a lecture: P. Baufreton (Snecma, France), B.H. Krogh (CMU, USA), and N. Lynch (MIT, USA). The tutorial lecture will be presented by C. Miller (Ohio State University, USA). The contributed lectures are based on the papers in this proceedings. In addition, there will be demonstrations of software tools for the design, analysis, and simulation of hybrid systems.

The program committee, chaired by the editors, has selected the 19 contributed papers out of 44 submitted papers. The editors are grateful to the members of the program committee for their generous support with the reviewing and the selection process.

The editors are grateful to the invited speakers, the contributed lecture speakers, the participants for their willingness to participate in the workshop, and the sponsoring institutions whose support has made this event possible. Finally, they would like to thank Ansgar Fehnker for making such excellent web pages, and Mirèse Willems and Lieke Schultze for efficient secretarial support.

January 1999
Nijmegen
Amsterdam

Frits W. Vaandrager
Jan H. van Schuppen

Steering Committee

Panos Antsaklis (University of Notre Dame)
Tom Henzinger (University of California, Berkeley)
Nancy Lynch (MIT, Cambridge)
Oded Maler (Verimag, Gières)
Amir Pnueli (Weizmann Institute, Rehovot)
Alberto Sangiovanni-Vincentelli (University of California, Berkeley)
Shankar Sastry (University of California, Berkeley)
Frits Vaandrager (University of Nijmegen)
Jan H. van Schuppen (CWI, Amsterdam)

Program Committee

Rajeev Alur (University of Pennsylvania)
Eugene Asarin (Institute for Information Transmission Problems, Moscow)
Albert Benveniste (INRIA, Rennes)
Rene Boel (University of Ghent)
Mike Branicky (Case Western Reserve University)
Sebastian Engell (University of Dortmund)
Datta Godbole (University of California, Berkeley)
Bruce Krogh (CMU, Pittsburgh)
Kim Larsen (University of Aalborg)
Mike Lemmon (University of Notre Dame)
John Lygeros (University of California, Berkeley)
Simin Nadjm-Tehrani (Linköping University)
Ernst-Ruediger Olderog (University of Oldenburg)
Anuj Puri (Bell Labs)
Xu Qiwen (UNU/IIST, Macau)
Anders Ravn (DTU, Lyngby)
Arjan van der Schaft (University of Twente)
Henny Sipma (Stanford University)
Howard Wong-Toi (Cadence, Berkeley)
Frits Vaandrager (co-chair) (University of Nijmegen)
Jan H. van Schuppen (co-chair) (CWI, Amsterdam)
Sergio Yovine (Verimag, Gières)

Referees

Luca Aceto
Luca de Alfaro
Mireille Broucke
Pedro R. D'Argenio
Frank Hoffmann

Dang Van Hung
Karl Henrik Johansson
Mikhail Kourjanski
Kåre J. Kristoffersen
Peter Niebert
Paul Pettersson
Olaf Stursberg
Stavros Tripakis
Carsten Weise

Sponsoring Institutions

Koninklijke Nederlandse Akademie van Wetenschappen
(KNAW; Royal Netherlands Academy of Sciences, Amsterdam)

Instituut voor Programmatuur en Algoritmiek
(IPA; Research Institute for Programming and Algorithms, Eindhoven)

Stichting Advancement of Mathematics (Amstelveen)

Computing Science Institute
(CSI; University of Nijmegen)

Centrum voor Wiskunde en Informatica
(CWI; Centre for Mathematics and Computer Science, Amsterdam)

Table of Contents

SACRES: A Step Ahead in the Development of Critical Avionics Applications

Philippe Baufreton

Snecma systèmes, Centre de Villaroche, BP 42
77552 Moissy–Crayamel Cedex, France
philippe.baufreton@snecma.fr

Abstract. Basically, aircraft engines can be divided into a control embedded system and a controlled system with its environment. The behaviour of the controlled system is given a priori, while the control system still needs to be designed in a way guaranteeing the correct overall behaviour under all operational conditions depending on the flight domain. A large quantity of functions in control systems can be described by a formal system expressed in block diagrams and state-based representations. These representations can be translated to formal based tools relying on the synchronous languages Signal and Statecharts.

SACRES is a tool set supporting the design of safety-critical embedded control systems. It integrates the tools and specification techniques Statemate, Sildex, and Timing Diagrams with tool components for automatic code generation, formal verification based on model checking techniques, and an innovative approach for automatic code validation for target code generated from DC+. Technical achievements are
- Integration of dataflow and state-based specification styles
- Formal specification of safety-critical properties
- Integration of efficient symbolic model checking techniques with Statemate and Sildex
- Automatic generation of efficient distributed code
- Automated correctness proofs for the generated code

The main benefits of the SACRES approach are reduced risks for design errors and decreased design times and costs for the development of dependable (safety critical) embedded systems. SACRES is an attempt to avoid unpredictability (particularly that arising from late feedback from testing) associated with development of safety critical systems, through the use of the maximum degree of automation, especially in respect of code generation and verification.

An outstanding property of SACRES is the combination of specification styles and specification tools being applied in practice with automatic tools to establish provable correctness with respect to required properties. Both dependability and productivity are increased by automatic code generation from high-level specifications such that the generated code can be validated against higher levels by rigorous proofs.

These techniques allow traditional tests by sampling to be replaced by rigorous checking techniques which correspond to 100% coverage of test cases. In order to address a global innovative approach in the near future which match the whole software configuration, the SACRES tool set should be interfaced with asynchronous techniques matching the operating system development. This raises an open question in the software development future new process.

F.W. Vaandrager and J.H. van Schuppen (Eds.): HSCC'99, LNCS 1569, pp. 1–1, 1999.

Approximating Hybrid System Dynamics for Analysis and Control

Bruce H. Krogh

Department of Electrical and Computer Engineering, Carnegie Mellon University
Pittsburgh, PA 15213-3890 USA
krogh@ece.cmu.edu

Abstract. The objective of this lecture is to survey and assess the state of the research on methods for approximating hybrid system dynamics. For all but the most trivial dynamic systems, approximations are necessary to make analysis and controller synthesis tractable. This is true for both continuous-state systems and discrete-state systems, and theories have been developed in both domains to justify and guide model simplification and approximation (although ad hoc engineering judgment remains the method of choice in most applications). For hybrid systems, decidability results indicate that approximations will always be necessary to solve analysis and controller synthesis problems for even the simplest systems. These results will be reviewed briefly as a motivation for the work on approximation methods.

We will then consider two principal types of approximations that have been explored in the literature. The first set of methods approximates general hybrid system dynamics with simpler hybrid models for which some computational tools exist, such as linear or timed automata. The second set of methods generates finite-state discretizations of the continuous dynamics in a hybrid system so that tools for discrete-state systems can be applied. In both cases, the literature will be reviewed and success with applications will be assessed. Strengths and limitations of each approach will be summarized, and the types of problems that can be solved using each approach will be identified. Some software packages for building and approximating hybrid system models will also be reviewed and examples will be presented.

The final part of the lecture will discuss prospects for the future, both in terms of a theory for approximating hybrid systems dynamics and tools for computer-aided analysis and controller synthesis. Open problems and directions for future work will be identified.

F.W. Vaandrager and J.H. van Schuppen (Eds.): HSCC'99, LNCS 1569, pp. 2–2, 1999.
© Springer-Verlag Berlin Heidelberg 1999

High-Level Modeling and Analysis of an Air-Traffic Management System*

Nancy Lynch

MIT Laboratory for Computer Science, Cambridge, MA 02139, USA
lynch@theory.lcs.mit.edu

Abstract. This talk describes progress in a current project on modeling and analyzing the TCAS II aircraft collision-avoidance system.

The state of the art in formal methods applied to air traffic management systems involves specifying software behavior in detail, using formalisms such as Statecharts. Although such methods are precise, they do not help much in understanding the systems intuitively; nor do they enable analysis of high-level global requirements, such as "Under condition A, the planes will not crash."

To aid people in understanding such systems, and to enable such analysis, we advocate defining high-level mathematical models for the system, including not only the control software, but also the airplanes, sensors, and pilots—that is, high-level hybrid system models.

In a current demonstration project at MIT and Berkeley, we have defined abstract models for the key system components of the new TCAS II (version 7) system. These are based formally on the Hybrid I/O Automaton (HIOA) model [1]. We are using these models to formulate and prove theorems about the behavior of the system under particular assumptions. Our results are intended only as illustrations—the models provide a foundation for study of a wide range of properties of the system's behavior. We hope that this project will help to produce improved validation methods for air-traffic management systems.

References

1. N.A. Lynch, R. Segala, F.W. Vaandrager, and H.B. Weinberg. Hybrid I/O automata. In R. Alur, T.A. Henzinger, and E.D. Sontag, editors, *Hybrid Systems III*, volume 1066 of *Lecture Notes in Computer Science*, pages 496–510. Springer-Verlag, 1996.

---

* Based on joint work with Carl Livadas and John Lygeros.

F.W. Vaandrager and J.H. van Schuppen (Eds.): HSCC'99, LNCS 1569, pp. 3–3, 1999.

Geometric Categories, O-Minimal Structures and Control

Chris Miller

Department of Mathematics, The Ohio State University, 231 W. 18th Street,
Columbus, Ohio 43210, USA
miller@math.ohio-state.edu

Abstract. The theory of subanalytic sets is an excellent tool in various analytic-geometric contexts, including geometric control theory. (See [1], for example.)

One can axiomatize the notion of "behaving like the category of subanalytic sets (in manifolds)" by introducing the notion of "analytic-geometric category". (The category of subanalytic sets is the smallest analytic-geometric category.) The objects of such a category share many of the hereditary and geometric finiteness properties of subanalytic sets. Proofs of the more difficult results of this nature, like the Whitney-stratifiability of sets and maps in such a category, often involve the use of charts to reduce to the case of subsets of $\mathbb{R}^n$. For subsets of $\mathbb{R}^n$, the theory of o-minimal structures on the real field, an abstraction of the theory of semialgebraic sets, provides an elegant and efficient setting in which to work. (See [2] and [3].)

(Some reasonable sets—like $\{(x, x^r) : x > 0\}$ for irrational r, $\{(x, e^x) : x > 0\}$, and $\{(x, \Gamma(x)) : x > 0\}$—are not *globally* subanalytic in $\mathbb{R}^2$. Because there are o-minimal structures on the real field which include these sets, we now have available analytic-geometric categories which include these sets "at infinity" among their objects.)

In analogy with the semilinear, semialgebraic and subanalytic settings, one considers hybrid systems whose relevant data (guards, resets, flows and so on) all belong to some o-minimal structure. It can be shown, for example, that such hybrid systems admit finite bisimulations; see [4].

References

[1] *Differential Geometric Control Theory*, Progr. Math. **27**, Birkhäuser, Boston, 1983.

[2] L. van den Dries, *Tame Topology and O-minimal Structures*, London Math. Soc. Lecture Note Series **248**, Cambridge University Press, 1998.

[3] L. van den Dries and C. Miller, *Geometric categories and o-minimal structures*, Duke Math. J. **84**:497–540, 1996.

[4] G. Lafferriere, G. Pappas, and S. Sastry, *O-minimal hybrid systems*, Technical Report, UC Berkeley, 1998. Available at http://www.mth.pdx.edu/~gerardo.

F.W. Vaandrager and J.H. van Schuppen (Eds.): HSCC'99, LNCS 1569, pp. 4–4, 1999.
© Springer-Verlag Berlin Heidelberg 1999

Polyhedral Flows in Hybrid Automata

Rajeev Alur*, Sampath Kannan**, and Salvatore La Torre***

Department of Computer and Information Science
University of Pennsylvania
Philadelphia, PA 19104
{alur,kannan,latorre}@cis.upenn.edu

Abstract. A *hybrid automaton* is a mathematical model for hybrid systems, which combines, in a single formalism, automaton transitions for capturing discrete updates with differential constraints for capturing continuous flows. Formal verification of hybrid automata relies on symbolic fixpoint computation procedures that manipulate sets of states. These procedures can be implemented using boolean combinations of linear constraints over system variables, equivalently, using polyhedra, for the subclass of *linear hybrid automata*. In a linear hybrid automaton, the flow at each control mode is given by a rate polytope that constrains the allowed values of the first derivatives. The key property of such a flow is that, given a state-set described by a polyhedron, the set of states that can be reached as time elapses, is also a polyhedron. We call such a flow a *polyhedral flow*. In this paper, we study if we can generalize the syntax of linear hybrid automata for describing flows without sacrificing the polyhedral property. In particular, we consider flows described by *origin-dependent rate polytopes*, in which the allowed rates depend, not only on the current control mode, but also on the specific state at which the mode was entered. We establish that flows described by origin-dependent rate polytopes, in some special cases, are polyhedral.

1 Introduction

A hybrid system consists of a collection of digital programs that interact with each other and with an analog environment. The formal modeling and analysis of the mixed digital-analog nature of hybrid systems requires a mathematical model that incorporates the discrete behavior of computer programs with the continuous behavior of environment variables, such as time, position, and temperature. The model of our choice is the *hybrid automaton* [ACH+95]. A hybrid automaton consists of a finite *control graph* whose vertices correspond to *control modes* and edges correspond to *control switches*. The automaton has a finite

---

* Supported in part by Bell Laboratories, Lucent Technologies, and by the NSF CAREER award CCR-9734115 and by the DARPA grant NAG2-1214.

** Supported in part by the ARO grant DAAG55-98-1-0393 and the NSF award CCR-96-19910.

*** Visiting from Department of Computer Science, University of Salerno, Italy.

number of real-valued variables that can change discretely during the switches or continuously as the time elapses at control modes. The flow (i.e. continuous evolution) of the variables at a control mode is specified by differential equations and differential inequalities that constrain the allowed rates, and by invariant predicates that constrain the allowed durations.

For analyzing hybrid systems, we build on the model-checking technology, in which a formal model of the system is checked, fully automatically, for correctness with respect to a requirement expressed in temporal logic [CE81,CK96]. Model checking requires exploration of the entire state space of the system. For discrete finite-state systems, this can be done enumeratively, by considering each state individually, or symbolically, by computing with constraints over boolean variables that encode state sets. Because of its ability to deal with very large state spaces, symbolic model checking has been proven an effective technique for debugging of complex hardware [BCD+92]. For hybrid systems, the state space is infinite, so an enumerative approach is impossible, but the symbolic approach can be extended to a class of hybrid automata called *linear hybrid automata* [AHH96]. A linear hybrid automaton requires that

1. The predicates describing initialization of the system, the enabling of the control switches, the update of variables during switches, and the invariants at control modes, are boolean combinations of linear inequalities over automaton variables.
2. The flow at a control mode is given by a rate polytope that constrains the allowed values of the first derivatives of the variables.

With these restrictions, it can be established that all the state-sets computed by the symbolic model checking procedures can be described by boolean combinations of linear inequalities. This theorem provides the foundations of the verifier HYTECH [AHH96,HHW97], that implements the analysis procedures using packages to manipulate polyhedra. Even though termination of the model checking procedure is not guaranteed— the model checking problem for linear hybrid automata is undecidable, the method has been shown to be useful in analysis of protocols of practical interest (cf. [HW95]).

The dynamics allowed by linear hybrid automata is very restrictive, and hence, modeling in HYTECH requires the user to approximate the system dynamics by identifying lower and upper bounds on the rates. Previous research has focused on identifying decidable subclasses [AD94,HKPV95,LPS98], and on general strategies for approximating the given dynamics using linear hybrid automata (cf. [HW96]). In this paper, our objective is to identify classes of hybrid automata which are more general than linear hybrid automata, but still admit symbolic analysis using polyhedra.

Of the two restrictions imposed by linear hybrid automata, the first one seems natural and tight, given that we don't want non-polyhedral sets. We investigate ways to generalize the second one in this paper. Given a set σ of states, let $post(\sigma)$ denote the set of states that can be reached starting from a state in σ as time elapses and variables evolve subject to the specified flow constraints. To be able to do reachability analysis using polyhedra, we want that *post* of a polyhedron must also be a polyhedron. We call such flows, which map polyhedra to polyhedra, *polyhedral flows*. For instance, the (2-dim) flow described by the equations

$$1 \leq \dot{x} \leq 2 \text{ and } \dot{x} = \dot{y},$$

describes motion diagonally upwards at a bounded speed. This flow is polyhedral, and is admitted by the syntax of linear hybrid automata. On the other hand, the flow described by the equations

$$\dot{x} = x \text{ and } \dot{y} = 1,$$

describes motion along the exponential curve, and is not polyhedral.

To generalize the flow constraints allowed by the syntax of linear hybrid automata, we consider flows that are described by *origin-dependent rate polytopes*, in which the allowed rates depend, not only on the current control mode, but also on the specific state at which the mode was entered. A sample 2-dimensional origin-dependent rate polytope is described by

$$\dot{x} = x_0 + 3 \text{ and } \dot{y} = y_0,$$

where (x_0, y_0) is the initial state upon entering the mode. Note that this particular flow is polyhedral. Unfortunately, not all origin-dependent rate polytopes specify polyhedral flows. We prove that flows are polyhedral in the following two special cases.

1. Flows describing motion towards a fixed target cannot be specified by rate polytopes, but can be specified by origin-dependent rate polytopes, and are polyhedral.
2. Flows described by the origin-dependent rate function $\dot{X} = \lambda \cdot X_0 + \Gamma$, where the vector X_0 denotes the initial values, λ is a constant, and Γ is a constant vector, are polyhedral.

In fact, we establish that these flows are *strongly polyhedral*: for any time δ, and for any polyhedron σ, the set $post_{<\delta}(\sigma)$ of states that can be reached at time δ or before, is a polyhedron. We also establish the tightness of our results by examples that are slight generalizations, but lead to non-polyhedral flows.

Our results are useful in two ways. First, they establish that symbolic fixpoint computation using polyhedra can be used for analysis of classes that are more general than linear hybrid automata. Second, they suggest new ways to approximate the system dynamics.

2   Hybrid Automata

A hybrid automaton [ACH+95] is a formal model to describe reactive systems with discrete and continuous components. Formally, a hybrid automaton H consists of the following components.

- A finite directed multi-graph (V, E). The vertices are called the *control modes* while the edges are called the *control switches*.
- A finite set X of real valued variables. A *valuation* ν is a function that assigns a real value $\nu(x)$ to each variable $x \in X$. The set of all valuations is denoted Σ_X. A *state* q is a pair (v, ν) consisting of a mode v and a valuation ν. The set of all states is denoted Σ. A *region* is a subset of Σ.
- A function *init*, that assigns to each mode v, a set $init(v) \subseteq \Sigma_X$ of valuations. This describes the initialization of the system: a state (v, ν) is initial if $\nu \in init(v)$. The region containing all initial states is denoted σ^I.
- A function *flow*, that assigns to each mode v, a set $flow(v)$ of C^∞-functions from R^+ to Σ_X. This describes the way variables can evolve in a mode.
- A function *inv*, that assigns to each mode v, a set $inv(v) \subseteq \Sigma_X$ of valuations. The system can stay in mode v only as long as the state is within $inv(v)$, and a switch must be taken before the invariant gets violated.
- A function *jump*, that assigns to each switch e, a set $jump(e) \subseteq \Sigma_X \times \Sigma_X$. This describes the enabling condition for a switch, together with the discrete update of the variables as a result of the switch.

The hybrid automaton H starts in an initial state. During its execution, its state can change in one of the two ways. A *discrete* change causes the automaton to change both its control mode and the values of its variables. Otherwise, a *continuous* activity causes only the values of variables to change according to the specified flows while maintaining the invariants. The operational semantics of the hybrid automaton is captured by defining transition relations over the state space Σ. For a switch $e = (v, v')$, we write $(v, \nu) \to_e (v', \nu')$ if $(\nu, \nu') \in jump(e)$. For a mode v and a time increment $\delta \in R^+$, we write $(v, \nu) \to_\delta (v, \nu')$ if there exists a function $f \in flow(v)$ such that $f(0) = \nu$, $f(\delta) = \nu'$, and $f(\delta') \in inv(v)$ for all $0 \leq \delta' \leq \delta$. The transition relations $\to_e$ capture the discrete dynamics, while the transition relations $\to_\delta$ captures the continuous dynamics.

Algorithms for symbolic reachability analysis of hybrid automata manipulate regions. Let σ be a region of H. For a switch e, $post_e(\sigma)$ contains states q' such that $q \to_e q'$ for some $q \in \sigma$. Similarly, define

$$post_\infty(\sigma) = \{ q' \mid q \to_\delta q' \text{ for some } q \in \sigma \text{ and some } \delta \in R^+ \}.$$

Then,

$$post(\sigma) = \cup_{e \in E} \, post_e(\sigma) \, \cup \, post_\infty(\sigma)$$

denotes the successor region of σ. A state q is said to be reachable if $q \in post^i(\sigma^I)$ for some natural number i.

In many cases, the automaton has a special timer variable which controls the time spent in different control modes. For a region σ and a time increment $\delta \in R^+$,

$$post_{\leq \delta}(\sigma) = \{ q' \mid q \to_{\delta'} q' \text{ for some } q \in \sigma \text{ and some } 0 \leq \delta' \leq \delta \}.$$

Thus, $post_{\leq \delta}(\sigma)$ contains all the states that can be reached up to time δ starting in σ.

The switches of a hybrid automaton can be labeled with events. When different components of a complex system are described individually by hybrid automata, the event-labels on switches of different components are used for synchronization [ACH+95]. We have omitted this aspect of the definition since it has no bearing on the analysis problem studied in this paper.

3   Polyhedral Analysis

The central problem in algorithmic formal verification of hybrid systems is to compute the set of reachable states of a given hybrid automaton. The set of all reachable states of a hybrid automaton can be computed by repeatedly applying *post* to the initial region. Model checking of more complex requirements specified in the temporal logics such as CTL [CE81] and TCTL [ACD93], involves symbolic fixpoint computation procedures with the computation of *post* as the basic building block.

If a hybrid automaton has n variables, then a valuation can be viewed as a point in the n-dimensional Euclidean space. A region σ of H is a *polytope* if there exists a mode v and a polytope σ' in the n-dimensional space such that σ equals $\{(v, \nu) \mid \nu \in \sigma'\}$. The mode v is called the mode of the polytope σ.

Let $H = (V, E, X, init, flow, inv, jump)$ be a hybrid automaton.

1. The automaton has *polyhedral initialization* if for each mode v, the region $init(v)$ is a polytope.
2. The automaton has *polyhedral switches* if for each polytope σ and each switch e, the region $post_e(\sigma)$ is a polytope.
3. The automaton has *polyhedral flows* if for each polytope σ, the region $post_\infty(\sigma)$ is a polytope.
4. The automaton has *strongly polyhedral flows* if for each polytope σ and time increment $\delta \in \mathbb{R}^+$, the region $post_{\leq \delta}(\sigma)$ is a polytope.

The hybrid automaton H is *(strongly) polyhedral* if it has polyhedral initialization, polyhedral switches, and (strongly) polyhedral flows. For a polyhedral hybrid automaton H, assuming the applications of *post* operators to polytopes are effectively computable, the set of reachable states can be effectively computed in an iterative manner.

The definition of polyhedral hybrid automata is semantic. The requirements about polyhedral initialization and polyhedral switches naturally lead to syntax:

1. For each mode v, $init(v)$ is described by a conjunction of linear inequalities over the variables X.
2. For each switch e, $jump(e)$ is described by a conjunction of linear inequalities over the variables $X \cup X'$, where the unprimed variables X refer to the values before the switch, and the primed variables X' refer to the values after the switch.

Note that, when the *jump* function is described in the above manner, for a polytope σ and a switch e, the polytope $post_e(\sigma)$ can be computed effectively using quantifier elimination (or projection) (see, for instance, [AHH96]).

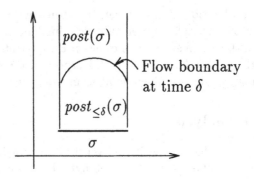

Fig. 1. Polyhedral, but not strongly polyhedral, flow

The requirement that the flows be polyhedral is more tricky. Since the invariant describes the boundary at which switches are triggered, it is natural to require, for each mode v, the region $inv(v)$ to be a polytope. As before, this can be ensured syntactically:

- For each mode v, $inv(v)$ is described by a conjunction of linear inequalities over the variables X.

In a *linear hybrid automaton*, flow at a mode v is given by a *rate polytope*, denoted $rate(v)$, an n-dimensional polytope that constrains the allowed values of the first derivates with respect to time. That is, a C^∞-function f belongs to $flow(v)$ iff the first derivative $\dot{f}$ of f with respect to time belongs to the polytope $rate(v)$ for all times $\delta \in \mathbf{R}^+$. Rate polytopes are specified as a conjunction of linear inequalities over the set $\dot{X}$ of dotted variables representing the first derivates of the corresponding variables in X. For instance, if the variables x and y denote the position of a robot on a plane, then the rate polytope

$$\dot{x} = \dot{y}, \ 1/\sqrt{2} \leq \dot{x} \leq \sqrt{2}$$

specifies the dynamics in which the robot moves diagonally at a speed between 1 to 2. Flows described by rate polytopes are polyhedral, and furthermore, allow effective computation of the *post* operator [AHH96].

The strong polyhedral property is useful in the following special case. Suppose the automaton has a special timer variable t that is used only to ensure an upper bound on the time spent in each mode. That is, t is reset to 0 on every switch, and for every mode v, the invariant $inv(v)$ has a conjunct $t \leq \delta_v$, for some constant δ_v. In this case, during analysis, we need not consider the timer variable as one of the dimensions. Reducing dimension by one can be computationally important. If σ is a region with mode v, then we must compute $post_{\leq \delta_v}(\sigma)$. For such analysis to be performed using polyhedra, the flows must be strongly polyhedral.

Observe that strongly-polyhedral flows are polyhedral, but as the illustration in Figure 1 indicates, the reverse need not hold.

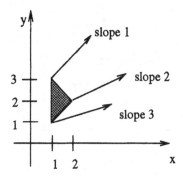

Fig. 2. Illustration of origin-dependent rates: $\dot{x} = x_0 + 2$ and $\dot{y} = y_0$

4   Rates dependent on initial state

Our objective is to understand if there are polyhedral flows that are more general than those specified by rate polytopes. The dynamics of a linear hybrid automaton can be understood as follows. After switching to a mode v, the system chooses a rate belonging to the polytope $rate(v)$, and then evolves at this fixed rate[1]. In our generalization, the rate polytope depends, not only on the mode, but also on the state upon entering the mode. More precisely, the flow is specified by a function which maps a mode v and a valuation ν to an n-dimensional polytope $rate(v, \nu)$. After switching to a mode v, if the valuation is ν_0, then the system chooses a rate belonging to the polytope $rate(v, \nu_0)$, and then evolves at this fixed rate. That is, a C^∞-function f belongs to $flow(v)$, for a mode v, iff $\dot{f} \in rate(v, f(0))$ for all times $\delta \in \mathrm{R}^+$. We call such flows *origin-dependent rate polytopes*.

We will use X_0 to denote the values of the corresponding variables in X at the beginning of a flow. Then, origin-dependent rate polytopes will be described by constraints involving the dotted variables $\dot{X}$ and the variables X_0. A sample 2-dimensional origin-dependent rate polytope is described by

$$\dot{x} \;=\; x_0 + 2 \text{ and } \dot{y} \;=\; y_0.$$

Figure 2 illustrates how different points evolve according to this flow.

Observe that for a flow function to be polyhedral, a necessary condition is that, for every state q, $post_\infty(q)$ is a polytope. This condition is not satisfied by flows described by differential equations in which the first derivative depends on the state in a continuous manner, but holds for origin-dependent rate polytopes. Origin-dependent rate polytopes do not guarantee polyhedral flows. In the sequel, we will identify some restrictions that ensure polyhedral flows.

---

[1] This is because there is a trajectory f starting in a valuation ν and ending in a valuation ν' with $\dot{f}$ belonging to the rate polytope at all times iff the rate of the straight-line trajectory from ν to ν' is in the rate polytope.

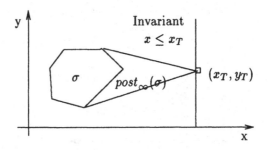

Fig. 3. Motion towards a fixed target

For a mode v, the automaton has polyhedral flows at v if for every polytope σ with mode v, $post_\infty(\sigma)$ is a polytope. For a flow to be polyhedral, it must be polyhedral at every mode.

4.1   Case 1: Motion towards a fixed target

As an example of an origin-dependent rate polytope, consider a robot whose position on a plane is described by two variables x and y. Suppose $\sigma \subseteq \mathbf{R}^2$ is a polytope that describes the set of possible current positions of the robot. Consider the dynamics in which the robot moves directly towards a fixed target located at (x_T, y_T) at fixed speed, say λ. Assume that the initial polytope σ lies entirely within the invariant $x \le x_T$. Then, if the original position of the robot is (x_0, y_0), the dynamics is described by the equation

$$\dot{x} = (x_T - x_0) \cdot \lambda / \sqrt{(x_T - x_0)^2 + (y_T - y_0)^2}, \quad \dot{y} \cdot (x_T - x_0) = \dot{x} \cdot (y_T - y_0)$$

Observe that for fixed values of the initial position (x_0, y_0), the above equations describe a polytope over $(\dot{x}, \dot{y})$. It is also easy to verify that $post_\infty(\sigma)$ is a polytope (see Figure 3). It turns out that this flow is also strongly-polyhedral: for any time δ, the boundary of the possible positions of the robot at time δ is also a polytope.

The polyhedral nature observed in this example holds for motion towards a fixed target in arbitrary dimensions, and even if the speed is not fixed. For two valuations ν and ν', let $dist(\nu, \nu')$ denote the Euclidean distance between them.

Theorem 1. *Let v be a mode of a hybrid automaton H. Suppose there exists a target valuation ν_T such that the flow is a origin-dependent rate polytope described by*

$$(\nu_T(x) - \nu(x)) \cdot \lambda_1 / dist(\nu, \nu_T) \le \dot{x} \le (\nu_T(x) - \nu(x)) \cdot \lambda_2 / dist(\nu, \nu_T),$$

for constants λ_1 and λ_2, and for all $y \in X$,

$$\dot{y} \cdot (\nu_T(x) - \nu(x)) = \dot{x} \cdot (\nu_T(y) - \nu(y)).$$

Furthermore, $inv(v)$ *is a polytope contained in a half-space defined by a hyper-plane passing through* ν_T. *Then, the automaton has strongly-polyhedral flow at the mode* v.

Let us note that it is straightforward to establish that the above flow is polyhedral: for a given polytope σ, $post_\infty(\sigma)$ is simply the convex hull of σ at the target. The following proof establishes the less obvious fact that the flow is strongly-polyhedral. The role of the invariant is to ensure that the flow does not go past the target point.

Proof. Let σ be a polytope with mode v, and let δ be a time increment. We first make a few simple observations:

If p is a point in $post_{\leq\delta}(\sigma)$ then there exists a point $\nu_0 \in \sigma$, a *constant* rate vector α and a time $\delta' \leq \delta$ such that the point ν_0 evolving at rate α for time δ' reaches the point p. The rate vector α is pointed towards the target, and has magnitude between λ_1 and λ_2. This observation follows from the convexity of the rate interval.

The evolution of a point is monotone towards the target. In other words, if $\delta_1 < \delta_2$, the position of any initial point ν at time δ_2 is on the ray joining its position at time δ_1 with ν_T.

Finally, if we consider any ray from ν_T, its intersection with $post_{\leq\delta}(\sigma)$ is an interval. Again this follows from the convexity of the rate polytope.

With these observations we will prove the theorem by proving that $post_{\leq\delta}(\sigma)$ is convex and has a finite number of vertices.

Lemma 1. $post_{\leq\delta}(\sigma)$ *is convex.*

Proof. Let p and q be arbitrary points in $post_{\leq\delta}(\sigma)$. For arbitrary c such that $0 \leq c \leq 1$ we have to establish that $\nu = cp + (1-c)q$ is in $post_{\leq\delta}(\sigma)$ (See Figure 4). By the previous observations there exist points $r, s \in \sigma$, rate vectors α_1, α_2, and times $t_1 \leq t_2 \leq \delta$ (without loss of generality) such that $p = r + \alpha_1 t_1$ and $q = s + \alpha_2 t_2$.

Consider the points $p' = r + \alpha_1 t_2$ (if this is outside the invariant region we choose $p' = \nu_T$) and $q' = s + \alpha_2 t_1$. By the fact that the rays from r and s meet at ν_T we have that the points p, q, p', and q' are co-planar. Consider the quadrilateral (possibly a triangle) of these four points. The ray from ν_T to ν intersects this quadrilateral twice, once on the line segment between p and q' and once on the line segment between q and p'. Let a and b be these points respectively. Let $a = dp + (1-d)q'$ for some $0 \leq d \leq 1$. Then the point $dr + (1-d)s$ evolving at rate $d\alpha_1 + (1-d)\alpha_2$ reaches point a at time t_1 and point b at time t_2. By the observation that the intersection of a ray from ν_T with $post_{\leq\delta}(\sigma)$ is an interval, we find that the point ν of interest is in $post_{\leq\delta}(dr + (1-d)s)$, and hence in $post_{\leq\delta}(\sigma)$. ∎

Lemma 2. *The set of vertices of* $post_{\leq\delta}(\sigma)$ *is finite.*

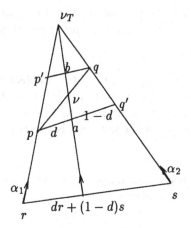

Fig. 4. Proof of convexity of $post_{\leq\delta}(\sigma)$ (Lemma 1)

Proof. Consider first a point $p \in post_{\leq\delta}(\sigma)$ which is in $post_{\leq\delta}(\nu)$ for $\nu \in \sigma$ where ν is not a vertex of σ. We claim that p is not a vertex in $post_{\leq\delta}(\sigma)$. Let $\delta' \leq \delta$ be the time and α be the rate vector such that $p = \nu + \alpha\delta'$. Since ν is not a vertex of σ, there exist $a, b \in \sigma$ and c such that $0 < c < 1$ such that $\nu = ca + (1-c)b$. Let $s, t \in post_{\leq\delta}(\sigma)$ be obtained respectively by allowing a and b to evolve along rate vector α for time δ'. Then $p = cs + (1-c)t$ proving that p is not a vertex.

Next consider a point $p \in post_{\leq\delta}(\nu)$ where ν is a vertex of σ. Suppose there is a rate vector α and a time $\delta' < \delta$ such that $p = \nu + \delta'\alpha$ and further suppose that α is not a vector representing the maximum or minimum speed allowed. Then it is easy to see again that p is not a vertex and can be expressed as the convex combination of two points in $post_{\leq\delta}(\sigma)$, one obtained by traveling at greater speed and the other obtained by traveling at lesser speed.

Thus the only potential vertices are those obtained by vertices of σ traveling at maximum speed for time δ or vertices of σ traveling at minimum speed for time 0. This is at most twice the number of vertices as in σ and hence finite. ∎

Note that if for a mode v, the rate polytope $rate(v, \nu)$ is independent of ν, then the automaton has polyhedral flows at v. The automaton can be polyhedral at different modes for different reasons, and still, we have polyhedral flows. However, as the next example shows, if different variables exhibit different types of evolution at the same mode, then the resulting flows are not polyhedral.

Consider a 3-dimensional automaton with three variables x, y, and z. Suppose the initial region is the line segment: $x = 0, z = 0, 0 \leq y \leq 1$. The variable z evolves at the constant rate 1 (independent of the state). The other two variables evolve together at a fixed speed along the direction towards the fixed target $(1, 0)$: for the initial state $(0, y_0, 0)$, $\dot{x} = 1/\sqrt{1 + y_0^2}$ and $\dot{y} = -y_0/\sqrt{1 + y_0^2}$. Suppose the invariant is $x \leq 1$, and let σ be the successor-region of the initial line segment under the specified flow. Verify that, while for a particular value of z, the corresponding projection of σ onto x and y is a polytope (in fact, a line segment), however, σ itself is curved.

4.2 Case 2: Separable linear dependence

Now we consider the case when the rate of change of a variable x_i depends linearly only on the initial value of x_i. That is, for each variable x_i, there exist constants λ_i and γ_i such that origin-dependent rate polytope is specified by $\dot{x}_i = x_i(0) \cdot \lambda_i + \gamma_i$. Thus, different components evolve independently, and the flow function at a mode is completely specified by the two vectors Λ and Γ. We call such an origin-dependent rate function linear and separable.

Does a separable linear rate function guarantee polyhedral flows? The answer is positive only in the two dimensional case. In the general case, there is additional requirement, namely, that all the components of the vector Λ be the same. For instance, the 3-dimensional origin-dependent rate polytope described by $\dot{x} = x_0 + 2$, $\dot{y} = y_0$, and $\dot{z} = z_0 - 5$, is polyhedral.

We state a technical lemma that allows us to prove the main theorem about separable linear rate functions.

Lemma 3. *Let p and q be points in $post_\infty(\sigma)$ such that p arises from point $a \in \sigma$ at time δ_1 and q arises from point $b \in \sigma$ at time δ_2 under a separable linear rate function. Let p' be the image of a at time δ_2 and q' be the image of b at time δ_1. Then the flow is strongly-polyhedral if and only if for all p, q, the points p, p', q and q' are co-planar.*

Theorem 2. *Suppose the flow function at a mode v is described by the origin-dependent separable linear rate function specified by the vectors Λ and Γ. Then, the automaton has strongly-polyhedral flows at the mode v if either*

1. *the automaton is of dimension 2, or*
2. *for all variables x_i and x_j, $\lambda_i = \lambda_j$.*

Conversely, if neither of these two conditions are satisfied, there exists a polytope σ such that $post_\infty(\sigma)$ is not polyhedral.

Proof. In the case that the automaton has dimension 2, all points are co-planar, and by Lemma 3 the resulting flow is polyhedral.

Now consider dimension greater than 2. Let p and q be points in $post_\infty(\sigma)$ and let a and b be initial vectors such that p arises from a at time δ_1 and q arises from b at time δ_2. Without loss of generality, let $\delta_1 \leq \delta_2$. Letting p_i denote the ith component of p and similarly for the other points, we have the following relations:

$$p_i = a_i + (\lambda_i a_i + \gamma_i)\delta_1,$$
$$p'_i = a_i + (\lambda_i a_i + \gamma_i)\delta_2,$$
$$q_i = b_i + (\lambda_i b_i + \gamma_i)\delta_2,$$
$$q'_i = b_i + (\lambda_i b_i + \gamma_i)\delta_1$$

These 4 points are co-planar if and only if $q - p$ is linearly dependent on $p' - p$ and $q' - p$. In order to take a look at this we calculate the ith components of these difference vectors.

$$(p' - p)_i = (\lambda_i a_i + \gamma_i)(\delta_2 - \delta_1)$$
$$(q' - p)_i = (b_i - a_i) + \delta_1(\lambda_i(b_i - a_i))$$
$$(q - p)_i = (b_i - a_i) + (\lambda_i b_i + \gamma_i)\delta_2 - (\lambda_i a_i + \gamma_i)\delta_1$$

To prove the partial converse stated we simplify by letting $\delta_1 = 1$, $\delta_2 = 2$ and $\gamma = 0$. We then get the equations:

$$(p' - p)_i = \lambda_i a_i,$$
$$(q' - p)_i = (b_i - a_i)(1 + \lambda_i),$$
$$(q - p)_i = (y_i - a_i) + 2\lambda_i y_i - \lambda_i a_i$$

For coplanarity of the original points we must have constants c_1 and c_2 such that for all i,

$$c_1 \lambda_i a_i + c_2(y_i - a_i)(1 + \lambda_i) = (y_i - a_i) + 2\lambda_i y_i - \lambda_i a_i$$

Rewriting, we get

$$a_i(c_1 \lambda_i - c_2 - c_2 \lambda_i + \lambda_i + 1) = y_i(1 + 2\lambda_i - c_2 - c_2 \lambda_i)$$

Choosing $a_1 = y_1 \neq 0$ we find that $c_1 = 1$ and substituting this value for c_1 we get $c_2 = \frac{(1+2\lambda_i)}{(1+\lambda_i)}$ (assuming $\lambda_i \neq -1$, although this case can also be handled).

It is clear that c_2 will work out to be a fixed constant if and only if all the λ_i are equal. In other words, if there exist i, j such that $\lambda_i \neq \lambda_j$, then the four points p, q, p', q' are not co-planar for some choice of a and b and hence there exists some segment which includes the points a and b for which $post_\infty$ is not polyhedral.

The forward direction of the theorem, i.e., the proof that if all the λ_i's are equal then the points p, p', q, q' are co-planar can be carried out with similar algebra. ∎

For dimension 3, there is a choice of the vector Λ such that the above theorem does not hold. In fact, the successor-region of a line segment can be a curved surface which cannot be expressed as a union of finitely many polyhedra. This can be understood with an example. Consider the line segment $x = y$ in the plane $z = 0$. Suppose the dynamics is given by $\dot{x} = x_0$, $\dot{y} = 2y_0$, and $\dot{z} = 1$. Then, the line traces a curved surface as time progresses (e.g. at time 1, it is the line $y = 3x/2$ in the plane $z = 1$, and at time 2, it is the line $y = 5x/3$ in the plane $z = 2$).

The computation of $post_\infty$ when Λ has all identical components is easy. Suppose σ is described by the the equation $AX \leq B$, where A is a constant matrix, X is the vector of variables, and B is a constant vector. Suppose the dynamics is given by $\dot{X} = \lambda X_0 + \Gamma$. Then,

$$post_\infty(\sigma) = \{ Y \mid \exists \delta, X. \; Y = X + \delta(\lambda X + \Gamma) \wedge AX \leq B \}$$

Eliminating X, we get

$$post_\infty(\sigma) = \{ Y \mid \exists \delta. \ A[\frac{Y - \delta \Gamma}{1 + \lambda \delta}] \le B \}$$

Hence, $post_\infty(\sigma)$ can be written as

$$\exists \delta. \ AY - A\delta\Gamma \le (1 + \lambda\delta)B$$

This is a linear expression, and the constraints on Y can be found by eliminating the variable δ. It is easy to modify the above computation to compute $post_{\le \delta}(\sigma)$ for a fixed δ.

5 Conclusions

We have studied ways to generalize the syntax of linear hybrid automata without sacrificing the polyhedral property of the resulting flows. The specific generalizations we look at are motivated by an on-going study aimed at application of formal modeling and verification to problems in multi-robot coordination. In particular, we considered origin-dependent rate polytopes in which the dynamics depends, not only on the control mode, but also on the state upon entry to the mode. Origin-dependent rate polytopes offer a different way to approximate complex dynamics. We have identified the polyhedral subclass of origin-dependent rate polytopes. When the flow specified by an origin-dependent rate polytope is not polyhedral, $post_\infty(\sigma)$ can be approximated by the convex-hull of the $post_\infty$ of the corner-points of σ. Such an approximation seems appropriate for origin-dependent rate polytopes for two reasons. First, the $post_\infty$ of a corner point is guaranteed to be polyhedral, and easily computable. Second, the convex-hull of the $post_\infty$ of the corner-points of σ is guaranteed to include $post_\infty(\sigma)$, a property not true of arbitrary dynamics.

References

[ACD93] R. Alur, C. Courcoubetis, and D.L. Dill. Model-checking in dense real-time. *Information and Computation*, 104(1):2–34, 1993.

[ACH+95] R. Alur, C. Courcoubetis, N. Halbwachs, T.A. Henzinger, P. Ho, X. Nicollin, A. Olivero, J. Sifakis, and S. Yovine. The algorithmic analysis of hybrid systems. *Theoretical Computer Science*, 138:3–34, 1995.

[AD94] R. Alur and D.L. Dill. A theory of timed automata. *Theoretical Computer Science*, 126:183–235, 1994.

[AHH96] R. Alur, T.A. Henzinger, and P.-H. Ho. Automatic symbolic verification of embedded systems. *IEEE Transactions on Software Engineering*, 22(3):181–201, 1996.

[BCD+92] J.R. Burch, E.M. Clarke, D.L. Dill, L.J. Hwang, and K.L. McMillan. Symbolic model checking: 10^{20} states and beyond. *Information and Computation*, 98(2):142–170, 1992.

[CE81]   E.M. Clarke and E.A. Emerson. Design and synthesis of synchronization skeletons using branching time temporal logic. In *Proc. Workshop on Logic of Programs*, LNCS 131, pages 52–71. Springer-Verlag, 1981.

[CK96]   E.M. Clarke and R.P. Kurshan. Computer-aided verification. *IEEE Spectrum*, 33(6):61–67, 1996.

[HHW97] T.A. Henzinger, P. Ho, and H. Wong-Toi. HyTech: a model checker for hybrid systems. *Software Tools for Technology Transfer*, 1, 1997.

[HKPV95] T.A. Henzinger, P. Kopke, A. Puri, and P. Varaiya. What's decidable about hybrid automata. In *Proceedings of the 27th ACM Symposium on Theory of Computing*, pages 373–382, 1995.

[HW95]   P.H. Ho and H. Wong-Toi. Automated analysis of an audio control protocol. In *Proceedings of the Seventh Conference on Computer-Aided Verification*, LNCS 939, pages 381–394. Springer-Verlag, 1995.

[HW96]   T.A. Henzinger and H. Wong-Toi. Linear phase-portrait approximations of nonlinear hybrid systems. In *Hybrid Systems III: Verification and Control*, LNCS 1066, pages 377–388. Springer-Verlag, 1996.

[LPS98]   G. Lafferriere, G. Pappas, and S. Sastry. O-minimal hybrid systems. 1998.

As Soon as Possible:
Time Optimal Control for Timed Automata*

Eugene Asarin[1] and Oded Maler[2]

[1] Institute for Information Transmission Problems, 19
Bol. Karetnyi per. 101447 Moscow, Russia
asarin@iitp.ru
[2] VERIMAG, 2, av. de Vignate, 38610 Gières, France
Oded.Maler@imag.fr

Abstract. In this work we tackle the following problem: given a timed automaton, and a target set F of configurations, restrict its transition relation in a systematic way so that from every state, the remaining behaviors reach F *as soon as possible*. This consists in extending the controller synthesis problem for timed automata, solved in [MPS95,AMPS98], to deal with *quantitative* properties of behaviors. The problem is formulated using the notion of a *timed game automaton*, and an optimal strategy is constructed as a fixed-point of an operator on the space of value functions defined on state-clock configurations.

1 Introduction

Most of the research on verification and synthesis of discrete systems is based on the following approach: the set L of all possible behaviors of the system is partitioned into L_φ and $L_{\neg\varphi}$, the former consisting of behaviors satisfying a certain property φ. Verification is the task of checking whether $L_{\neg\varphi}$ is empty while synthesis is the restriction of the transition relation of the system in order to achieve this fact. A typical example would be the property that all behaviors of the system eventually reach a subset F of the state-space.

In many situations, however, this all-or-nothing classification of behaviors as good or bad is too crude, and we would like to distinguish, for example, between behaviors which reach F within one or million steps. This suggests a richer model where *quantitative* performance measures are associated with behaviors based on length, cumulative cost of states and transitions, probabilities, etc. Verification is then transformed into finding the worst-case performance measure, while synthesis is rephrased as the search for an optimal controller, minimizing the above quantity. Timed automata (TA) [AD94] are system models where quantitative timing constraints are added to discrete transition systems. So far most of the

---

* This work was partially supported by the European Community Esprit-LTR Project 26270 VHS (Verification of Hybrid systems), the French-Israeli collaboration project 970MAEFUT5 (Hybrid Models of Industrial Plants) and by Research Grants 97-01-00692 and 96-15-96048 of Russian Foundation of Basic Research.

F.W. Vaandrager and J.H. van Schuppen (Eds.): HSCC'99, LNCS 1569, pp. 19–30, 1999.

work on verification and synthesis for TA concentrated on qualitative all-or-nothing properties[1] and the only (important) role of the timing information was to constrain the set of possible behaviors. In this work we associate a most natural performance measure to behaviors of TA, namely the *time* which elapses until they reach a certain set F of configurations.

In order to treat the synthesis problem we use our previously-introduced [MPS95,AMPS98] model of *Timed Game Automaton* (TGA) which is nothing but the usual timed automaton of [AD94] where the actions are partitioned into those of the controller and those of the uncontrolled environment. On this model various controller synthesis problems for *qualitative* properties can be formulated and solved. One example is the eventuality problem: find the set of "winning" states from which the controller can enforce the TGA into a set F, and compute the "strategy" for these states. In this work we solve the following quantitative generalization: find for each state the *minimal* time in which the controller can enforce the automaton into F, regardless of uncontrolled events, and construct a controller which achieves this optimum for each state. Previous techniques could only tell whether or not this minimal time is finite.

As in [AMPS98] the solution is obtained by a backward fixed-point calculation on the state-space of the TGA. The main difference is that the iteration is performed on a function from the TGA state-space into $\mathbb{R}$ which denotes roughly the min-max of the temporal distance from the state to F (this is similar to value/policy iterations used in dynamic programming, e.g. [Ber95]). Since the state-space of a TGA is non-countable, these functions cannot be tabulated, as is done in shortest-path algorithms on finite graphs. Fortunately we prove that as in the restricted case of functions from clocks to $\{0, 1\}$, all the value functions encountered in the iteration of our synthesis algorithm belong to a special well-founded class of functions admitting a simple linear-algebraic representation. This guarantees the termination of our algorithm.

The rest of the paper is organized as follows: In section 2 we re-introduce the *Timed Game Automaton* model. In section 3 we formulate the controller synthesis problem and describe our synthesis algorithm. In section 4 we prove that the algorithm is effective and correct (partial correctness and termination).

2   Games on Timed Systems

2.1   The Context

Timed games are extensions of discrete games (or equivalently of plant models used in the theory of supervisory control for discrete event systems [RW87]) where the players' actions may take place anywhere on the physical time axis, subject to certain timing constraints. For such games we take the model of *timed*

---

[1] To be more precise, quantitative information can be expressed in various real-time logics, but not in its full richness: one can verify whether all behaviors of a system reach the target within a given fixed bound, but not determine the minimal constant for which such a property is true.

automata [AD94], in which automata are equipped with auxiliary continuous variables called *clocks* which grow uniformly when the automaton is in some state. The clocks interact with the transitions by participating in pre-conditions (guards) for certain transitions and they are possibly reset when some transitions are taken.

In this continuous-time setting, a player might choose at a given moment to wait some time t and *then* take a transition. In this case, it should consider not only what the adversary can do *after* this action but also the possibility that the latter *will not wait for t time*, and perform an action at some $t' < t$.

For a more elaborate description of the game-theoretic approach to untimed and timed synthesis, the reader is encouraged to look at the lengthy introductions of [AMP95,AMPS98] as well as [TLS98].

2.2 Timed Game Automata

Let Q be a finite set of *states* and let $X = \mathbb{R}_+^d$ for some integer d be the *clock space*. We denote elements of X as $\boldsymbol{x} = (x_1, \ldots, x_d)$ and use $\boldsymbol{x} + t$ for $\boldsymbol{x} + (t, t, \ldots, t)$. Elements of $Q \times X$ are called *configurations*. A subset of X is called a *k-zone* if it can be obtained as a boolean combination of inequalities of the form $x_i \leq c$, $x_i < c$, $x_i - x_j \leq c$, where $c \in \{0, 1, \ldots, k\}$. The set of zones is denoted by $\mathcal{Z}(X)$. A zone Z is *right-open* if for any $\boldsymbol{x} \in Z$ there is a $\tau > 0$ such that the interval $(\boldsymbol{x}, \boldsymbol{x} + \tau)$ is also included in Z. These properties of subsets of X extend naturally to subsets of $Q \times X$, e.g. we say that $P \subseteq Q \times X$ is a zone if it can be written as finite union of sets of the form $\{q_i\} \times P_i$ such that every P_i is a zone. A function $\rho : X \to X$ is a *reset* function if it sets some of the coordinates of its argument to 0 and leaves the others intact. The set of all such functions is denoted by $\mathcal{J}(X)$. We assume two finite sets of actions A and B, a special empty action ε and let $A^\varepsilon = A \cup \{\varepsilon\}$ and $B^\varepsilon = B \cup \{\varepsilon\}$. The set A represents our actions (i.e. possible actions of the controller), while B stands for uncontrollable actions of the environment. The action ε stands for "wait and see".

Definition 1 (Timed Game Automaton).
A Timed game automaton (TGA) is a tuple $\mathcal{A} = (Z, A, B, T^A, T^B, \delta, \rho)$ where $Z \subseteq Q \times X$ is a zone, Q and X are the state and clock spaces, A and B are two distinct action alphabets, $T^A \subseteq Q \times X \times A^\varepsilon$ and $T^B \subseteq Q \times X \times B^\varepsilon$ are timing constraints for the two types of actions, the functions $\delta : Q \times A^\varepsilon \times B^\varepsilon \to Q$ and $\rho : Q \times A^\varepsilon \times B^\varepsilon \to \mathcal{J}(X)$ indicate which state is reached when performing a (possibly joint) action and which clocks are reset in that occasion.

Further requirements are the following: for every state q and action $a \in A^\varepsilon$, the set $T^A(q, a) = \{\boldsymbol{x} : (q, \boldsymbol{x}, a) \in T^A\}$ is a k-zone. We assume k to be fixed throughout the paper — it is the largest constant in the definition of the TGA. Similar requirements hold for T^B. We assume that $\delta(q, \varepsilon, \varepsilon) = q$ and that $\rho(q, \varepsilon, \varepsilon)$ is the identity function (if both sides refrain from action nothing happens). We require that the automaton is *strongly non-Zeno*, that is, in every cycle in the transition graph of the automaton (induced by δ), there is at least one transition which resets a clock variable x_i to zero, and at least one transition which can be taken only if $x_i \geq 1$. This is a very important condition as it prevents the controller and the environment to achieve their goals using unrealistic tricks that stop time.

The last requirement is that the zone $T^A(q, \varepsilon)$ is *right-open* for any q. This means that if player A is allowed to wait in a configuration (q, x), then it can really wait for a small additional positive amount of time. This requirement is important for preventing infeasible zero-time idling in A's strategy

Intuitively, when the automaton is at a configuration (q, x), time can progress as long as both players agree, that is, $(q, x, \varepsilon) \in T^B \cap T^A$. As soon as one of them can take an action, i.e. $(q, x, a) \in T^A$ for some $a \in A$ or $(q, x, b) \in T^B$ for some $b \in B$, or both, a transition can be taken. This can be formalized as follows:

Definition 2 (Steps and Runs). *A joint step of a TGA A is $(q, x) \longrightarrow (q', x')$ which is either*

1. *a time step (of duration t):*

$$(q, x) \xrightarrow{\ t\ } (q, x + t)$$

   such that $t > 0$, and for every $t' < t$,
   $(q, x + t', \varepsilon) \in T^A \cap T^B$.
2. *a discrete step*

$$(q, x) \xrightarrow{(a, b)} (q', x')$$

   such that $(a, b) \neq (\varepsilon, \varepsilon)$, $(q, x, a) \in T^A$, $(q, x, b) \in T^B$, $q' = \delta(q, a, b)$, and $x' = \rho(q, a, b)(x)$.

A run of a TGA A starting from (q_0, x_0) is a sequence of joint steps

$$\xi : (q_0, x_0) \longrightarrow (q_1, x_1) \longrightarrow \cdots$$

Note that $(q, x, \varepsilon) \in T^A \cap T^B$ means that both A and B agree to let time progress by a *positive* amount. On the other hand it is possible to reach a state where ε is not permitted by one or more of the two players: in such a situation the only thing that can happen is a discrete step.

For ease of notation we introduce the total transition function $\bar{\delta} : Q \times X \times A^\varepsilon \times B^\varepsilon \to Q \times X \cup \{\top, \bot\}$. Put

$$\bar{\delta}(q, x, a, b) = \begin{cases} (q, x) & \text{when } a = b = \varepsilon, (q, x, \varepsilon) \in T^A \cap T^B \\ \bot & \text{when } (q, x, a) \notin T^A \\ \top & \text{when } (q, x, a) \in T^A, (q, x, b) \notin T^B \\ (\delta(q, a, b), \rho(q, a, b)(x)) & \text{otherwise.} \end{cases}$$

Notice that the last line corresponds to the "normal" case when $(a, b) \neq (\varepsilon, \varepsilon)$, $(q, x, a) \in T^A$, and $(q, x, b) \in T^B$.

Clearly, replacing T^A by any subset of it, will restrict the range of action that the player A can chose at certain configurations, and hence decrease the set of behaviors of the TGA. We formulate various controller synthesis problems as finding appropriate restrictions of T^A, that we call *strategies*. These strategies are not necessary deterministic as they might allow A more than one action at a given configuration.

3   The Problem and the Algorithm

Definition 3 (Brachystochronic Problem). *Given a TGA $\mathcal{A}$ and a set $F \subset Q \times X$ find a strategy $T_*^A \subseteq T^A$ for player A which allows him or her to reach the target set F as fast as possible whatever player B does.*

The following algorithm is a generalization of the synthesis method for eventuality games on timed automata. The algorithm iterates over the value function which gives an upper estimate of arrival times to the set F. When we have a value function $f : Q \times X \cup \{\top, \bot\} \to \mathbb{R}_+ \cup \{0, \infty\}$ at any step of the iteration it means that we have a controller which allows to reach F from $(q, \boldsymbol{x})$ in no more than $f(q, \boldsymbol{x})$ time. The algorithm for finding the value function has the following form:

Algorithm 1 (Value Iteration for TGA).

{Initialization}
$$f_0(q, \boldsymbol{x}) = \begin{cases} 0 \text{ when } (q, \boldsymbol{x}) \in F \cup \{\top\} \\ \infty \qquad \text{otherwise;} \end{cases}$$
$$n := 0;$$
{Iteration}
> **repeat**
> $$n := n + 1;$$
> $$f_n := \pi(f_{n-1});$$
> **until** $f_n = f_{n-1};$

{Strategy extraction}
$$f_* := f(n);$$
$$T_*^A = \alpha(f_*).$$

The operators used in the algorithm are as follows:

$$\pi(f) = \min\{f, \pi_{\text{Act}}(f), \pi_{\text{Idle}}(f)\}, \tag{1}$$

where

$$\pi_{\text{Act}}(f)(q, \boldsymbol{x}) = \min_{a \in A} \max_{b \in B^{\varepsilon}} f(\bar{\delta}(q, \boldsymbol{x}, a, b)) \tag{2}$$

and

$$\pi_{\text{Idle}}(f)(q, \boldsymbol{x}) = \inf_{t \in \mathbb{R}_+} v(q, \boldsymbol{x}, t) \tag{3}$$

where

$$v(q, \boldsymbol{x}, t) = \max(\sup_{\tau < t} g(q, \boldsymbol{x}, \tau), t + f(q, \boldsymbol{x} + t)) \tag{4}$$

and

$$g(q, \boldsymbol{x}, \tau) = \max_{b \in B}(\tau + f(\bar{\delta}(q, \boldsymbol{x} + \tau, \varepsilon, b))) \tag{5}$$

Intuitively $g(q, \boldsymbol{x}, \tau)$ specifies the worst (largest) f of the configuration in which the game might be after B has performed an action at τ while A was idling. Similarly $v(q, \boldsymbol{x}, t)$ is the worst thing that can happen to A after deciding to idle for t time at $(q, \boldsymbol{x})$: it includes all the possible outcomes of B's actions at $\tau < t$. The best waiting time for A is the t which minimizes $v(q, \boldsymbol{x}, t)$ and its outcome π_{Idle} is compared with π_{Act} which denotes the best outcome of an immediate action.

The strategy extraction operator is given by the formula

$$\alpha(f)(q, \boldsymbol{x}) = \begin{cases} \alpha_{Act}(f)(q, \boldsymbol{x}) \text{ when } \pi_{\text{Act}}(f)(q, \boldsymbol{x}) < \pi_{\text{Idle}}(f)(q, \boldsymbol{x}) \\ \{\varepsilon\} \text{ when } \pi_{\text{Idle}}(f)(q, \boldsymbol{x}) < \pi_{\text{Act}}(f)(q, \boldsymbol{x}) \\ \alpha_{Act}(f)(q, \boldsymbol{x}) \cup \{\varepsilon\} \text{ when } \pi_{\text{Act}}(f)(q, \boldsymbol{x}) = \pi_{\text{Idle}}(f)(q, \boldsymbol{x}) < \infty \\ \emptyset \text{ when } \pi_{\text{Act}}(f)(q, \boldsymbol{x}) = \pi_{\text{Idle}}(f)(q, \boldsymbol{x}) = \infty \end{cases}$$

where

$$\alpha_{Act}(f)(q, \boldsymbol{x}) = \{a \in T^A(q, \boldsymbol{x}) | \max_{b \in B^\varepsilon} f(\bar{\delta}(q, \boldsymbol{x}, a, b)) = \pi_{\text{Act}}(f(q, \boldsymbol{x}))\}.$$

Unfortunately, the subtleties of real-valued time complicate the situation a bit. The problem is that the strategy extracted by α might violate the right-openness requirement. This can happen in a situation where player A has to take a transition the sooner the better but *after* time t_0. This is reflected in the strategy $\alpha(f)$ as follows: player A has action ε enabled until t_0 *including* t_0 and a "good" discrete transition enabled *after* t_0. But formally speaking, this strategy is blocking at time t_0.

To overcome this problem we introduce non-blocking approximations $\alpha_\varsigma(f)$ for small $\varsigma > 0$, which satisfy the right-openness conditions (but use non-integers in the guards). As we will show in the next section, taking this strategy for ς small enough, player A can reach F in time arbitrarily close to $f_*(q, \boldsymbol{x})$.

The ς-relaxed strategy is obtained from $\alpha(f)$ by enabling the ε action on a narrow stripe: $\alpha_\varsigma(f) = \alpha(f) \cup \{(q, \boldsymbol{x}, \varepsilon) : (q, x) \in S_\varsigma(f)\}$, where

$$S_\varsigma(f) = \text{interior}\{(q, \boldsymbol{x}) : \pi_{\text{Idle}}(f)(q, \boldsymbol{x}) < \pi_{\text{Act}}(f)(q, \boldsymbol{x}) + \varsigma\}.$$

4 Correctness Proof

4.1 Partial correctness

Our first aim is to prove that if the algorithm converges then it gives the optimal least-restrictive solution of the problem.

Lemma 1. *If player A uses the strategy $T_{n,\varsigma}^A = \alpha_\varsigma(f_{n-1}))$ from an initial position in (q, x) with $v = f_n(q, x) < \infty$, then the TGA reaches the target set F in no more then $v + n\varsigma$ time with no more then n transitions whatever the adversary does.*

Proof. Straightforward induction over n.   □

Let $\lambda > 0$ be any given (small) positive number.

Corollary 1. *Suppose that algorithm 1 converges in N iterations. Let $\varsigma = \lambda/N$ If player A uses the strategy $T_{*,\varsigma}^A = \alpha_\varsigma(f_*)$ from an initial position (q, x) such that $v = f_*(q, x) < \infty$ then the TGA reaches the target set F in no more then $v + \lambda$ time whatever the adversary does.*

Lemma 2. *For every (q, x) and n, player B can prevent the TGA from reaching F during at least n steps or $f_n(q, x) - \lambda$ time, whatever player A does.*

Proof. Straightforward induction over n.   □

Corollary 2. *Whatever player A does from an initial position (q, x) player B can prevent the TGA from reaching the target set F during at least $f_*(q, x) - \lambda$ time.*

Corollary 3. *If from an initial position (q, x) player A makes a transition not in T_*^A, then the adversary B can prevent the TGA from reaching the target set F during strictly more than $f_*(q, x)$ time.*

These corollaries imply the partial correctness of the algorithm.

Claim. Suppose the algorithm 1 converges. Then

1. The function $f_*(q, x)$ gives the infimal time necessary for A to drive the TGA from (q, x) to F whatever B does;
2. The strategy $T_{*,\varsigma}^A = \alpha_\varsigma(f_*)$ guarantees arrival to F in time $f_*(q, x) + \lambda$;
3. It is least restrictive: any strategy which guarantees arrival to F in time $f_*(q, x)$ is a sub-strategy of T_*^A.

4.2   Simple functions

Now we need to prove two properties of Algorithm 1. The first one is effectiveness: the algorithm manipulates functions from $Q \times \mathbb{R}^n$ to $\mathbb{R}$ and we need to find a finite representation of these functions in order to implement the algorithm. The second issue is to prove that the algorithm converges. Both effectiveness and finite convergence follow from the fact that all the functions used in the algorithm belong to the following class which is closed under the operator π.

Definition 4. *A function* $f : X \to \mathbb{R}_+ \cup \{0, \infty\}$ *is referred to as a k-simple one iff it can be represented as*

$$f(\boldsymbol{x}) = \begin{cases} c_i \text{ when } \boldsymbol{x} \in D_i \\ d_j - x_{l_j} \text{ when } \boldsymbol{x} \in E_j \end{cases}$$

where $D_i, i = 1, \ldots, M$ *and* $E_j, j = 1, \ldots, N$ *are k-zones,* $E_j \subseteq \{\boldsymbol{x} | x_{l_j} \leq k\}$ *and* $c_i, d_j \in \mathbb{N} \cup \{\infty\}$. *A function* $f : Q \times X \to \mathbb{R}_+ \cup \{0, \infty\}$ *is k-simple if* $f(q, \cdot)$ *is k-simple for any fixed* $q \in Q$

The additional boundedness condition on the E_j's means that the value of x_{l_j} can influence f only when $x_{l_j} < k$, the largest constant in the definition of the TGA. Clearly, any k-simple function admits a finite representation as a list of zones D_i and E_j and of constants c_i and d_j. It is easy to see that equality of k-simple functions is decidable, and that they satisfy some closure properties:

Lemma 3. *If* $f(\boldsymbol{x}), g(\boldsymbol{x})$ *are k-simple and* $\rho(\boldsymbol{x})$ *a reset function, then* $\max(f, g)$, $\min(f, g)$ *and* $f(\rho(\boldsymbol{x}))$ *are k-simple as well. The set* $\{\boldsymbol{x} | f(\boldsymbol{x}) < g(\boldsymbol{x})\}$ *is a k-zone.*

Notice that all closure results mentioned in this subsection are effective in the sense that it is possible to transform algorithmically representations of the original functions into representations of the resulting functions.

Another property of k-simple functions will be crucial for the proof of convergence of the algorithm.

Lemma 4 (Well-foundedness). *Any decreasing sequence of k-simple functions is finite.*

4.3   Effectiveness and convergence

The key step in the proof of convergence and effectiveness of Algorithm 1 is the closure of simple function under the operators used in the algorithm. Let $\Delta(\boldsymbol{x}, Z)$ be the largest backward distance from $\boldsymbol{x}$ to a zone Z, i.e. $\Delta(\boldsymbol{x}, Z) = \sup\{t \geq 0 | \boldsymbol{x} + t \in Z\}$. It is easy to see that Δ is simple in the first argument.

Lemma 5 (Closure of k-simple functions under π). *If* $f(q, \boldsymbol{x})$ *is k-simple and the TGA is k-bounded, then* $\pi_{\mathrm{Act}}(f)$, $\pi_{\mathrm{Idle}}(f)$ *and* $\pi(f)$ *are k-simple. Their representations can be found effectively.*

Proof. For $\pi_{\text{Act}}(f)$ it follows immediately from lemma 3 and the fact that for any fixed a, b and q the function $\bar{\delta}(q, \cdot, a, b)$ is a reset function (restricted to a zone).

The case of $\pi_{\text{Idle}}(f)$ is more difficult. Looking at equations (3), (4), and (5) one can see that $\pi_{\text{Idle}}(f)$ is obtained by elimination (via inf and sup) of the t variable from functions whose domain includes state-space and time. For example $g(q, \boldsymbol{x}, \tau)$ of (5) is defined as

$$g(q, \boldsymbol{x}, \tau) = \max_{b \in B}\{\tau + f(\bar{\delta}(q, \boldsymbol{x} + \tau, \varepsilon, b))\} = \tau + \max_{b \in B}\{\bar{\delta}(q, \boldsymbol{x} + \tau, \varepsilon, b)\}$$

and from the closure of $k-$simple function under reset, we can conclude that $g(q, \boldsymbol{x}, t)$ is of the form $t + f'(q, \boldsymbol{x} + t)$ for a k-simple function f'. This motivates the definition of the following class of functions:

Definition 5 (Bizones and Nice Functions). *A k-bizone is a union of sets of the form $\{(\boldsymbol{x}, t) | \boldsymbol{x} \in C \wedge \boldsymbol{x} + t \in D\}$ where C and D are k-zones. A function $f : X \times \mathbb{R}_+ \to \mathbb{R}_+ \cup \{0, \infty\}$ is referred to as k-nice iff it can be represented as*

$$g(\boldsymbol{x}, t) = \begin{cases} c_i + t \text{ when } (\boldsymbol{x}, t) \in D_i \\ d_j - x_{l_j} \text{ when } (\boldsymbol{x}, t) \in E_j, \end{cases}$$

where $D_i, i = 1, \ldots, M$ and $E_j, j = 1, \ldots, N$ are k-bizones, $E_j \subseteq \{\boldsymbol{x} | x_{l_j} \leq k\}$ and $c_i, d_j \in \mathbb{N} \cup \{\infty\}$.

Sublemma 6 (Properties of k-Nice Functions).

1. *If $f(\boldsymbol{x})$ is k-simple, then $t + f(\boldsymbol{x} + t)$ is k-nice.*
2. *If g and h are k-nice so are $\min\{g, h\}$ and $\max\{g, h\}$.*
3. *If $g(\boldsymbol{x}, \tau)$ is k-nice, then $h(\boldsymbol{x}, t) = \sup_{\tau < t} g(\boldsymbol{x}, \tau)$ is k-nice.*
4. *If $g(\boldsymbol{x}, t)$ is k-nice, then $\inf_{t \in \mathbb{R}_+} g(\boldsymbol{x}, t)$ is k-simple.*

Proof.

1. Immediate from definitions.
2. Let

$$g(\boldsymbol{x}, t) = \begin{cases} c_i + t \text{ when } (\boldsymbol{x}, t) \in D_i \\ d_i - x_{l_i} \text{ when } (\boldsymbol{x}, t) \in E_i, \end{cases}$$

and

$$h(\boldsymbol{x}, t) = \begin{cases} r_m + t \text{ when } (\boldsymbol{x}, t) \in P_m \\ s_m - x_{l_n} \text{ when } (\boldsymbol{x}, t) \in Q_m, \end{cases}$$

and let

$$u(\boldsymbol{x}, t) = \max\{g(\boldsymbol{x}, t), h(\boldsymbol{x}, t)\} = \begin{cases} g(\boldsymbol{x}, t) \text{ when } g(\boldsymbol{x}, t) > h(\boldsymbol{x}, t) \\ h(\boldsymbol{x}, t) \text{ when } g(\boldsymbol{x}, t) \leq h(\boldsymbol{x}, t), \end{cases}$$

To obtain a representation for $u(\boldsymbol{x}, t)$ we combine each line of the formula for g with each line of the formula for h and verify that the following types of sets, which appear when comparing lines of these formulae, are bizones.

$\{(\boldsymbol{x},t)|r+t \leq c+t\}$: In fact it is either $\mathbb{R}_+^{n+1}$ or $\emptyset$, hence a bizone.

$\{(\boldsymbol{x},t)|s-x_i \leq c+t\}$: It can be written as $\{(\boldsymbol{x},t)|x_i+t \geq s-c\}$, which is a bizone.

$\{(\boldsymbol{x},t)|s-x_i \leq d-x_i\}$: It is either $\mathbb{R}_+^{n+1}$ or $\emptyset$, hence a bizone.

$\{(\boldsymbol{x},t)|s-x_i \leq d-x_j\}$: It can be written as $\{(\boldsymbol{x},t)|x_i-x_j \geq s-d\}$, which is a bizone.

When intersected with $D_i \cap P_m$, $D_i \cap Q_m$, $E_i \cap P_m$, and $E_i \cap Q_m$ these sets produce bizones participating in the definition of the function $u(\boldsymbol{x},t)$. The boundedness condition is inherited from the bizones E_i and Q_m. The reasoning for min is identical.

3. Let

$$g(\boldsymbol{x},\tau) = \begin{cases} c_i + \tau \text{ when } (\boldsymbol{x},\tau) \in D_i \\ d_j - x_{l_j} \text{ when } (\boldsymbol{x},\tau) \in E_j, \end{cases}$$

and $h(\boldsymbol{x},t) = \sup\limits_{\tau<t} g(\boldsymbol{x},\tau)$.

Observe that the set $[\boldsymbol{x},\boldsymbol{x}+t] = \{\boldsymbol{x}+\tau|t \in [0,t]\}$ intersects finitely many D_i and E_j sets from the definition of g. We associate with them the following (partial) functions:

$$u_j(\boldsymbol{x},t) = d_j - x_{l_j} \qquad \text{when } [\boldsymbol{x},\boldsymbol{x}+t] \cap E_j \neq \emptyset$$

$$v_i(\boldsymbol{x},t) = c_i - t \qquad \text{when } \boldsymbol{x}+t \in D_i$$

$$w_i(\boldsymbol{x},t) = c_i + \Delta(\boldsymbol{x},D_i) \text{ when } [\boldsymbol{x},\boldsymbol{x}+t] \cap D_i \neq \emptyset \wedge \boldsymbol{x}+t \notin D_i$$

All the defining conditions are bizones, the functions are k-nice in $\boldsymbol{x}$ and t and so is h which is their max.

4. Similar to statement 3.

This concludes the proof of the sublemma.   □

It follows immediately from the sublemma that $\pi_{\text{Idle}}(f)$ is k-simple. The case of $\pi(f)$ is immediate from two previous cases and Lemma 3.   □

Corollary 4. *Algorithm 1 is effective.*

Corollary 5. *Algorithm 1 always converges.*

Proof. The sequence f_n is a decreasing sequence of k-simple functions. By virtue of Lemma 4 it stabilizes.   □

Together with Claim 4.1, this concludes the correctness proof of the algorithm.

Theorem 1 (Main Result). *Algorithm 1 always converges and produces the least-restrictive optimal strategy for the brachystochronic problem for Timed Automata.*

5   Example

Consider the 1-clock TGA of figure 1 where the adversary B is trivial and the target set F is $\{q_4\}$. From q_0 player A can choose between waiting, going to q_1 (a losing sink), to q_2 (while resetting the clock) or to q_3. Intuitively, for smaller values of x it might be better to do b and go to q_2 because A does not lose much time by resetting the clock and can benefit from the smaller transition guard. The value function and strategy obtained by our algorithm are depicted below.

q	x	f	T_*^A
q_4	$[0, \infty)$	0	$\{\varepsilon\}$
q_1	$[0, \infty)$	∞	$\emptyset$
q_2	$[0, 2)$	$2 - x$	$\{\varepsilon\}$
	$[2, \infty)$	0	$\{a\}$
q_3	$[0, 5)$	$5 - x$	$\{\varepsilon\}$
	$[5, \infty)$	0	$\{a\}$
q_0	$[0, 3)$	2	$\{b\}$
	$[3, 3]$	2	$\{\varepsilon, b\}$
	$(3, 4)$	$5 - x$	$\{\varepsilon\}$
	$[4, 5]$	$5 - x$	$\{\varepsilon, c\}$
	$(5, \infty)$	0	$\{c\}$

As one can see, q_4 is a winning state right from the start while from q_1 you can never reach q_4. The clock spaces of q_2 and q_3 are partitioned into two intervals: one after the clock values reach their respective guards, where the value of f is 0, and the interval before that, where f measures the time until the satisfaction of the guards. Finally $f(q_0, x)$ is obtained as $\min\{f(q_1, x), f(q_2, 0), f(q_3, x)\}$ and one can observe that $x = 3$ is the breakpoint after which it is better to take the c transition to q_3. In the absence of an adversary, not all the complexities of the algorithm are demonstrated in this example.

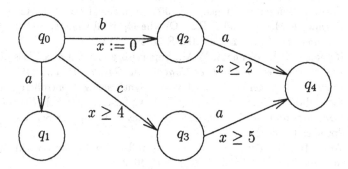

Fig. 1. A TGA with a trivial adversary. Missing guards and invariants are *true*.

6   Concluding Remarks

This work constitutes yet another step in the process of lifting classical results from automata to timed automata, and the main result can be rephrased as finding (min-max) shortest paths in non-countable (but well-behaving) graphs. The algorithm can be easily extended to models where integer costs are associated with transitions. However, assigning different costs to the passage of time at different states will transform the model into a more general hybrid system (such as the ones whose synthesis problem is treated in [W97]), and will probably make the problem harder, if not undecidable.

From the application point of view, many problems related to digital circuit design or scheduling problems in manufacturing and telecommunication, can be formulated as optimal control problems in the framework we have introduced. We believe that unifying the qualitative "property-based" approach, which is dominant in verification, with optimization-oriented approaches used elsewhere[2] is important for the development of hybrid systems research.

References

[AD94]   Alur, R., Dill, D.L.: A Theory of Timed Automata. *Theoretical Computer Science* **126** (1994) 183–235

[AMP95]   Asarin, E., Maler, O., Pnueli, A.: Symbolic Controller Synthesis for Discrete and Timed Systems. In: Antsaklis, P., Kohn, W., Nerode, A., Sastry, S. (eds.): *Hybrid Systems II.* Lecture Notes in Computer Science, Vol. 999. Springer (1995) 1–20

[AMPS98] Asarin, E., Maler, O., Pnueli, A., Sifakis, J.: Controller Synthesis for Timed Automata. In: *Proc. IFAC Symposium on System Structure and Control.* Elsevier (1998) 469–474

[Ber95]   Bertsekas, D.P.: *Dynamic Programming and Optimal Control.* Athena Scientific (1995)

[MPS95]   Maler, O., Pnueli, A., Sifakis, J.: On the Synthesis of Discrete Controllers for Timed Systems. In: Mayr, E.W., Puech, C. (eds.): *Proc. STACS'95.* Lecture Notes in Computer Science, Vol. 900. Springer (1995) 229–242

[PA89]   Passino, K.M., Antsaklis, P.J.: On the Optimal Control of Discrete Event Systems. In: *Proc. CDC'89.* IEEE (1989) 2713–2718

[RW87]   Ramadge, P.J., Wonham, W.M.: Supervisory Control of a Class of Discrete Event Processes. *SIAM J. of Control and Optimization* **25** (1987) 206–230

[TLS98]   Tomlin, C., Lygeros, J., Sastry, S.: Synthesizing Controllers for Nonlinear Hybrid Systems, in Henzinger, T.A., Sastry, S. (eds.): *Hybrid Systems: Computation and Control.* Lecture Notes in Computer Science, Vol. 1386. Springer (1998) 360–373

[W97]   Wong-Toi, H.: The Synthesis of Controllers for Linear Hybrid Automata. In: *Proc. CDC'97.* IEEE (1997) 4607–4612

---

[2] In [PA89], for example, transition costs are added to the discrete event models and an optimal controller synthesis problem is formulated and solved.

Verification of Hybrid Systems via Mathematical Programming

Alberto Bemporad and Manfred Morari

Institut für Automatik, Swiss Federal Institute of Technology
ETH Zentrum - ETL I29, CH 8092 Zürich, Switzerland
tel. +41-1-632 7626, fax +41-1-632 1211
{bemporad,morari}@aut.ee.ethz.ch
http://control.ethz.ch

Abstract. This paper proposes a novel approach to the verification of hybrid systems based on linear and mixed-integer linear programming. Models are described using the Mixed Logical Dynamical (MLD) formalism introduced in [5]. The proposed technique is demonstrated on a verification case study for an automotive suspension system.

1   Introduction

Hybrid models describe processes evolving according to dynamics and logic rules. The adjective "hybrid" stems from the fact that both continuous and discrete quantities are needed to describe the behaviour of the process at hand. Hybrid systems have recently grown in interest not only for being theoretically challenging, but also for their impact on applications. Although many physical phenomena are hybrid in nature, the main interest is directed to real-time systems, where physical processes are controlled by embedded controllers. For this reason, it is important to have available tools to guarantee that this combination behaves as desired. Verification algorithms for hybrid systems are aimed at providing such a certification.

Most of the literature about verification of hybrid systems originates from the artificial intelligence realm, and solvers rely on symbolic computation. In this paper, we propose an approach stemming from system science and propose a solver based on mathematical programming. As an example application, we report a case study on an automotive suspension system.

2   Mixed Logic Dynamic (MLD) Systems

The mixed logic dynamic (MLD) form has been introduced in [5]. It is a modeling framework that allows to describe various classes of systems, like finite state machines interacting with dynamic systems, piecewise linear systems, systems with mixed discrete/continuous inputs and states, systems with qualitative outputs, and so on. Physical constraints, constraint prioritization, and heuristics

F.W. Vaandrager and J.H. van Schuppen (Eds.): HSCC'99, LNCS 1569, pp. 31–45, 1999.

can also be included in the description of the system. For details, we defer to [5]. Here we only give the general MLD form

$$x(t+1) = Ax(t) + B_1w(t) + B_2\delta(t) + B_3z(t) \tag{1a}$$

$$y(t) = Cx(t) + D_1w(t) + D_2\delta(t) + D_3z(t) \tag{1b}$$

$$E_2\delta(t) + E_3z(t) \leq E_1w(t) + E_4x(t) + E_5 \tag{1c}$$

where $x = \begin{bmatrix} x_c \\ x_\ell \end{bmatrix}$, is the state of the system, whose components are distinguished between continuous $x_c \in \mathbb{R}^{n_c}$ and logical $x_\ell \in \{0,1\}^{n_\ell}$; $y = \begin{bmatrix} y_c \\ y_\ell \end{bmatrix}$, $y_c \in \mathbb{R}^{p_c}$ $y_\ell \in \{0,1\}^{p_\ell}$, is the output vector collecting quantities of interest, $w = \begin{bmatrix} w_c \\ w_\ell \end{bmatrix}$ is a vector of disturbances entering the system, collecting both continuous disturbances $w_c \in \mathbb{R}^{m_c}$, and binary disturbances $w_\ell \in \{0,1\}^{m_\ell}$ (e.g. faults [4]); $\delta \in \{0,1\}^{r_\ell}$ and $z \in \mathbb{R}^{r_c}$ represent auxiliary logical and continuous variables respectively. The auxiliary variables are introduced whenever logic propositions are translated into linear inequalities. The key idea of the approach is in fact to transform logic statements into mixed-integer linear inequalities. For instance: "$X_1 \wedge X_2$ FALSE" becomes "$\delta_1 + \delta_2 \leq 1$", or "if X_1 TRUE then Pressure $P_1 \leq P_0$" becomes "$P_1 - P_0 \leq M(1 - \delta_1)$", where M is a large number. All these constraints are summarized in the inequality (1c). Note that the description (1) is only apparently linear because of the integrality constraints. Also, the form (1) involves linear discrete-time dynamics. One might formulate a continuous time version by replacing $x(t+1)$ by $\dot{x}(t)$ in (1a), or a nonlinear version by changing the linear equations and inequalities in (1) to more general nonlinear functions. We restrict the dynamics to be linear and discrete-time in order to obtain computationally tractable schemes. Nevertheless, we believe that this framework permits the description of a very broad class of systems.

3   Automatic Verification of MLD Systems

Consider a linear discrete-time hybrid system of the form (1). Given a set of initial states $\mathcal{X}(0)$ and a set of disturbances $\mathcal{W}$, consider the following *Verification Problems*:

VP1   Verify that $\forall w \in \mathcal{W}$ and $\forall x(0) \in \mathcal{X}(0)$ the state $x(t) \in \mathcal{X}_s$, where $\mathcal{X}_s$ is an assigned set of *safe* states

VP2   Find the maximum range for $y(t)$

$$\max_{t \geq 0, w(t) \in \mathcal{W}, x(0) \in \mathcal{X}(0)} \{C_c x_c(t) + C_\ell x_\ell(t) + D_1 w(t) + D_2\delta(t) + D_3z(t)\}$$

3.1   A Simple but Numerically Impractical Solution

In principle, problem **VP1** can be addressed by solving $\forall T \geq 0$ the following Mixed-Integer Feasibility Test (MIFT)

$$\left\{ \begin{array}{l} x(0) \in \mathcal{X}(0) \\ x(T) \notin \mathcal{X}_s \\ w(t) \in \mathcal{W} \\ x(t+1) = Ax(t) + B_1 w(t) + B_2 \delta(t) + B_3 z(t) \\ E_2 \delta(t) + E_3 z(t) \leq E_1 u(t) + E_4 x(t) + E_5 \\ \qquad 0 \leq t \leq T \end{array} \right. \tag{2}$$

and **VP2** through the Mixed-Integer Program (MIP)

$$\max_{x(0), \{w(t), \delta(t), z(t)\}_{t=0}^T} Cx(T) + D_1 w(T) + D_2 \delta(T) + D_3 z(T)$$

$$\text{subj. to} \left\{ \begin{array}{l} x(0) \in \mathcal{X}(0) \\ w(t) \in \mathcal{W}, \ 0 \leq t \leq T \\ x(t+1) = Ax(t) + B_1 w(t) + B_2 \delta(t) + B_3 z(t) \\ E_2 \delta(t) + E_3 z(t) \leq E_1 u(t) + E_4 x(t) + E_5 \end{array} \right. \tag{3}$$

Even in the case of polyhedral sets $\mathcal{W}$, $\mathcal{X}_s$, $\mathcal{X}(0)$, solving the MIFT (2) and the Mixed Integer Linear Program (MILP)–(3) for large T becomes prohibitive. In fact, each problem (2) is $\mathcal{NP}$-complete because of the presence of integer variables [5], which means that in the worst case the required computation time grows exponentially with T [12].

3.2   A General Procedure for Verification of MLD Systems

The numerical complexities discussed above are due to the presence of free integer variables in the optimization problems. Note, however, that the binary variables δ are related to conditions on the continuous states x_c. Therefore, a trajectory of the hybrid system (1) can be partitioned in subtrajectories with $\delta(t) \equiv$ const, and analogously with $x_\ell(t) \equiv$ const. With this idea in mind, we consider a hypothetic partition of the continuous state space $\mathbb{R}^{n_c}$ in subregions $\mathcal{C}_i$ where system (1) evolves with $\delta(t) \equiv \delta_i$, $x_\ell(t) \equiv x_{\ell i}$, namely

$$\mathcal{C}_i = \{x_c \in \mathcal{X}_c : \exists z \in \mathbb{R}^{r_c}, w \in \mathbb{R}^m, \text{ such that}$$
$$E_3 z \leq E_1 w + E_4 \left[\begin{smallmatrix} x_c \\ x_{\ell i} \end{smallmatrix} \right] + E_5 - E_2 \delta_i \}$$

The number of these subregions is at most $2^{r_\ell + n_\ell}$, corresponding to all 0,1 combinations of each component of vector $\left[\begin{smallmatrix} \delta \\ x_\ell \end{smallmatrix} \right]$. However, in general the number of nonempty sets $\mathcal{C}_i$ is much smaller, as most combinations will not fulfill the constraints stemming from translations of logical propositions (for instance, the logic proposition "$[\delta_1 = 1] \wedge [\delta_2 = 1]$ FALSE" becomes $\delta_1 + \delta_2 \leq 1$, which rules out the combination $(\delta_1, \delta_2) = (1, 1)$). Without loss of generality, we assume that the logical components x_ℓ of the state are of the form

$$x_\ell(t) = e_J \delta(t)$$

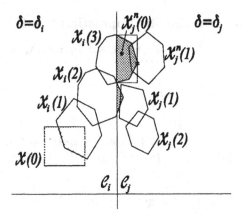

Fig. 1. Algorithm for verification.

where $e_J = [0, \ldots, 0, 1, 0, \ldots 0]$ is the J-th row of the $r_\ell \times r_\ell$ identity matrix. In fact, the state transition of logical states derives in general from a logic predicate involving literals associated with components of $\delta(t)$ and $x_\ell(t)$, and the latter can be expressed again as additional auxiliary variables $\delta_\ell(t)$, by simply adding the constraints $\delta_\ell(t) \leq x_\ell(t)$, $-\delta_\ell(t) \geq -x_\ell(t)$ in (1c).

The main idea underlying the algorithm is sketched in Fig. 1 and can be summarized as follows. Assume that $\mathcal{X}(0) \subseteq \mathcal{C}_i$ for some $i \in \{0, \ldots, 2^{r_\ell - 1}\}$. Then, consider the set $\mathcal{X}_i(t)$ of all possible evolutions from $\mathcal{X}(0)$ driven by $w(t) \in \mathcal{W}$ and such that $\delta(t) \equiv \delta_i$,

$$\mathcal{X}_i(t) = \{x : \exists x_c(0) \in \mathcal{X}(0), \{w(k)\}_{k=0}^t \text{ such that } x(k) \in \mathcal{C}_i, \forall k \leq t\}.$$

Note that $\mathcal{X}_i(t) = \text{Reach}(t, \mathcal{X}(0)) \bigcap \mathcal{C}_i$, where $\text{Reach}(t, \mathcal{X}(0))$ denotes the *reach set* from $\mathcal{X}(0)$. When at a certain time $\bar{t}$ there exists some state $x_c(\bar{t}) \in \text{Reach}(t, \mathcal{X}(0))$ such that $x_c(\bar{t}) \notin \mathcal{C}_i$, say $x_c(\bar{t}) \in \mathcal{C}_j$ (i.e. $\delta(\bar{t}) = \delta_j \neq \delta_i$ satisfies the constraints in (1) for some $w(\bar{t}) \in \mathcal{W}$, $z(\bar{t}) \in \mathbb{R}^{r_c}$), then $\mathcal{X}_j(0) \triangleq \text{Reach}(t, \mathcal{X}(0)) \bigcap \mathcal{C}_j$ is used to start the exploration of a new region with $\delta(t) \equiv \delta_j$. A certain evolution $\mathcal{X}_i(t)$ is considered explored or *fathomed* at time T_i when $\mathcal{X}_i(T_i) \subseteq \mathcal{X}_i(T_i - 1)$, i.e. the set $\mathcal{X}_i(T_i)$ has shrunk or is invariant, or is empty (i.e. all trajectories escape from $\mathcal{X}(T_i - 1)$). It may happen that during the exploration inside $\mathcal{C}_j$ at time t some states $x \in \mathcal{X}^* \triangleq \text{Reach}(t, \mathcal{X}_j(0)) \bigcap \mathcal{C}_i$ enter again a region $\mathcal{C}_i$ where an exploration has already been performed[1]. In this case, if $\mathcal{X}^* \not\subseteq \bigcup_{t=0}^{T_i} \mathcal{X}_i(t)$, $\forall t \leq T_i$, a new exploration is performed from $\mathcal{X}_i^1(0) = \mathcal{X}^*$; otherwise no action is taken. The procedure stops only when all evolutions $\mathcal{X}_i^h(\cdot)$ have been explored. Problem **VP2** is solved by maximizing $C_c x_c(t) + D_1 w(t) + D_3 z(t)$ over $\mathcal{X}_i^h(t) \times \mathcal{W} \times \mathbb{R}^{r_c}$. When the aim is to solve **VP1**, the procedure stops when $\mathcal{X}_i^h(t) \times \{e_J \delta_i\} \bigcap \mathcal{X}_u \neq \emptyset$, where $\mathcal{X}_u$ is the complementary set of $\mathcal{X}_s$, i.e. the set of "unsafe" states. This algorithm is formulated in Table 1. Note that there is no

---

[1] If the system where in continuous time, then $\mathcal{X}^*$ would be a point or, at most, a polyhedron of dimension $n_c - 1$.

need to investigate a priori all possible regions C_i, corresponding to all possible combinations of δ. The algorithm will explore only those combinations which are reachable from the assigned initial state $\mathcal{X}(0)$.

Algorithm 1.

0. **Push problem** $P_0^0 = \{i = 0,\ h = 0,\ \delta_i = 0,\ T_i^h = 0,\ \mathcal{X}_i^h(0) = \mathcal{X}(0) \subseteq C_0\}$ **on STACK.** $Y_{\max} = [-\infty, \ldots, -\infty]'$.
1. **While STACK nonempty,**
   1.1. **Pop problem** P_i^h **from STACK.**
   1.2. $t \leftarrow 0$.
   1.3. **If** $\mathcal{X}_i^h(0) \subseteq \bigcup_{\tau=1}^{T_j^k} \mathcal{X}_j^k(\tau)$, **for some fathomed problem** P_j^k, **go to 1.**
   1.4. $Y_{\max} \leftarrow \max\{Y_{\max}, \max_{x_c \in \mathcal{X}_i^h(t), w \in W, z \in \mathbb{R}^{r_c}} C_c x_c + C_\ell x_{\ell i} + D_1 w + D_2 \delta_i + D_3 z\}$.
   1.5. **If** $\mathcal{X}_i^h(t) \times \{e_J \delta_i\} \bigcap \mathcal{X}_u \neq \emptyset$, **system is unsafe. Stop.**
   1.6. $t \leftarrow t + 1$.
   1.7. **For all** $C_j \neq C_i$ **such that** $\text{Reach}(t, \mathcal{X}_i^h(0)) \bigcap C_j \neq \emptyset$:
      1.7.1. $n \leftarrow \max\{k : P_j^k$ **is on STACK**$\} + 1$.
      1.7.2. $P_j^n \leftarrow \{j,\ n,\ \delta_j,\ T_j^n = 0,\ \mathcal{X}_j^n(0) = \text{Reach}(t, \mathcal{X}_i^h(0)) \bigcap C_j\}$.
      1.7.3. **Push** P_j^n **on STACK.**
   1.8. **Compute** $\mathcal{X}_i^h(t)$.
   1.9. **If** $\mathcal{X}_i^h(t) \subseteq \mathcal{X}_i^h(t-1)$ **or** $\mathcal{X}_i^h(t) = \emptyset$, **fathom** P_i^h **and go to 1.**
   1.10. **Go to 1.4.**
2. **Stop.**

Table 1. Basic algorithm for verification of hybrid systems.

As the problem of verification of hybrid systems is undecidable [1, 11], there is no guarantee that Algorithm 1 will terminate. However from a practical point of view decidability would not be enough, as there is no difference between a non-terminating procedure and a procedure which runs out of time or memory [2]. Nevertheless, when the proposed algorithm terminates, it provides an answer to **VP1** and **VP2** respectively.

4 Verification of MLD Systems Based on Mathematical Programming

Below we describe how Algorithm 1 can be implemented using Linear Programming (LP) and Mixed Integer Linear Programming (MILP). We assume that the set of disturbances W and the set of initial states $\mathcal{X}(0)$, as well as the safe set $\mathcal{X}_s \triangleq \{x : K_1 x \leq K_2\}$, are polyhedra, and that relations between continuous and logic variables have the form $[\delta = 1] \leftrightarrow [x_c \leq 0]$. The latter assumption, which is rather general and satisfied in many applications, allows one to search for new subregions by simply minimizing and maximizing the components of the continuous part of the state [2].

Because of the linear form of MLD system (1), the evolution $x(t)$ can be expressed as

$$x(t) = A^t x_0 + \sum_{i=0}^{t-1} A^i \left[B_1 w(t-1-i) + B_2 \delta(t-1-i) + B_3 z(t-1-i) \right] \quad (4)$$

subject to (1c). Therefore, optimization of linear functions $f(x(t), w(t), \delta(t), z(t))$ results in an MILP if the variables $\delta(t)$ are free, or LP if $\delta(t)$ are fixed (for instance by setting $\delta(t) = \delta_i$ to enforce $x(t) \in C_i$). The same holds for feasibility problems of the form $x(t) \in \mathcal{X}$, where $\mathcal{X}$ is a polyhedron. These considerations allow one to rewrite Algorithm 1 as follows.

Algorithm 2.

0. Push problem $P_0^0 = \{i = 0, \ h = 0, \ \delta_i = 0, \ T_i^h = 0, \ \mathcal{X}_i^h(0) = \mathcal{X}(0) \subseteq C_0\}$ on STACK. $Y_{\max} = [-\infty, \dots, -\infty]'$.
1. While STACK nonempty,
   1.1. Pop problem P_i^h from STACK.
   1.2. $t \leftarrow 0$,

$$M_j(0) \leftarrow \max_{x \in \mathcal{X}_i^h(0)} \begin{bmatrix} x(t) \\ -x(t) \end{bmatrix}$$

   1.3. For all fathomed problems P_i^k:
       1.3.1. For $\tau = 0, \dots, T_i^k$
           1.3.1.1. For $v \in \{v_1, \dots, v_n\}$=vertices of $\mathcal{X}_i^h(0)$ solve the feasibility problem

$$\begin{cases} x_c(\tau) = v \\ (4)+(1c), \ t = 0, \dots, \tau \\ w(t) \in W \\ \delta(t) = \delta_i \\ x_\ell(0) = x_{\ell i} \\ x_c(0) \in \mathcal{X}_i^k(0) \end{cases}$$

           1.3.1.2. If feasible $\forall v$, fathom P_i^h and go to 1.
   1.4. Solve

$$\bar{Y} \leftarrow \begin{cases} \max_{\{w(k), z(k)\}_{k=0}^t, x_c(0)} C_c x_c(t) + D_1 w(t) + D_3 z(t) + [D_2 \delta_i + C_\ell x_{\ell i}] \\ \text{subj. to} \begin{cases} (4)+(1c), \ k = 0, \dots, t \\ w(k) \in W, \ k = 0, \dots, t \\ x_c(0) \in \mathcal{X}_i^h(0) \\ x_\ell(0) = x_{\ell i} \\ \delta(k) = \delta_i, \ k = 0, \dots, t \end{cases} \end{cases}$$

   and set $Y_{\max} \leftarrow \max\{Y_{\max}, \bar{Y}\}$

---

[2] The algorithm can be extended for $[\delta = 1] \leftrightarrow [Cx_c \leq 0]$ by minimizing/maximizing $C^{[j]}x$, where $C^{[j]}$ denotes the j-th row of C. Equivalently, the algorithm described below can be used by defining new state components $x_{\text{aug}} = Cx_c$.

1.5. Solve the feasibility problem defined by the constraints in 1.4.
and $K_1^{[j]}x(t) \geq K_2^{[j]}$, $j = 1, \ldots$, number of rows of K_1. If any is
feasible, system is unsafe. Stop.

1.6. $t \leftarrow t + 1$

1.7. For $j = 1, \ldots, n_c$:

  1.7.1. Solve

$$M_j(t) \leftarrow \begin{cases} \max_{\{w(k),z(k)\}_{k=0}^{t}, x_c(0), \delta(t)} \begin{bmatrix} x(t) \\ -x(t) \end{bmatrix} \\ \text{subj. to} \begin{cases} (4) + (1c), \ k = 0, \ldots, t \\ w(k) \in W, \ k = 0, \ldots, t \\ \delta(k) = \delta_i, \ k = 0, \ldots, t-1 \\ \delta(t) \in \{0,1\}^{r_\ell} \\ x_c(0) \in \mathcal{X}_i^h(0) \\ x_\ell(0) = x_{\ell i} \end{cases} \end{cases}$$

  1.7.2. For each optimization in 1.7.1., if $\delta(t) = \delta_j \neq \delta_i$,

    1.7.2.1. $n \leftarrow \max\{k : P_j^k \text{ is on STACK}\} + 1$.

    1.7.2.2. If $\mathcal{X}_j^n(0) \not\subseteq \mathcal{X}_j^k(0)$, $\forall P_j(k)$ on STACK, push $P_j^n = \{j, n, \delta_j, T_j^n = 0, \mathcal{X}_j^n(0) \subseteq \mathcal{C}_j\}$ on STACK

    1.7.2.3. Recompute 1.7.1. with the additional constraint $\delta(t) = \delta_i$

  1.8. If $M_j(t) \leq M_j(t-1)$ or the problems in 1.7.1. or 1.7.2.3. are
  infeasible, fathom P_i^h and go to 1.1.

  1.9. Goto 1.4.

2. Stop.

At step (1.7.2.2.), one must define the new set $\mathcal{X}_j^n(0)$. According to Algorithm 1,
one should define $\mathcal{X}_j^n(0) = \text{Reach}(t, \mathcal{X}_i^h(0)) \cap \mathcal{C}_j$, where the reach set at time t
$\text{Reach}(t, \mathcal{X}_i^h(0))$ is implicitly defined by Eqs. (4)+(1c). This definition of $\mathcal{X}_j^n(0)$,
although exact, has the disadvantage that the number of constraints defining
$\mathcal{X}_j^n(0)$ might keep growing during the execution of Algorithm 2. In this paper
we propose two alternatives, leading to inner and outer approximations of $\mathcal{X}_j^n(0)$
respectively. The inner approximation $\mathcal{X}_j^n(0) = \{x_c(t)\}$ ($x_c(t)$ corresponds to the
point marked as '$\star$' in Fig. 1). The second consists of approximating $\mathcal{X}_j^n(0)$ with
a hyper-rectangle (dashed rectangle in Fig. 1), which is computed by covering the
points obtained as in step (1.7.2.3.) by letting $\delta(t) = \delta_j$. Other approximations
are possible, and will be investigated in the future. For instance, a better inner
approximation consists of taking the convex hull of these points. Ellipsoidal approximations seem to be not appropriate, as they would result in quadratic constraints in the optimization problems. Parallelotopic or higher order polyhedra
can be better approximations. Another technique can consist in approximating
recursively the reach sets during the exploration, so that $\text{Reach}(t, \mathcal{X}(0))$ always
has a number of faces which is less than a specified limit. Finally, inner and
outer approximations can be run in parallel, by increasing their complexity until
lower and upper bounds to the verification problem converge within a desired
threshold. These ideas will be investigated in future research.

At step (1.3.1.1.) the algorithm checks the stronger condition $\mathcal{X}_i^h(0) \subseteq \mathcal{X}_i^k(t)$
for some $t \leq T_i^k$, instead of $\mathcal{X}_i^h(0) \subseteq \bigcup_{i=0}^{T_i^k} \mathcal{X}_i^k(t)$. Because of convexity of the
sets $\mathcal{X}_i^k(t)$, the first condition is easier to test, as only inclusion of the vertices

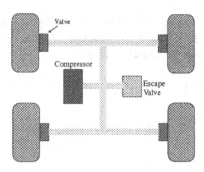

Fig. 2. Suspension system.

of $\mathcal{X}_i^h(0)$ must be checked. Note that although the results of the verification algorithm are not affected, the number of explored regions might increase.

4.1   Complexity of Algorithm 2

The optimization problems have the following complexity: (1.2) solves $2n_c$ LPs; (1.3.1.1.) one feasibility problem over linear constraints (LP); (1.4.) p_c LPs; (1.5.) m feasibility tests over linear constraints (LP), where m=number of rows in matrix K_1, i.e. number of inequalities defining the safe set $\mathcal{X}_s$; (1.7.1.) $2n_c$ MILPs with r_ℓ integer variables; (1.7.2.2) depends on the complexity of the sets $\mathcal{X}_j^k(0)$, for instance requires very little computation when $\mathcal{X}_j^k(0)$ are hyper-rectangles; (1.7.2.3.) $\leq 2n_c$ LPs. Note that compared to the simple approach described in Sect 3.1 where the number of integer variables involved in the optimization grows with time, in each optimization at step (1.7.1.) the number of integer variables remains equal to r_ℓ.

Although the number of continuous variables adds only minor computational complexity when compared to the number of integers, Algorithm 2 solves linear problems with a number of variables proportional to T_i^k. Our experience is that the problems are usually fathomed after a small number of time steps, i.e. T_i^k do not increase much. However, one might get large T_i^k when fast sampling is applied to slow dynamics, for instance, because of the presence of different time constants in the systems.

5   Verification of an Automotive Electronic Height Control System

In this section we describe an application of Algorithm 2 to the case study proposed first in [13], and reconsidered in [7] and [8]. The aim is to verify that an automotive control system satisfies certain driving comfort requirements.

5.1   Description of the System

The chassis level of the car is controlled by a pneumatic suspension system. The level is raised by pumping air into the system, and lowered by opening an escape valve. The configuration can be seen in Fig. 2. For the sake of simplicity, as in [13,7], we consider an abstract model including only one wheel. The suspension system is commanded by a logic controller, whose behavior is represented in Fig. 3. In short, the controller switches the compressor on when the level of the chassis is below a certain outer tolerance OTl, off when it reaches again an inner tolerance ITl, while it opens the valve when the level is above OTh, and closes it again when the level decreases below ITh. Because of high-frequency disturbances due to irregularities of the surface of the road, the controller switches based on a filtered version $f(t) = \frac{1}{1+as}h(t)$ of the measured level h of the chassis. The filter is reset to $f = 0$ each time f returns within the inner range [ITl, ITh]. The compressor can lift the chassis at a rate $cp(t) \in [cp_{min} \ cp_{max}]$, and the escape valve can lower it at a rate $ev(t) \in [ev_{min} \ ev_{max}]$. These parameters are reported in Table 2. We model this uncertainty as unmeasured disturbances, by letting $cp(t) = \overline{cp} + \Delta cp(t)$, $ev(t) = \overline{ev} + \Delta ev(t)$, where $\overline{cp} = \frac{cp_{max}+cp_{min}}{2}$, $\overline{ev} = \frac{ev_{max}+ev_{min}}{2}$, and $\Delta cp(t)$, $\Delta ev(t)$ range within $[\frac{cp_{min}-cp_{max}}{2}, \frac{cp_{max}-cp_{min}}{2}]$ and $[\frac{ev_{min}-ev_{max}}{2}, \frac{ev_{max}-ev_{min}}{2}]$ respectively.

Symbol	OTh	OTl	ITh	ITl	$1/a$	T_s	cp_{min}	cp_{max}	ev_{min}	ev_{max}	d_{min}	d_{max}	f_{sp}
Value	20	-40	16	-6	2	1	1	2	-2	-1	-1	1	0
Unit	mm	mm	mm	mm	s	s	mm s^{-1}	mm s^{-1}	mm s^{-1}	mm s^{-1}	mm s^{-1}	mm s^{-1}	mm

Table 2. Model parameters

6   Modeling the Automotive Hybrid System in MLD Form

The Electronic Height Controller is represented by the automaton depicted in Fig. 3. We introduce auxiliary binary variables in order to translate the automaton into the MLD form (1). We will use a shortened notation by writing δ instead of $[\delta = 1]$, and $\bar{\delta}$ instead of $[\delta = 0]$. Define

$$\delta_1 \leftrightarrow [f(t) \leq ITh], \quad \delta_2 \leftrightarrow [f(t) \leq ITl]$$
$$\delta_3 \leftrightarrow [f(t) \geq OTh], \quad \delta_4 \leftrightarrow [f(t) \leq OTl]$$

As OTl<ITl<ITh<OTH, it is easy to see that

$$\bar{\delta}_1 \to \delta_1, \delta_2; \quad \delta_2 \to \delta_1, \bar{\delta}_3; \quad \delta_3 \to \bar{\delta}_3$$

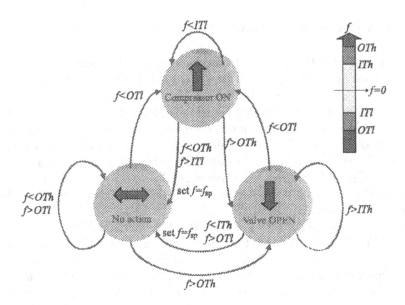

Fig. 3. Hybrid automaton for the controller.

The three states of the automaton in Fig. 3 are represented by a two-dimensional logic state

$$x_\ell = \begin{cases} \begin{bmatrix} 0 \\ 0 \end{bmatrix} & \text{if compressor OFF, valve CLOSED} \\ \begin{bmatrix} 0 \\ 1 \end{bmatrix} & \text{if compressor ON, valve CLOSED} \\ \begin{bmatrix} 1 \\ 0 \end{bmatrix} & \text{if compressor OFF, valve OPEN} \end{cases} \tag{5}$$

By using Karnough map techniques, (5) can be rewritten as

$$x_{\ell 1}(t+1) = \bar{x}_{\ell 1}(t)\delta_3(t) + x_{\ell 1}(t)\bar{\delta}_1(t) \tag{6}$$
$$x_{\ell 2}(t+1) = x_{\ell 2}(t)\delta_2(t) + \bar{x}_{\ell 1}(t)\bar{\delta}_4(t) \tag{7}$$

(note that typically (6)–(7) are available from the team which designed the logic controller). By defining $\delta_5 = \bar{x}_{\ell 1}\delta_3$, $\delta_6 = x_{\ell 1}\bar{\delta}_1$, $\delta_7 = x_{\ell 2}\delta_2$, $\delta_8 = \bar{x}_{\ell 1}\bar{\delta}_4$, $\delta_9 = \delta_5 \vee \delta_6$, $\delta_{10} = \delta_7 \vee \delta_8$, Eqs. (6)–(7) are equivalent to $x_{\ell 1}(t+1) = \delta_9(t)$, $x_{\ell 2}(t+1) = \delta_{10}(t)$. As the state $\begin{bmatrix} 1 \\ 1 \end{bmatrix}$ does not exist, the additional constraints $x_{\ell 1} + x_{\ell 2} \leq 1$, $\delta_9 + \delta_{10} \leq 1$ are included[3].

The input $u(t)$ of the system is redefined as

$$u(t) = \begin{cases} \overline{cp} + \Delta cp(t) & \text{if } x_\ell(t) = \begin{bmatrix} 0 \\ 1 \end{bmatrix} \\ \overline{ev} + \Delta ev(t) & \text{if } x_\ell(t) = \begin{bmatrix} 1 \\ 0 \end{bmatrix} \\ 0 & \text{if } x_\ell(t) = \begin{bmatrix} 0 \\ 0 \end{bmatrix} \end{cases} \tag{8}$$

By letting $\delta_{11} = \bar{x}_{\ell 1}x_{\ell 2}$, $\delta_{12} = x_{\ell 1}\bar{x}_{\ell 2}$, with $\delta_{11} + \delta_{12} \leq 1$, and $z_1 = cp\,\delta_{11}$, $z_2 = ev\,\delta_{12}$, one gets $u(t) = z_1(t) + z_2(t)$. The continuous dynamics of the car

---

[3] Contrary to optimization over continuous variables, in mixed-integer programming constraints involving integer variables can help the solver significantly.

and the filter are sampled by exact discretization (by introduction of zero-order holders), namely

$$f(t+1) = e^{-aT_s} f(t) + (1 - e^{-aT_s}) h(t) \tag{9}$$

$$h(t+1) = h(t) + T_s [d(t) + u(t)] \tag{10}$$

As the filter is reset to a set point value f_{sp} during the transitions $\begin{bmatrix} 0 \\ 1 \end{bmatrix} \to \begin{bmatrix} 0 \\ 0 \end{bmatrix}$ and $\begin{bmatrix} 1 \\ 0 \end{bmatrix} \to \begin{bmatrix} 0 \\ 0 \end{bmatrix}$, i.e.

$$[[x_\ell(t) = \begin{bmatrix} 1 \\ 0 \end{bmatrix} \wedge x_\ell(t+1) = \begin{bmatrix} 0 \\ 0 \end{bmatrix}] \vee [x_\ell(t) = \begin{bmatrix} 1 \\ 0 \end{bmatrix} \wedge x_\ell(t+1) = \begin{bmatrix} 0 \\ 0 \end{bmatrix}]] \to [f(t) = f_{sp}]$$

we introduce the variables $\delta_{13} = \overline{(\bar{\delta}_9 \bar{\delta}_{10} \delta_{14})}$, where $\delta_{14} = x_{\ell 1} \vee x_{\ell 2}$, and modify (9) in the form

$$f(t) = z_3(t), \ z_3(t) = [e^{-aT_s} f(t) + (1 - e^{-aT_s}) h(t)] \delta_{13}(t) + f_{sp}[1 - \delta_{13}(t)]$$

In summary, the system has $x_c = [f, h]'$, $x_\ell = [x_{\ell 1}, x_{\ell 2}]'$, $w = [\Delta cp, \Delta ev, d]'$, $\delta = [\delta_1, \ldots, \delta_{14}]'$, and $z = [z_1, z_2, z_3]'$.

Matrix	A	B_1	B_2	B_3	E_1	E_2	E_3	E_4	E_5
Sparsity (%)	93.75	92.67	96.43	75.00	97.74	89.46	93.22	85.59	47.45

Table 3. Sparsity of MLD matrices

6.1 MLD Translation & HYSDEL List

The system described above is translated into the MLD form (1) by using the language HYSDEL (HYbrid System DEscription Language) currently developed at the Automatic Control Lab, ETH Zürich. The description of the system in HYSDEL is reported in Table 4. The HYSDEL compiler automatically generates the matrices of the system. The sparsity is reported in Table 3. The number of constraints in (1c) is 59.

6.2 Numerical Results

Algorithm 2 has been implemented in Matlab using rectangular approximations of new regions to explore, and provides a maximum range $-44.54149 \leq h(t) \leq 25.00000$. The rectangular approximations $\mathcal{X}_i^h(0)$ in the (f, h) plane are shown in Fig. 4. The algorithm uses interpreted m-code and terminates in 761 s on a PC Pentium II 300 MHz with 96 Mb RAM. In [7], where this verification problem is solved analytically, the authors report the exact range $-43 \leq h(t) \leq 23$. In [13], the authors use HyTech [2] for symbolic verification, and obtain the range $-47 \leq h(t) \leq 27$. These bounds are slightly conservative because, in order

```
% Description of variables and constants

state f,h,xl1,xl2;
input d,dc,dev;

const OTh, OT1, ITh, IT1;
const M1,M2,M3,M4,m1,m2,m3,m4;
const e;
const Ts,cbar,evbar,eats,cmax,cmin,evmax,evmin;

% Variable types

real f,h,z1,z2,z3,d,dc,dev;
logic d1,d2,d3,d4,d5,d6,d7,d8,d9,d10,d11,d12,d13,d14;

% Relations

d1 = {f-ITh <= 0, M1, m1, e};
d2 = {f-IT1 <= 0, M2, m2, e};
d3 = {f-OTh >= 0, M3, m3, e};
d4 = {f-OT1 >= 0, M4, m4, e};

d5 = ~xl1 & d3;    % Should be accepted also: d5=(1-xl1)&d3, d5=(1-xl1)*d3
d6 = xl1 & ~d1;
d7 = xl2 & d2;
d8 = ~xl2 & ~d4;
d9 = d5 | d6;
d10 = d7 | d8;
d11 = ~xl1 & xl2;
d12 = xl1 & ~xl2;
d13 = ~(~d9 & ~d10 & d14);
d14 = xl1 | xl2;

z1 = d11 * (cbar + dc) {cmax,cmin,e};
z2 = d12 * (evbar + dev) {evmax,evmin,e};
z3 = (eats * f + (1 - eats) * h)*d13 {10*OTh,10*OT1,e};

% Other constraints

must xl1 + xl2 <= 1;
must ~(d9 & d10);
must ~(d11 & d12);

% Update

update f = z3;
update h = h + Ts * (d + z1 + z2);
update xl1 = d9;
update xl2 = d10;
```

Table 4. HYSDEL description of the automotive active leveler.

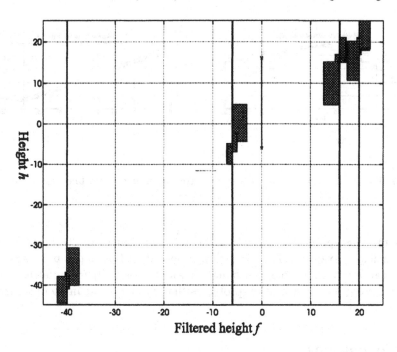

Fig. 4. Evolution of Algorithm 2 in the (f, h) plane using rectangular approximations of the new regions. The vertical thick lines represent the guard lines, where the logic conditions switch.

to fit the model used in HyTech, the dynamics is described as $k_1 \leq \dot{x} \leq k_2$. The authors report a computation time of 62 m on a Sun SparcStation 20 with 128 Mb RAM. In [8], the author reports computation times up to one day.

Although some conservativeness arises from the rectangular approximations, the algorithm presented in this paper provides a larger range mainly because of the discrete-time filter dynamics. In fact, one can easily check that the sequence of disturbances $d(t) \equiv 1$, $\Delta cp(t) \equiv 0$, $\Delta ev(t) \equiv 0$ leads the initial state $f(0) = 0$, $h(0) = 14.76537$ to $h(16) = 24.76537$ (as $f(8) = 19.99999$, the transition from the logic state $(0, 0)$ to $(1, 0)$ happens between $t = 9$ and $t = 10$. See Fig. 5(a)). Note that by using a point-wise inner approximation, one gets $h(t) \leq 24.35318$. Concerning the lower bound, by applying $d(t) \equiv -1$, $\Delta cp(t) \equiv 0$, $\Delta ev(t) \equiv 0$ from the initial state $f(0) = 0$, $h(0) = -5.54149$, one reaches exactly the computed lower bound $h(39) = -44.54149$ (in fact for $f(37) = -39.99999$, the transition from the logic state $(0, 0)$ to $(0, 1)$ happens between $t = 38$ and $t = 39$. See Fig. 5(b)).

These different approaches to the solution of the verification problem have their benefits and disadvantages. The method proposed in [2] uses symbolic computation, can handle a wide class of verification problems, but requires approximation of the dynamics and is computationally expensive. In [8], there is no approximation of the dynamics, but the author uses parallelotopic approximations of the reach sets in order to compute the solution of the verification

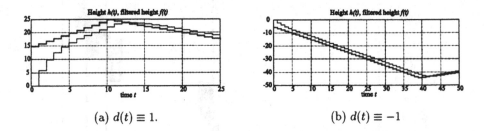

(a) $d(t) \equiv 1$.   (b) $d(t) \equiv -1$

Fig. 5. Worst-case simulation with step disturbances $d(t)$ (thin line: $f(t)$; thick line: $h(t)$).

problem. The approach suggested in [7] is exact, but seems to be tailored to the particular example. The algorithm proposed in this paper promises to be computationally affordable, can handle a wider class of hybrid models, and allows the solution of verification problems that can be recast in an optimization framework.

7   Conclusions

In this paper we have presented a novel approach to the verification of hybrid systems. It is based on linear/mixed-integer linear optimization and relies on the modeling formalism introduced in [5]. Computational feasibility of the approach has been shown in a non-trivial case study. Future research will be devoted to examine different inner/outer approximation techniques, iterative approximation of the reach sets, and improving the efficiency of the computer codes used to test the proposed algorithms.

Acknowledgments

The authors thank Prof. Sanjoy Mitter, Dr. Nicola Elia, and Fabio Torrisi for fruitful discussions.

References

1. R. Alur, C. Courcoubetis, T.A. Henzinger, and P.-H. Ho: Hybrid automata: an algorithmic approach to the specification and verification of hybrid systems. In A.P. Ravn R.L. Grossman, A. Nerode and H. Rischel, editors, *Hybrid Systems*, volume 736 of *Lecture notes in computer Science*, pages 209–229. Springer Verlag, 1993.
2. R. Alur, T.A. Henzinger, and P.-H. Ho: Automatic symbolic verification of embedded systems. *IEEE Trans. on Software Engineering*, 22:181–201, 1996.
3. A. Asarin, O. Maler, A. Pnueli: On the Analysis of Dynamical Systems having Piecewise-Constant Derivatives. *Theoretical Computer Science*, 138:35–65, 1995.

4. A. Bemporad, D. Mignone, and M. Morari: Moving horizon estimation for hybrid systems and fault detection. *Submitted American Control Conference*, 1999.
5. A. Bemporad and M. Morari: Control of systems integrating logic, dynamics, and constraints. *Automatica*, 35(3), March 1999.
6. T. M. Cavalier, P. M. Pardalos, and A. L. Soyster: Modeling and integer programming techniques applied to propositional calculus. *Computers Opns Res.*, 17(6):561–570, 1990.
7. N. Elia and B. Brandin: Verification of an automotive active leveler. Technical report, Dept. of Electrical Engineering and Computer Science, M.I.T., 1999.
8. A. Fehnker: Automotive control revised - linear inequalities as approximation of reachable sets. In *Hybrid Systems: Computation and Control*, volume 1386 of *Lecture notes in Computer Science*, pages 110–125. Springer Verlag, 1998.
9. R. Fletcher and S. Leyffer: A mixed integer quadratic programming package. Technical report, University of Dundee, Dept. of Mathematics, Scotland, U.K., 1994.
10. R. Fletcher and S. Leyffer: Numerical experience with lower bounds for miqp branch-and-bound. Technical report, University of Dundee, Dept. of Mathematics, Scotland, U.K., 1995. submitted to SIAM Journal on Optimization, http://www.mcs.dundee.ac.uk:8080/ sleyffer/miqp_art.ps.Z.
11. Y. Kesten, A. Pnueli, J. Sifakis, and S. Yovine: Integration graphs: a class of decidable hybrid systems. In A.P. Ravn R.L. Grossman, A. Nerode and H. Rischel, editors, *Hybrid Systems*, volume 736 of *Lecture notes in computer Science*, pages 179–208. Springer Verlag, 1993.
12. R. Raman and I. E. Grossmann: Relation between milp modeling and logical inference for chemical process synthesis. *Computers Chem. Engng.*, 15(2):73–84, 1991.
13. T. Stauner, Olaf Müller, and M. Fuchs: Using HYTECH to verify an automotive control system. In O. Maler, editor, *Hybrid and Real-Time Systems*, volume 1201 of *Lecture notes in Computer Science*, pages 139–153. Springer Verlag, 1997.
14. M.L. Tyler and M. Morari: Propositional logic in control and monitoring problems. *Automatica*, in print, 1999.
15. H.P. Williams: *Model Building in Mathematical Programming*. John Wiley & Sons, Third Edition, 1993.

Orthogonal Polyhedra: Representation and Computation*

Olivier Bournez[1], Oded Maler[1], and Amir Pnueli[2]

[1] VERIMAG, Centre Equation, 2, av. de Vignate, 38610 Gières, France
{bournez,maler}@imag.fr
[2] Dept. of Computer Science, Weizmann Inst. Rehovot 76100, Israel
and Univ. Joseph Fourier, Grenoble, France
amir@wisdom.weizmann.ac.il

Abstract. In this paper we investigate *orthogonal polyhedra*, i.e. polyhedra which are finite unions of full-dimensional hyper-rectangles. We define representation schemes for these polyhedra based on their vertices, and show that these compact representation schemes are canonical for all (*convex and non-convex*) polyhedra in *any* dimension. We then develop efficient algorithms for membership, face-detection and Boolean operations for these representations.

1 Introduction and Motivation

Traditionally, most of the applications of computational geometry are concerned with low-dimensional spaces, motivated mainly by problems in graphics, vision and robotics. On the other hand, the analysis of dynamical systems is often done in state-spaces of higher dimension. Since geometry plays an important role in the analysis of such dynamical systems, one would expect that computational geometry will be used extensively in computer-aided design tools for control systems. Although applied mathematicians write algorithms that operate in such spaces (optimization, ODEs, PDEs) the point of view and the concerns are sometimes different from those of mainstream computational geometry. The only notable exception is the treatment of convex polyhedra in linear programming where the points of view of applied mathematics and computational geometry coincide.

Recently, attempts have been made to re-approach computer science and control theory in order to build a theory of *hybrid systems*. These are dynamical systems, defined over both discrete and continuous state variables, intended to model the interaction of computerized controllers with their physical environments (see [AKNS95,M97,HS98] for a representative sample) and to extend the scope of program verification techniques toward continuous systems. One fundamental problem in this domain is the following: *Given a dynamical system*

---

* This work was partially supported by the European Community Esprit-LTR Project 26270 VHS (Verification of Hybrid systems) and the French-Israeli collaboration project 970MAEFUT5 (Hybrid Models of Industrial Plants).

F.W. Vaandrager and J.H. van Schuppen (Eds.): HSCC'99, LNCS 1569, pp. 46–60, 1999.
© Springer-Verlag Berlin Heidelberg 1999

defined by $\dot{x} = f(x)$, where x takes its values in the state-space $\mathbb{R}^d$, and given $P \subseteq \mathbb{R}^d$, calculate (or approximate) the set of points in the state-space reached by trajectories (solutions) starting in P. In [DM98] a method called *face lifting* was proposed, based on previous work of [KM91] [G96]. It consists of restricting P to the class of polyhedra, and iteratively "lifting" the faces of the polyhedra outward according the the maximal value of the normal component of f along the face (see [DM98] for a detailed description and Fig. 1 for an illustration).

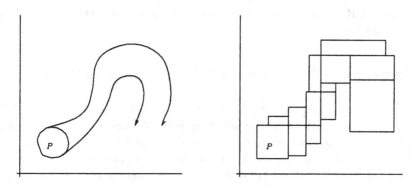

Fig. 1. A dynamical system and trajectories starting at a set P (left) and an approximation of the reachable states by polyhedra (right).

The main computational-geometric burden associated with this approach is related to the representation of intermediate polyhedra (non-convex in general), identifying their faces, decomposing them into convex subsets, and performing face lifting as well as other set-theoretic operations. Due to the complicated structure of high-dimensional non-convex polyhedra, the approach taken in [DM98] consists in restricting the class of subsets to contain only *orthogonal* (axis-parallel, isothetic) polyhedra which can be written as finite unions of full-dimensional hyper-rectangles. A special case of these polyhedra are what was called in [DM98] *griddy* polyhedra, which are generated from unit *hypercubes* with integer-valued vertices. Since arbitrary orthogonal polyhedra can be obtained from griddy ones by appropriate stretching and translation, we restrict our attention to the latter, and use the term orthogonal in order not to introduce additional terminology.

The main contribution of the paper is the definition of several canonical representation schemes for *non-convex* orthogonal polyhedra in *any* dimension. All these schemes are vertex-based and their sizes range between $O(nd)$ and $O(n2^d)$ where n is the number of vertices and d is the dimension. Based on these representations we develop relatively-efficient algorithms for membership, face detection, and Boolean operations on arbitrary orthogonal polyhedra of any dimension. The generalization of these results to more general classes of polyhedra, in particular to *timed polyhedra* used in the verification of timed automata will be reported elsewhere.

Beyond the original motivation coming from computer-aided control system design, we believe that orthogonal polyhedra and subsets of the integer grid are fundamental objects whose computational aspects deserve a thorough investigation.

The rest of the paper is organized as follows: in section 2 we define orthogonal polyhedra and their representation schemes. Section 3 is devoted to algorithms for deciding membership of a point in a polyhedron. In section 4 we discuss face detection and Boolean operations. Finally, in section 5 we mention some future research directions.

2   Orthogonal Polyhedra and Their Representation

We assume that all our polyhedra live inside a bounded subset $X = [0, m]^d \subseteq \mathbb{R}^d$ (in fact, the results will hold also for $X = \mathbb{R}^d_+$). We denote elements of X as $\mathbf{x} = (x_1 \ldots, x_d)$ and the zero and unit vector by $\mathbf{0}$ and $\mathbf{1}$. A d-dimensional grid is a product of d subsets of $\mathbb{N}$. In particular, the *elementary grid* associated with X is $\mathbf{G} = \{0, 1 \ldots, m-1\}^d \subseteq \mathbb{N}^d$. For every point $\mathbf{x} \in X$, $\lfloor \mathbf{x} \rfloor$ is the grid point corresponding to the integer part of the components of $\mathbf{x}$. The grid admits a natural partial order with $(m-1, \ldots, m-1)$ on the top and $\mathbf{0}$ as bottom. The set of subsets of the elementary grid forms a Boolean algebra $(2^{\mathbf{G}}, \cap, \cup, \sim)$ under the set-theoretic operations.

Definition 1 (Orthogonal Polyhedra). *Let $\mathbf{x} = (x_1, \ldots, x_d)$ be a grid point. The* elementary box *associated with $\mathbf{x}$ is a closed subset of X of the form $B(\mathbf{x}) = [x_1, x_1+1] \times [x_2, x_2+1] \times \ldots [x_d, x_d+1]$. The point $\mathbf{x}$ is called the leftmost corner of $B(\mathbf{x})$. The set of boxes is denoted by $\mathbf{B}$. An orthogonal polyhedron P is a union of elementary boxes, i.e. an element of $2^{\mathbf{B}}$.*

One can see that $2^{\mathbf{B}}$ is closed[1] under the following operations:

$$A \sqcup B = A \cup B$$

$$A \sqcap B = cl(int(A) \cap int(B))$$

$$\neg A = cl(\sim A)$$

(where cl and int are the topological closure and interior operations[2]) and that the bijection B between $\mathbf{G}$ and $\mathbf{B}$ which associates every box with its leftmost corner generates an isomorphism between $(2^{\mathbf{G}}, \cap, \cup, \sim)$ and $(2^{\mathbf{B}}, \sqcap, \sqcup, \neg)$. In the sequel we will switch between point-based and box-based terminology according to what serves better the intuition.

---

[1] It is not closed under usual complementation and intersection.

[2] See [Bro83] for definitions.

Definition 2 (Color Function). *Let P be an orthogonal polyhedron. The color function $c : X \rightarrow \{0,1\}$ is defined as follows: If $\mathbf{x}$ is a grid point than $c(\mathbf{x}) = 1$ iff $B(\mathbf{x}) \subseteq P$; otherwise, $c(\mathbf{x}) = c(\lfloor \mathbf{x} \rfloor)$.*

We say that a grid point $\mathbf{x}$ is black (resp. white) and that $B(\mathbf{x})$ is full (resp. empty) when $c(\mathbf{x}) = 1$ (resp. 0). Note that c almost coincides with the characteristic function of P as a subset of X. It differs from it only on right-boundary points (see figure 2).

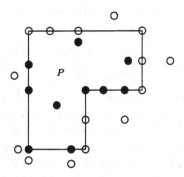

Fig. 2. An orthogonal polyhedron and a sample of the values of the color function it induces.

Definition 3 (Facets and Vertices). *In the following we consider z to be an integer in $[0, m)$, $\mathbf{x} = (x_1, \ldots, x_d)$ and a polyhedron P with a color function c.*

- *The i-predecessor of a point $\mathbf{x}$ is $\mathbf{x}^{i-} = (x_1, \ldots, x_i - 1, \ldots, x_d)$. We use $\mathbf{x}^{ij-}$ as a shorthand for $(\mathbf{x}^{i-})^{j-}$.*
- *An i-hyperplane is a $(d-1)$-dimensional subset $\mathcal{H}_{i,z}$ of X consisting of all points satisfying $x_i = z$.*
- *An i-facet of P is $F_{i,z}(P) = cl\{\mathbf{x} \in \mathcal{H}_{i,z} : c(\mathbf{x}) \neq c(\mathbf{x}^{i-})\}$. We say that elements of $F_{i,z}(P)$ are i-traversed.[3] Note that a facet is an orthogonal polyhedron in $\mathbb{R}^{d-1}$ rather than in $\mathbb{R}^d$.*
- *A vertex is a (non-empty) intersection of d distinct facets. The set of vertices of P is denoted by $V(P)$.*
- *An i-vertex-predecessor of $\mathbf{x}$ is a vertex of the form $(x_1, \ldots, x_{i-1}, z, \ldots, x_d)$, $z \leq x_i$. The first i-vertex-predecessor of $\mathbf{x}$, denoted by $\mathbf{x}^{i\leftarrow}$, is the one with the maximal z.*

When $\mathbf{x}$ has no i-predecessor (resp. i-vertex-predecessor) we write $\mathbf{x}^{i-} = \perp$ (resp. $\mathbf{x}^{i\leftarrow} = \perp$).

---

[3] A facet can be decomposed into two parts according to the orientation, that is, $F_{i,z} = F_{i,z}^+ \cup F_{i,z}^-$ where $F_{i,z}^+ = F_{i,z} \cap \{\mathbf{x} : c(\mathbf{x}) = 0\}$ and $F_{i,z}^- = F_{i,z} \cap \{\mathbf{x} : c(\mathbf{x}) = 1\}$. We call any such F^+ or F^- an *oriented facet*.

One can check that these definitions capture the intuitive meaning of a facet and a vertex and, in particular, that the boundary of an orthogonal polyhedron is the union of its facets. Another useful concept is that of neighborhood:

Definition 4 (Neighborhood). *The neighborhood of a grid point* $\mathbf{x}$ *is the set*

$$\mathcal{N}(\mathbf{x}) = \{x_1 - 1, x_1\} \times \ldots \times \ldots \{x_d - 1, x_d\}$$

(*the vertices of a box lying between* $\mathbf{x} - \mathbf{1}$ *and* $\mathbf{x}$). *For every* i, $\mathcal{N}(\mathbf{x})$ *can be partitioned into left and right* i-*neighborhoods*

$$\mathcal{N}^{i-}(\mathbf{x}) = \{x_1 - 1, x_1\} \times \ldots \times \{x_i - 1\} \times \ldots \times \{x_d - 1, x_d\}$$

and

$$\mathcal{N}^{i}(\mathbf{x}) = \{x_1 - 1, x_1\} \times \ldots \times \{x_i\} \times \ldots \times \{x_d - 1, x_d\}$$

A *representation scheme* for $2^{\mathbf{B}}$ (or $2^{\mathbf{G}}$) is a set $\mathcal{E}$ of syntactic objects such that there is a surjective function ψ from $\mathcal{E}$ to $2^{\mathbf{B}}$ (i.e. every syntactic object represents at most one polyhedron and every polyhedron has at least one corresponding object). If ψ is also an injection we say that the representation scheme is *canonical* (every polyhedron has a unique representation). There are two obvious representation schemes for orthogonal polyhedra. One is the trivial *explicit* representation consisting of an enumeration of the values of c on every grid point, i.e. a d-dimensional zero-one array with m^d entries. The other is the *Boolean representation* based on all the formulae generated from inequalities of the form $x_i \geq z$ via Boolean operations. Clearly this is a representation but not a canonical one even if we restrict formulae to be in disjunctive normal form (a union of hyper-rectangles).

The *vertex representation*, around which this paper is built, consists of the set $\{(\mathbf{x}, c(\mathbf{x})) : \mathbf{x} \text{ is a vertex}\}$, namely the vertices of P along with their color. One of the main results of the paper is that this is indeed a representation scheme for $2^{\mathbf{B}}$ (canonicity is evident). Note that the set of vertices alone is *not* a representation due to ambiguity (see Fig. 3). Also notice that not every set of points and colors is a valid representation of a polyhedron.

We will also use the *neighborhood representation* in which additional information is attached to each vertex, namely the color of all the 2^d points in its neighborhood. Transforming a vertex representation into this one (whose size is $O(n2^d)$) can be performed as a pre-processing stage. Finally we extend the *extreme vertex representation*, which was proposed independently by Aguilera and Ayala in [AA97,AA98] for 3-dimensional orthogonal polyhedra, and show that it is a representation for *any* dimension.

3   Deciding Membership

In this section we show that all the abovementioned representation schemes are valid by providing decision procedures for the *membership* problem: *Given a representation of a polyhedron* P *and a grid point* $\mathbf{x}$, *determine* $c(\mathbf{x})$, *that is, whether* $B(\mathbf{x}) \subseteq P$.

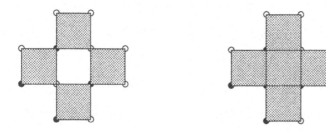

Fig. 3. Two orthogonal polyhedra and their corresponding vertex representations. Note that they have the same set of vertices and only the color of one of the vertices distinguishes one from the other.

3.1 Vertex Representation

Observation 1 (Vertex Rules).
1) A point $\mathbf{x}$ is on an i-facet iff

$$\exists \mathbf{x}' \in \mathcal{N}^i(\mathbf{x}) \text{ s.t. } c(\mathbf{x}'^{i-}) \neq c(\mathbf{x}') \tag{1}$$

2) A point $\mathbf{x}$ is a vertex iff

$$\forall i \in \{1,\ldots,d\} \exists \mathbf{x}' \in \mathcal{N}^i(\mathbf{x}) \text{ s.t. } c(\mathbf{x}'^{i-}) \neq c(\mathbf{x}') \tag{2}$$

3) A point $\mathbf{x}$ is not a vertex iff

$$\exists i \in \{1,\ldots,d\} \forall \mathbf{x}' \in \mathcal{N}^i(\mathbf{x}) \ c(\mathbf{x}'^{i-}) = c(\mathbf{x}') \tag{3}$$

Example: Take $d = 2$ and $\mathbf{x} = (x_1, x_2)$. Then:
$\mathbf{x}$ is on a 1-facet iff $c(x_1 - 1, x_2 - 1) \neq c(x_1, x_2 - 1) \lor c(x_1 - 1, x_2) \neq c(x_1, x_2)$.
It is on a 2-facet iff $c(x_1 - 1, x_2 - 1) \neq c(x_1 - 1, x_2) \lor c(x_1, x_2 - 1) \neq c(x_1, x_2)$.
It is a vertex if both of the above hold and a *non-vertex* if
$c(x_1 - 1, x_2 - 1) = c(x_1, x_2 - 1) \land c(x_1 - 1, x_2) = c(x_1, x_2) \lor$
$c(x_1 - 1, x_2 - 1) = c(x_1 - 1, x_2) \land c(x_1, x_2 - 1) = c(x_1, x_2)$.
This is illustrated in Fig. 4.

Lemma 1 (Color of a Non-Vertex). *Let $\mathbf{x}$ be a non-vertex and let j be a direction such that for every $\mathbf{x}' \in \mathcal{N}^j(\mathbf{x}) - \{\mathbf{x}\}$, $c(\mathbf{x}') = c(\mathbf{x}'^{j-})$. Then $c(\mathbf{x}) = c(\mathbf{x}^{j-})$.*

Proof. Since $\mathbf{x}$ is not a vertex there exists i such that for every $\mathbf{x}' \in \mathcal{N}^i(\mathbf{x})$ $c(\mathbf{x}') = c(\mathbf{x}'^{i-})$. If $j = i$ we are done and $c(\mathbf{x}) = c(\mathbf{x}^{i-})$. Otherwise, we know that not being on an i-facet implies, in particular, $c(\mathbf{x}^{ij-}) = c(\mathbf{x}^{j-})$. In the j-direction we have $c(\mathbf{x}^{ij-}) = c(\mathbf{x}^{i-})$ and using $c(\mathbf{x}^{i-}) = c(\mathbf{x})$ and the transitivity of equality we get $c(\mathbf{x}) = c(\mathbf{x}^{j-})$ (see Fig. 5).   □

Consequently we can calculate the color of a non-vertex $\mathbf{x}$ based on the color of all points in $\mathcal{N}(\mathbf{x}) - \{\mathbf{x}\}$: just find some j satisfying the conditions of Lemma 1 and let $c(\mathbf{x}) = c(\mathbf{x}^{j-})$. This gives immediately a decision procedure for the membership problem:

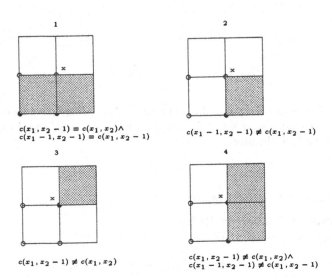

1

$$c(x_1, x_2 - 1) = c(x_1, x_2) \wedge$$
$$c(x_1 - 1, x_2 - 1) = c(x_1, x_2 - 1)$$

2

$$c(x_1 - 1, x_2 - 1) \neq c(x_1, x_2 - 1)$$

3

$$c(x_1, x_2 - 1) \neq c(x_1, x_2)$$

4

$$c(x_1, x_2 - 1) \neq c(x_1, x_2) \wedge$$
$$c(x_1 - 1, x_2 - 1) \neq c(x_1, x_2 - 1)$$

Fig. 4. Some examples of the vertex and facet conditions for a point $x = (x_1, x_2)$: 1) x is not on a 1-facet. 2) and 3) x is on a 1-facet (for different reasons). In these cases it is also on a 2-facet and hence a vertex. 4) The point is on 1-facet but not on a 2-facet

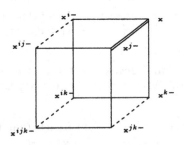

Fig. 5. An illustration of the proof of Lemma 1: horizontal lines indicate equalities in the i direction and dashed lines equalities in the j direction. The equality between $c(x)$ and $c(x^{j-})$ is derived.

Theorem 1 (Membership for Vertex Representation). *The membership problem for vertex representation can be solved in time $O(n^d d 2^d)$ using space $O(n^d)$.*

Proof. We start at $\mathbf{x}$ and call recursively the membership procedure of all the $2^d - 1$ point in $\mathcal{N}(\mathbf{x}) - \{\mathbf{x}\}$. Termination is guaranteed because we go down in the partial-order on $2^{\mathbf{G}}$ and either encounter vertices or reach the origin. We can avoid duplicate calls to the same point by memorizing the visited points and thus visit every point in the grid at most once. This gives an $O(N d 2^d)$ algorithm where N is the size of the grid.

This algorithm is not very efficient because in the worst-case one has to calculate the color of all the grid points between $\mathbf{0}$ and $\mathbf{x}$. We can improve it using the notion of an *induced grid*: let the *i-scale* of P be the set of the i-coordinates of the vertices of P and let the induced grid be the Cartesian product of its i-scales (see Fig. 6). One can see that the induced grid is the smallest (coarsest) grid containing all the vertices, that every rectangle in this grid has a uniform color and that the size of the grid is $O(n^d)$. Hence, calculating the color of a point reduces to finding its closest "dominating" point on the induced grid and applying the algorithm to that grid in $O(n^d d 2^d)$ time.   □

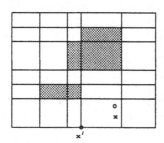

Fig. 6. A polygon, its induced grid, and a point $\mathbf{x}$ dominated by $\mathbf{x}'$.

Corollary 1 (Main Result). *The vertex representation is a canonical representation for orthogonal polyhedra.*

3.2   Neighborhood Representation

By fixing d we now suggest an $O(n \log n)$ membership algorithm for the neighborhood representation, based on successive projections of P into polyhedra of a smaller dimension.

Definition 5 (*i-Slice and i-Section*). *Let P be an orthogonal polyhedron and z an integer in $[0, m)$.*

- An *i-slice* of P, is the d-dimensional orthogonal polyhedron $J_{i,z}(P) = P \sqcap \{\mathbf{x} : z \leq x_i \leq z + 1\}$.
- An *i-section* of P, is the $d-1$-dimensional orthogonal polyhedron $\mathcal{J}_{i,z}(P) = J_{i,z}(P) \cap \mathcal{H}_{i,z}$.

These notions are illustrated in Fig. 7.

Clearly, the membership of $\mathbf{x} = (x_1, \ldots, x_d)$ in P can be reduced into membership in $\mathcal{J}_{i,x_i}(P)$, which is a $(d-1)$-dimensional problem. By successively reducing dimensionality for every i we obtain a point whose color is that of $\mathbf{x}$. We show how the main computational activity, the calculation of i-sections, can be done using the neighborhood representation.

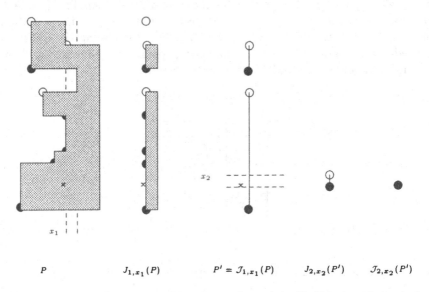

P $\qquad\qquad$ $J_{1,x_1}(P)$ $\qquad\qquad$ $P' = \mathcal{J}_{1,x_1}(P)$ $\qquad$ $J_{2,x_2}(P')$ $\qquad$ $\mathcal{J}_{2,x_2}(P')$

Fig. 7. Calculating the membership of $\mathbf{x} = (x_1, x_2)$ in P: P is transformed via its 1-slice $J_{1,x_1}(P)$, into a 1-section $P' = \mathcal{J}_{1,x_2}(P)$. Then P' is transformed, via its 2-slice $J_{2,x_2}(P')$, into its 2-section $\mathcal{J}_{2,x_2}(P')$ which is a point. The vertices of P which are $\mathbf{x}^{i\leftarrow}$ for some $\mathbf{x} \in \mathcal{H}_{1,x_1}$ are indicated.

Lemma 2 (Vertex of a Section). *Let P be a polyhedron and let P' be its i-section at $x_i = z$. A point $\mathbf{x}$ is a vertex of P' iff $\mathbf{y} = \mathbf{x}^{i\leftarrow} \neq \perp$ and for every $j \neq i$, there exists $\mathbf{x}' \in \mathcal{N}^i(\mathbf{y}) \cap \mathcal{N}^j(\mathbf{y})$ such that $c(\mathbf{x}'^{j-}) \neq c(\mathbf{x}')$. Moreover, when this condition is true, the neighborhood of $\mathbf{x}$ relative to $\mathcal{J}_{i,z}(P)$ is given by $\mathcal{N}^i(\mathbf{y})$.*

Proof. First, observe that $\mathbf{x}$ is a vertex of P' if it satisfies that condition itself, i.e. for every $j \neq i$, there exists $\mathbf{x}' \in \mathcal{N}^i(\mathbf{x}) \cap \mathcal{N}^j(\mathbf{x})$ such that $c(\mathbf{x}'^{j-}) \neq c(\mathbf{x}')$.[4]

Assume $\mathbf{x}$ satisfies the condition. There exists $\mathbf{y} = (x_1, \ldots, x_{i-1}, z, \ldots, x_d)$ such that $c(\mathcal{N}^i(\mathbf{y})) = c(\mathcal{N}^i(\mathbf{x}))$ and $c(\mathcal{N}^{i-}(\mathbf{y})) \neq c(\mathcal{N}^i(\mathbf{y}))$ with z maximal with this property. Since $c(\mathcal{N}^i(\mathbf{y})) = c(\mathcal{N}^i(\mathbf{x}))$, $\mathbf{y}$ satisfies the condition as well and since $c(\mathcal{N}^{i-}(\mathbf{y})) \neq c(\mathcal{N}^i(\mathbf{y}))$, $\mathbf{y}$ is a vertex of P. Since z is maximal with this property we have $\mathbf{y} = \mathbf{x}^{i\leftarrow}$.

Conversely assume $\mathbf{y} = \mathbf{x}^{i\leftarrow}$ exists and it satisfies the condition. Then $c(\mathcal{N}^i(\mathbf{x})) = c(\mathcal{N}^i(\mathbf{y}))$, because otherwise, by the above reasoning, there would be a vertex between $\mathbf{x}$ and $\mathbf{y}$. Hence $\mathbf{x}$ satisfies the condition as well.   □

Theorem 2 (Membership for Neighborhood Represe ıtation). *The membership problem for the neighborhood representation can be solved in time* $O(nd^2(\log n + 2^d))$.

Proof. First observe that it takes $O(nd \log n)$ to determine the vertices $\mathbf{y}$ which are $\mathbf{x}^{i\leftarrow}$ for some $\mathbf{x} \in \mathcal{H}_{i,z}$. There are at most n such points. Using the previous lemma it is possible to determine, using $O(d2^d)$ time, whether each of the corresponding points on $\mathcal{H}_{i,z}$ are vertices of the section. Hence it takes $O(nd(\log n + 2^d))$ to get rid of one dimension, and this is repeated d times until P is contracted into a point.   □

Remark: A similar algorithm with the same complexity can be used to calculate the color of all the points in a *neighborhood* of $\mathbf{x}$ which we describe informally. The algorithm takes double slices (which are d-dimensional thick sections of width two) of P, as illustrated in Fig. 8, and successively reduces P into the neighborhood of $\mathbf{x}$. This variation on the algorithm is used for doing Boolean operations.

3.3 Extreme Vertex Representation

The next representation scheme, inspired by the representation proposed by Aguilera and Ayala [AA97,AA98] for 3-dimensional polyhedra, can be viewed as a compaction of the neighborhood representation. Instead of maintaining all the neighborhood of each vertex, it suffices to keep only the *parity* of the number of black points in that neighborhood — in fact, it suffices to keep only vertices with odd parity. We use $\pi(\mathbf{x})$, $\pi^i(\mathbf{x})$ and $\pi^{i-}(\mathbf{x})$ to denote the parity of the number of black points in $\mathcal{N}(\mathbf{x})$, $\mathcal{N}^i(\mathbf{x})$ and $\mathcal{N}^{i-}(\mathbf{x})$, respectively. We will use the convention $\pi(\bot) = 0$.

Definition 6 (Extreme Points). *A point $\mathbf{x}$ is said to be* extreme *if* $\pi(\mathbf{x}) = 1$.

By enumerating all the possible configurations in dimension 1, 2 and 3, it can be checked that this definition coincides with the geometrical definition presented in [AA97,AA98] for these dimensions.

---

[4] Note the difference from the condition for being a vertex of P: there, the i-coordinates of the $\mathbf{x}'$s can be either z or $z - 1$ but here we insist on z. This is the reason some vertices of P disappear after making a section (see Fig. 7).

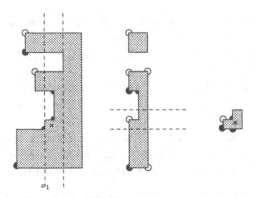

Fig. 8. Calculating the color of a neighborhood of a point.

Observation 2. *Any extreme point* $\mathbf{x}$ *is a vertex.*

Proof. By induction on the dimension d. The assertion for $d = 1$ is immediate. Now in dimension d, choose an arbitrary direction $i \in \{1, \ldots, d\}$. Exactly one of the neighborhoods $\mathcal{N}^{i-}(\mathbf{x})$ and $\mathcal{N}^{i}(\mathbf{x})$ contains an odd number of black points. Assume without loss of generality that it is $\mathcal{N}^{i}(\mathbf{x})$. By induction hypothesis such a neighborhood implies that $\mathbf{x}$ is a vertex in $\mathcal{J}_{i,x_i}(P)$. This means that for every $j \neq i$, there exists $\mathbf{x}' \in \mathcal{N}^{j}(\mathbf{x})$ such that $c(\mathbf{x}'^{j-}) \neq c(\mathbf{x}')$. Since one cannot have $c(\mathbf{x}') = c(\mathbf{x}'^{i-})$ for all $\mathbf{x}' \in \mathcal{N}^{i}(\mathbf{x})$, $\mathbf{x}$ is a vertex of P.   □

The converse is not true and vertices need not be extreme as one can see in Fig. 9.

The *extreme vertex representation* consists in representing an orthogonal polyhedron by the set of its extreme vertices.[5] Note that in dimension 1 all vertices are extreme and hence the vertex and extreme vertex representations practically coincide.

In order to do successive projections on this representation we need a rule, similar to Lemma 2, for determining which points are extreme vertices of an i-section. The following is a corollary of Lemma 2:

Corollary 2. *Let* $\mathbf{x} = (x_1, \ldots, x_{i-1}, z, x_i, \ldots, x_d)$ *be a point and let* $\mathbf{y} = (\mathbf{x}^{i-})^{i\leftarrow}$ *be its (strict) i-vertex-predecessor. Then* $\pi^{i-}(\mathbf{x}) = \pi^{i}(\mathbf{y})$.

Proof. Observation 2 implies that if $\pi^{i-}(\mathbf{x}) = 1$ then $\mathbf{x}^{i-}$ must be a vertex of $\mathcal{J}_{i,z-1}(P)$. By Lemma 2, $\mathcal{N}^{i-}(\mathbf{x}) = \mathcal{N}^{i}(\mathbf{y})$.

Conversely, Observation 2 implies that if $\pi^{i}(\mathbf{y}) = 1$ then for every $j \neq i$, there exists $\mathbf{x}' \in \mathcal{N}^{i}(\mathbf{y}) \cap \mathcal{N}^{j}(\mathbf{y})$ such that $c(\mathbf{x}'^{j-}) \neq c(\mathbf{x}')$. By applying Lemma 2 to $\mathbf{x}^{i-}$ one gets that $\mathcal{N}^{i-}(\mathbf{x})$ must be equal to $\mathcal{N}^{i}(\mathbf{y})$.   □

---

[5] To be more precise, an additional bit for the color of the origin is needed. From this information, the color of all extreme vertices can be inferred.

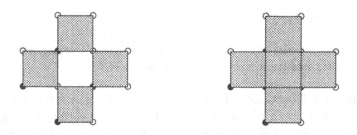

Fig. 9. All the vertices of the two polyhedra are extreme except vertices A, B, C and D.

Note that when $\pi^{i-}(\mathbf{x}) = \pi^i(\mathbf{x}) = 0$, $\mathcal{N}^{i-}(\mathbf{x}) = \mathcal{N}^i(\mathbf{y})$ need not hold.

Lemma 3 (Extreme Vertices of a Section). *Let P be a polyhedron and let $P' = \mathcal{J}_{i,z}(P)$. A point $\mathbf{x}$ is an extreme vertex of P' iff it has an odd number of extreme i-vertex-predecessors.*

Proof. First note that $\mathbf{x}$ is extreme iff $\pi^{i-}(\mathbf{x}) \neq \pi^i(\mathbf{x})$. We prove by induction on the number of vertex predecessors of $\mathbf{x}$. Assume $\mathbf{x}$ has no vertex predecessors. In this case $\pi^{i-}(\mathbf{x}) = 0$ and $\pi^i(\mathbf{x}) = 1$ iff $\mathbf{x}$ is extreme. Suppose it is true for $n - 1$ vertex predecessors and let $\mathbf{x}$ have n strict vertex predecessors $\mathbf{y}^1, \ldots, \mathbf{y}^n$. By the induction hypothesis $\pi^i(\mathbf{y}^n)$ is equal to the number of extreme vertices among $\mathbf{y}^1, \ldots, \mathbf{y}^n$. By Corollary 2, $\pi^i(\mathbf{y}^n) = \pi^{i-}(\mathbf{x})$ and we have $\mathbf{x}$ not extreme if $\pi^i(\mathbf{x}) = \pi^{i-}(\mathbf{x}) = \pi^i(\mathbf{y}^n)$ and $\mathbf{x}$ extreme if $\pi^i(\mathbf{x}) \neq \pi^{i-}(\mathbf{x}) = \pi^i(\mathbf{y}^n)$. In both cases $\pi^i(\mathbf{x})$ coincides with the parity of the number of extreme vertices.   □

One gets immediately:

Corollary 3. *Given an orthogonal polyhedron P and two integers i, z, one can compute in time $O(dn \log n)$ an extreme vertex representation of $\mathcal{J}_{i,z}(P)$.*

Applying the successive projection technique we get:

Theorem 3 (Membership for Extreme Vertex Representation). *The extreme vertex representation is canonical for orthogonal polyhedra in arbitrary dimension and the membership problem for this representation can be solved in time $O(nd^2 \log n)$.*

4   Other Operations

While our representations might be very compact, their usefulness will be measured by how much can algorithms operate on them *without* retrieving the color of every point. As it turns out, face detection is rather simple, and Boolean operations can be performed on neighborhoods of vertices and potential vertices whose number is quadratic in the number of vertices.

4.1   Face Detection

The problem of face detection is the the following: *Given a orthogonal polyhedron P, a direction i and an integer z, calculate the facet $F_{i,z}$.*

Observation 3 (Vertices of Facets). *Let $c_{i,z}$ be the color function of the facet $F_{i,z}(P)$, i.e. $c_{i,z}(\mathbf{x}) = 1$ iff $c(\mathbf{x}^{i-}) \neq c(\mathbf{x})$. Then, $\mathbf{x}$ is a vertex of $F_{i,z}$ iff it is a vertex of P with $x_i = z$ and it satisfies the vertex condition relative to $c_{i,z}$, that is, for every $j \neq i$ there exists $\mathbf{x}' \in \mathcal{N}^j(\mathbf{x}) \cap \mathcal{N}^i(\mathbf{x})$ such that $c_{i,z}(\mathbf{x}'^{j-}) \neq c_{i,z}(\mathbf{x})$.*

For the neighborhood representation one just needs to check the above condition for every vertex of P. Extreme vertices always satisfy the condition and hence one gets:

Theorem 4 (Face Detection). *The face detection problem for orthogonal polyhedra can be done in $O(nd2^d)$ using neighborhood representation and in $O(n)$ using the extreme vertex representation.*

4.2   Boolean Operations

Complementation is trivial for all our representations. Intersection and union are similar and we discuss the first (the second can be performed anyway via de-Morganization). We assume two orthogonal polyhedra P_1 and P_2 with n_1 and n_2 vertices respectively. After intersection some vertices disappear and some new vertices are created (see Fig. 10). However not every point is a candidate to be a vertex of the intersection.

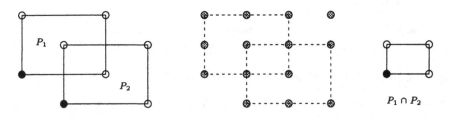

Fig. 10. Intersection of two polyhedra. In the middle one can see all the candidates for being vertices of the intersection.

Lemma 4. *A point $\mathbf{x}$ is a vertex of $P_1 \cap P_2$ only if for every i, $\mathbf{x}$ is on an i-facet of P_1 or on an i-facet of P_2.*

Proof. If there where some i such that $\mathbf{x}$ was not i-traversed in both polyhedra, it remains so after intersection.   □

Lemma 5. *Let* $\mathbf{x}$ *be a vertex of* $P_1 \cap P_2$ *which is not an original vertex, and let* I_1 *(resp.* I_2*) be the set of directions i for which* $\mathbf{x}$ *is on an i-facet of* P_1 *(resp.* P_2*). Then there exists a vertex* $\mathbf{y}^1$ *of* P_1 *and a vertex* $\mathbf{y}^2$ *of* P_2*, such that* $\mathbf{x} = \max(\mathbf{y}^1, \mathbf{y}^2)$ *where* max *is applied coordinate-wise.*

Proof. First we observe that if $\mathbf{x}$ is traversed at directions I_1 in P_1 then there is a vertex $\mathbf{y}$ such that it agrees with $\mathbf{x}$ on all the I_1 coordinates and is smaller than $\mathbf{x}$ in the remaining directions. The same reasoning applies to P_2.   □

From this we can conclude that the candidates for being vertices of $P_1 \cap P_2$ are restricted to the following set:

$$V(P_1) \cup V(P_2) \cup \{\mathbf{x} : \exists \mathbf{y}^1 \in V(P_1)\, \exists \mathbf{y}^2 \in V(P_2) \text{ s.t. } \mathbf{x} = \max(\mathbf{y}^1, \mathbf{y}^2)\}$$

whose number is not greater then $n_1 + n_2 + n_1 n_2$. Combining this with the slicing results we have (assuming $n_1 n_2 >> n_1 + n_2$):

Theorem 5 (Boolean Operations). *The intersection of two orthogonal polyhedra with* n_1 *and* n_2 *vertices can be calculated in time* $O(n_1 n_2 d^2 2^d (n_1 + n_2))$ *using the extreme vertex representation.*

Proof. For every pair of vertices calculate their max as a potential vertex of the intersection. Then compute the color of its neighborhood (if it was not a vertex of P_1 and P_2). Finally calculate point-wise the intersection of the neighborhoods of each point and determine whether or not it is a vertex of $P_1 \cap P_2$ using the standard vertex rules. Note that when the vertices of a given polyhedron are sorted in a lexicographical order as a preprocessing step, it takes $O(nd^2)$ time to determine the color of an arbitrary point.   □

5   Past and Future Directions

Orthogonal polyhedra were studied intensively by research communities such as *Computer Graphics, Solid Modeling, Computational Geometry*, etc. An elaborate survey of these disciplines, their results and methodologies is outside the scope of this paper, but it is fair to say that at least the first two, for obvious reasons, rarely look at dimensions higher than 3. The work reported in [AA97,AA98], which we extended to arbitrary dimension, is the only one we have found relevant to our approach.

We have investigated a representation scheme for orthogonal polyhedra and devised algorithms for the basic operations on them. These algorithms have been implemented and will be integrated into the system described in [DM98]. In this direction, it will be interesting to give a characterization of "typical" orthogonal polyhedra arising from continuous operations, and evaluate the average case complexity of the representation and algorithms on these. Applications of this technique to the analysis of programs with integer variables should be examined as well.

We are currently extending our results to the more general class of polyhedra manipulated by verification and synthesis algorithms for *timed automata*, generated by the (finitely many) elements of the "region graph" [AD94].

References

[AD94]   R. Alur and D.L. Dill, A Theory of Timed Automata, *Theoretical Computer Science* 126, 183–235, 1994.

[AA97]   A. Aguilera and D. Ayala, Orthogonal Polyhedra as Geometric Bounds, in Constructive Solid Geometry, *Proc. Solid Modeling 97*, 1997

[AA98]   A. Aguilera and D. Ayala, Domain Extension for the Extreme Vertices Model (EVM) and Set-membership Classification, *Proc. Constructive Solid Geometry 98*, 1998.

[AKNS95] P. Antsaklis, W. Kohn, A. Nerode and S. Sastry (Eds.), *Hybrid Systems II*, LNCS 999, Springer, 1995.

[Bro83]   A. Brondsted, *An Introduction to Convex Polytopes*, Springer, 1983.

[DM98]   T. Dang, O. Maler, Reachability Analysis via Face Lifting, in T.A. Henzinger and S. Sastry (Eds), *Hybrid Systems: Computation and Control*, LNCS 1386, 96-109, Springer, 1998.

[G96]   M.R. Greenstreet, Verifying Safety Properties of Differential Equations, in R. Alur and T.A. Henzinger (Eds.), *Proc. CAV'96*, LNCS 1102, 277-287, Springer, 1996.

[HS98]   T.A. Henzinger and S. Sastry (Eds), *Hybrid Systems: Computation and Control*, LNCS 1386, Springer, 1998.

[KM91]   R.P. Kurshan and K.L. McMillan, Analysis of Digital Circuits Through Symbolic Reduction, *IEEE Trans. on Computer-Aided Design*, 10, 1350-1371, 1991.

[M97]   O. Maler (Ed.), *Hybrid and Real-Time Systems, Int. Workshop HART'97*, LNCS 1201, Springer, 1997.

A Geometric Approach to Bisimulation and Verification of Hybrid Systems

Mireille Broucke

Deparment of Electrical Engineering and Computer Sciences
University of California, Berkeley CA 94720, USA
mire@eecs.berkeley.edu

Abstract. An approximate verification method for hybrid systems in which sets of the automaton are over-approximated, while leaving the vector fields intact, is presented. The method is based on a geometrically-inspired approach, using tangential and transversal foliations, to obtain bisimulations. Exterior differential systems provide a natural setting to obtain an analytical representation of the bisimulation, and to obtain the bisimulation under parallel composition. We define the symbolic execution theory and give applications to coordinated aircraft and robots.

1 Introduction

We consider a hybrid system which is viewed as a two level system with a finite automaton at the top level and a dynamical system corresponding to each location at the lower level. Former approaches to reachability problems for hybrid systems have taken the view that the initial and final regions, and the enabling conditions and reset conditions are fixed by the problem specification. To obtain a computationally tractable algorithm, a bisimulation is formed with respect to these constraints. Here we are interested in obtaining bisimulations for verification of the safety problem for multiple autonomous agents modeled by their kinematics.

A primary focus of research is to extend the class of hybrid systems that have a finite bisimulation. Two fundamental and potentially compelling questions are: *can a bisimulation of a hybrid system be found analytically?* and, *what geometric structure should the continuous dynamics of the hybrid system possess in order to have a finite bisimulation?* We provide some results on these questions, but our approach is, in general, approximate. In particular, the initial and final regions, enabling conditions, and reset conditions will be approximated so that they are compatible with the bisimulation. This work has been inspired by the papers by Caines [2,3] and the groundbreaking paper of Alur and Dill [1].

1.1 Notation

x' refers to the updated value of a variable x after a transition is taken, and $\dot{x}$ refers to the time derivative. d_H is the Hausdorff metric. $\overline{\sigma} \in \Sigma^*$ refers to a

F.W. Vaandrager and J.H. van Schuppen (Eds.): HSCC'99, LNCS 1569, pp. 61–75, 1999.

finite string of events $\sigma_i \in \Sigma$. All manifolds, vector fields, curves and maps are of class C^∞. Manifolds are assumed to be connected, paracompact, and Hausdorff. $C^\infty(M)$, $\mathcal{X}(M)$, and $\Omega^k(M)$ denote the sets of smooth real-valued functions, smooth vector fields, and k-forms defined on a manifold M. The wedge product of $\alpha, \beta \in \Omega(M)$ is denoted $\alpha \wedge \beta$. $\Omega(M) = \oplus_{k=0}^\infty \Omega^k(M)$ with the wedge product is the exterior algebra on M. $d : \Omega^k(M) \to \Omega^{k+1}(M)$ is the exterior derivative. $\omega \in \Omega^k(M)$ is *exact* if there exists an $\alpha \in \Omega^{k-1}(M)$ such that $\omega = d\alpha$.

2 Hybrid automata

A hybrid automaton is a system $A = (Q, \Sigma, D, Q^0, I, E, J, Q^J)$ consisting of the following components:

State space $Q = L \times M$ consists of a finite set L of control locations and n continuous variables $x \in M$, where M is an n-dimensional differentiable manifold.

Events Σ is a finite observation alphabet.

Vector fields $D : L \to \mathcal{X}(M)$ is a function assigning an autonomous vector field to each location. We will use the notation $D(l) = f_l$. For location l, the dynamics are given by $\dot{x} = f_l(x)$, $f_l \in \mathcal{X}(M)$.

Initial conditions $Q^0 : L \to 2^M$ is a function assigning an initial set of states for each location. If the automaton is started in location l, then $x \in Q^0(l)$ at $t = 0$. We assume $Q^0(l) \subseteq I(l)$.

Invariant conditions $I : L \to 2^M$ is a function assigning for each location an invariant condition on the continuous states. The invariant condition $I(l) \subset M$ restricts the region on which the continuous states can evolve for location l.

Control switches E is a set of control switches. $e = (l, \sigma, l')$ is a directed edge between a source location l and a target location l' with observation $\sigma \in \Sigma$.

Jump conditions $J : E \to G \times R$ is a function assigning to each edge a guard condition and a reset condition. G is the set of guard conditions g on the continuous states, where $g \subset M$ is compact. R is the set of reset conditions r where $r : M \to 2^M$ is a compact set-valued map. We use the notation $G(e) = g_e$ and $R(e) = r_e$, and we assume for each $e = (l, \sigma, l') \in E$, $g_e \subseteq I(l)$, $r_e(g_e) \subseteq I(l')$.

Final condition $Q^J \subset Q$ is a set of final states. We will assume there is one final location so that $Q^J = \{l^J\} \times X^J$, $X^J \subset M$, and we assume $X^J \subseteq I(l^J)$.

Semantics A state is a pair (l, x) satisfying $x \in I(l)$. The invariant can be used to enforce edges from location l. In location l the continuous state evolves according to the vector field f_l. $\Sigma(l)$ will denote the set of events possible at $l \in L$ and $E(l)$ will denote the set of edges possible at $l \in L$. An edge is enabled when the discrete location is l and the continuous state satisfies $x \in g_e$, for $e \in E(l)$. When the transition $e = (l, \sigma, l')$ is taken, the event σ is recorded, the discrete location becomes l', and the continuous state is reset (possibly non-deterministically) to $x' := r_e(x)$.

For $\sigma \in \Sigma$ a σ-step is a tuple $\overset{\sigma}{\rightarrow} \subset Q \times Q$ and we write $q \overset{\sigma}{\rightarrow} q'$. Define $\phi_t^l(x)$ to be a trajectory of f_l at l, starting from x and evolving for time t. For $t \in \mathbb{R}^+$, define a t-step to be the tuple $\overset{t}{\rightarrow} \subset Q \times Q$. We write $(l, x) \overset{t}{\rightarrow} (l', x')$ iff (1) $l = l'$, (2) at $t = 0, x' = x$, and (3) for $t \geq 0, x' = \phi_t^l(x)$, where $\dot{\phi}_t^l(x) = f_l(\phi_t^l(x))$. We will use the label λ to represent a t-step with an arbitrary time passage.

A *trajectory* π of A is a finite or infinite sequence of the form $\pi : q_0 \overset{\tau_0}{\rightarrow} q_1 \overset{\tau_1}{\rightarrow} q_2 \overset{\tau_2}{\rightarrow} \ldots$ where $q_0 \in Q^0$, and for all $i \geq 0$, $q_i \in Q, \tau_i \in \Sigma \cup \mathbb{R}^+$. We assume throughout a *non-zeno* condition: every trajectory of A admits a finite number of σ-steps in any bounded time interval. Finally, given a set of initial states $Q^0 \subseteq Q$, the *reach set* of A, $Reach_A$, is the set of states that can be reached by any trajectory of A.

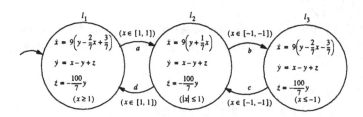

Fig. 1. Double scroll hybrid automaton.

Example 1. Consider the hybrid automata of Figure 1. The invariants for locations l_1, l_2, l_3 are $x \geq 1, |x| \leq 1, x \leq -1$, respectively. The dynamics in each location are either affine linear or linear. It has been shown that this hybrid automaton has a homoclinic orbit and by Shilnikov's theorem the system has a Smale horseshoe implying the existence of a chaotic attractor [4].

Bisimulation A *bisimulation* of A is a binary relation $\simeq \subset Q \times Q$ satisfying the condition that for all states $p, q \in Q$, if $p \simeq q$ and $\sigma \in \Sigma \cup \{\lambda\}$, then
(1) if $p \overset{\sigma}{\rightarrow} p'$, then there exists q' such that $q \overset{\sigma}{\rightarrow} q'$ and $p' \simeq q'$, and
(2) if $q \overset{\sigma}{\rightarrow} q'$, then there exists p' such that $p \overset{\sigma}{\rightarrow} p'$ and $p' \simeq q'$.

3   Verification

In this section we develop an approach to verification that approximates the enabling, reset, initial and final conditions, but leaves the vector fields intact. An equivalence relation that gives a bisimulation is defined, and its existence for a vector field will be shown, in a local sense.

Let K be a subset of an n-dimensional manifold M homeomorphic to the closed, unit n-cube in $\mathbb{R}^n$. For each $l \in L$ we construct a finite cover of K, denoted C_l, consisting of a finite collection of compact n-dimensional cells c_i such that $K = \cup_i^m c_i$. The boundary of each cell consists of a set of $2n$ faces of dimension $(n-1)$ and a collection of edges of dimension $n-2$ to 1 and a set of 2^n vertices. We require $int(c_i) \neq \emptyset$ and $int(c_i) \cap int(c_j) = \emptyset, \forall i \neq j, c_i, c_j \in C_l$.

Let C be such a cover of K. The diameter of $c \in C$ is $\rho(c) = \sup\{d(x,y)| x, y \in c\}$, where d is a Riemannian metric defined on M. The *mesh* of C is $\mu(C) = \sup\{\rho(c)|c \in C\}$. The *resolution* α of C is $\alpha(C) = \inf\{\rho(c)|c \in C\}$. A cover C' refines[1] cover C if $\mu(C') < \alpha(C)$.

Fact *If C has a resolution $\alpha(C) > 0$, then there exists a refinement of C, denoted C' with $\alpha(C') > 0$.*

If V is a closed subset of K, we say $(V)_\mu$ is a *μ-approximation* of V with respect to C with mesh μ, given by

$$(V)_\mu = \{c \in C \mid c \cap V \neq \emptyset, \mu(C) = \mu\}.$$

If $P = \{l\} \times U \subset Q$, then we write $(P)_\mu = \{l\} \times (U)_\mu$.

Fact $d_H(V, (V)_\mu) \leq \mu$.

Let $\mathcal{C}(K) = \{C_l \mid l \in L\}$ be the set of covers of K for automaton A. $\mathcal{C}(K)$ induces an equivalence relation $\simeq$ on Q. We say $q \simeq q'$, where $q = (l, x)$ and $q' = (l', x')$ iff

(1) $l = l'$,

(2) $x \notin K$ iff $x' \notin K$,

(3) if $x, x' \in K$, then $x \in c$ iff $x' \in c$, $\forall c \in C_l$.

We say cover C_l of K at l is a *stable partition of the flow* if for all l, x, x', y and $t \geq 0$, if $(l, x) \simeq (l, x')$ and $y = \phi_t(x)$, then there exists a y' and $t' \geq 0$ such that $y' = \phi_{t'}(x')$ and $(l, y) \simeq (l, y')$.

Suppose we are given a collection of stable partitions $\mathcal{C}(K)$ of $K \subset M$ for hybrid automaton A. We write $\mathcal{C}(K, \mu)$ if $\mu(C_l) = \mu > 0$ for all $l \in L$. We define the approximate hybrid automaton

$$A_\mu = (Q, \Sigma, D, Q_\mu^0, I_\mu, E, J_\mu, Q_\mu^f).$$

Q, Σ, D, and E are unchanged. Q_μ^0, I_μ, J_μ, and Q_μ^f are the μ-approximations of the respective sets. That is,

$$Q_\mu^0(l) = (Q^0(l) \cap K)_\mu,$$
$$I_\mu(l) = (I(l) \cap K)_\mu,$$
$$J_\mu(e) = ((g_e \cap K)_\mu, (r_e)_\mu)$$
$$Q_\mu^f = l^f \times (X^f \cap K)_\mu.$$

Let $e = (l, \sigma, l')$ and $m(x)$ be the number of cells having non-empty intersection with the point x. We define O_x to be the set of points that lie in the same intersection of cells as x. That is,

$$O_x = \left\{ y \in \bigcap_{i=1}^{m(x)} c_i \mid \forall c_i \in C_l \,.\, x \in c_i \right\}. \tag{1}$$

---

[1] We use a nonstandard definition of refinement of covers.

The set-valued map $(r_e)_\mu$ is defined point-wise by

$$(r_e)_\mu(x) = (r_e(O_x) \cap K)_\mu.$$

The modified reset map ensures that the points of O_x are "indistinguishable" after the reset. This operation introduces extra non-determinacy in the approximated model because the identity map is not preserved, in general.

We will say A_μ is an *over-approximation* of A on K if the following additional conditions are satisfied: (1) $Q^0(l) \subseteq K$, each $l \in L$, (2) $g_e, r_e(g_e) \subseteq K$, each $e \in E$, (3) $X^f \subseteq K$, and (4) $I(l) \subseteq K$.

Fact *If A_μ is an over-approximation of A on K, then $Reach_A|_K \subseteq Reach_{A_\mu}|_K$.*

Approximate Verification Problem:
Given hybrid automaton A, $C(K, \mu)$ with $\mu > 0$, and $P \subset L \times K$, determine if $(P)_\mu \cap Reach_{A_\mu} = \emptyset$.

Remarks:

1. If $(P)_\mu \cap Reach_{A_\mu} = \emptyset$ and A_μ is an over-approximation of A on K, then $P \cap Reach_A = \emptyset$. However, if either $(P)_\mu \cap Reach_{A_\mu} \neq \emptyset$ or A_μ is not an over-approximation of A on K, we have no conclusive answer about the original safety problem.

2. If $(P)_\mu \cap Reach_{A_\mu} = \emptyset$ for $\mu > 0$ then for all $\delta < \mu$, $(P)_\delta \cap Reach_{A_\delta} = \emptyset$. Therefore, we can find a coursest μ-approximation A_{μ^*} which verifies that the original system is safe.

Theorem 1 (Stable Partitions). *Given hybrid automaton A and $K \subset M$ homeomorphic to the closed, unit n-cube, suppose there exists $C(K, \mu)$, a collection of stable partitions of K. Then $\simeq$ is a bisimulation for A_μ.*

Proof. Consider first a t-step. Suppose $(l, x) \simeq (l, y)$. Suppose there exists $t_1 \geq 0$ such that $x' = \phi_{t_1}(x)$. By the stability of C_l, there exists $t_2 \geq 0$ and y' such that $y' = \phi_{t_2}(y)$ and $(l, x') \simeq (l, y')$. Next consider a σ-step. Let (l, x) be a state satisfying $x \in (g_e)_\mu$ for some $e = (l, \sigma, l') \in E(l)$ and suppose $(l, x) \simeq (l, y)$. After the σ-step x is reset to some $x' \in (r_e(O))_\mu$. $(l, x) \simeq (l, y)$ implies $y \in (g_e)_\mu$ and $O_y = O_y$. In particular, letting $y' = x'$, we have $y \xrightarrow{\sigma} y'$ and $(l', x') \simeq (l', y')$. Reversing the data in the above two steps provides the converse statements.

3.1 Local existence

Consider a point $q = (l, x), l \in L, x \in M$. We say q is a *regular point* of A if (1) $f_l(x) \neq 0$, and (2) $x \notin \partial g_e, \forall e \in E(l)$. For such a point we can show that "locally" a stable partition for A exists. That is, at regular point q, we can find $\{l\} \times U, U \subset M$, a neighborhood of q, and a partition C_l of $\{l\} \times U$ that gives a bisimulation on $L \times M$. Almost all the interesting behavior of the hybrid system is excluded here, but the intent is to show that locally vector fields have the right structure for bisimulation, and to give the reader a flavor of the more substantial result to come later.

A *first integral* of $\dot{x} = f(x), x \in M$ is a function $g : M \to \mathbb{R}$ satisfying $L_f g = 0$, where $L_f g$ is the Lie derivative of g along f. One can see that that if $\phi : I \to M$ is an integral curve, then $g \circ \phi = c, c \in \mathbb{R}$; that is, integral curves stay on level sets of g.

A bisimulation $\simeq$ of automaton A is said to be a *local bisimulation on* $P \subset L \times M$ *if* $p \notin P$ *and* $p' \notin P$ *together imply* $p \simeq p'$.

Theorem 2 (Local Existence). *Let* $q = (l, x_0)$ *be a regular point of hybrid automaton A. Then there exists* $\{l\} \times U$, *a neighborhood of q, with $U \subset M$ closed, such that if A satisfies*
1) $Q^0 \subseteq \{l\} \times U$,
2) any trajectory of A that leaves $\{l\} \times U$ never returns to it, unless it is reset, then there exists a local bisimulation of A on $\{l\} \times U$.

Proof. By the Flow Box Theorem [9], there exists a closed neighborhood U of x_0 and a diffeomorphism $h : U \to V \subset \mathbb{R}^n$, where $V = [-1, 1]^n$, such that $\dot{x} = f_l(x)$ expressed in $y = h(x)$ coordinates is

$$\dot{y}_1 = 0, \dot{y}_2 = 0, \dots \dot{y}_n = 1. \tag{2}$$

There exist $n - 1$ independent functions $y_1 = c_1, \dots, y_{n-1} = c_{n-1}$ that are first integrals of (2), and they define $(n-1)$ mutually transverse submanifolds, passing through each $y = (c_1, \dots, c_{n-1}, y_n)$. A submanifold transverse to the flow of (2) is given by $y_n = c_n$. Fix $N \in \mathbb{Z}^+$ and define $\Delta = \frac{1}{N} > 0$. Take the subcollection of submanifolds $y_1 = w_1, \dots, y_n = w_n$, where $w_i \in \{0, \pm\Delta, \pm 2\Delta, \dots, \pm 1\}$. Call this collection of submanifolds $S = \{s_\alpha\}$ and let $\overline{U} = U \setminus \cup_\alpha \{s_\alpha\}$. $\overline{U}$ is the union of $(2N)^n$ disjoint open sets $\{c_\beta\}$. Let $\tilde{s}_\alpha = h^{-1}(s_\alpha)$ and $\tilde{c}_\beta = h^{-1}(c_\beta)$.

We can define the equivalence relation $\simeq$ on $L \times M$. For $p = (l, x)$ and $q = (l', x')$, we say $p \simeq q$ iff
1) $l = l'$,
2) $x \notin U$ iff $x' \notin U$,
3) if $x, x' \in U$, then $x \in \tilde{s}_\alpha$ iff $x' \in \tilde{s}_\alpha$ and $x \in \tilde{c}_\beta$ iff $x' \in \tilde{c}_\beta$, $\forall \alpha, \beta$.

$\simeq$ is clearly a local bisimulation on U, using the fact that no trajectories of A enter U without either being initialized there or being reset there. Finally, $\simeq$ by construction, has a finite number of equivalence classes.

4   Construction of bisimulations

In this section we elaborate on the geometric construction suggested in the previous section to show how to derive an analytical representation of the bisimulation. The main geometric tool is foliations. The reader is referred to [7], [12] for background.

Given an n-dimensional manifold M a smooth *foliation* of dimension p or codimension $q = n - p$ is a collection of disjoint connected subsets $F = \{s_\alpha\}$ whose disjoint union forms a partition of M. The foliation satisfies the property that each point of M has a neighborhood U and a system of coordinates y :

$U \to \mathbb{R}^p \times \mathbb{R}^q$ such that for each s_α, the (connected) components of $(U \cap s_\alpha)$ are given by $y_{p+1} = c_1, \ldots, y_{p+q} = c_q$, where $c_i \in \mathbb{R}$. Each connected subset is called a *leaf* of the foliation. We are interested in foliations whose leaves are regular submanifolds of dimension p in M, and we construct the foliations using submersions. A foliation globally defined by a submersion is called *simple*.

Let $f \in \mathcal{X}(M)$. We will define two types of simple co-dimension one foliations with respect to f, called tangential and transversal foliations. For this we require a notion of transversality of foliations.

A map $h : M \to N$ is *transverse* to foliation F of N if either $h^{-1}(F) = \emptyset$, or if for every $x \in h^{-1}(F)$, $h_* T_x M + T_{h(x)} F = T_{h(x)} N$. A submanifold P on M is *transverse* to foliation F of M if the inclusion map $i : P \to M$ is transverse to F. A foliation F' is said to be *transverse* to F if each leaf of F' is transverse to F. A foliation in general does not admit a transversal foliation, but a local submanifold Σ_x of M such that Σ_x intersects every leaf in at most one point (or nowhere) and $T_x \Sigma_x + T_x F = T_x M$ can be found.

A *tangential foliation* F of M is a co-dimension one foliation that satisfies $f(x) \in T_x F, \forall x \in M$; that is, f is a cross-section of the tangent bundle of F. A *transversal foliation* $F_\perp$ of M is a co-dimension one foliation that satisfies $f(x) \notin T_x F, \forall x \in M$. A tangential foliation is therefore an invariant of the flow, whereas integral curves hit the leaves of a transversal foliation transversally.

We construct a collection F_i of $n - 1$ tangential foliations on $K \subset M$ and one transversal foliation $F_n := F_\perp$ on K. Additionally, we require a regularity condition on this collection of n foliations: *each pair of foliations $(F_i, F_j), i \neq j$ is transverse*. If the foliations are constructed via submersions, the following lemma provides an algebraic test for regularity.

Lemma 1. *Let M be an n-dimensional manifold and define $h_i : M \to \mathbb{R}, i = 1, \ldots n$, a collection of submersions on M. If dh_i are linearly independent on $K \subset M$, then the foliations defined by $h^{-1}(\mathbb{R})$ are mutually transverse on K.*

We will not use all of the leaves of a foliation, but only some finite subset of them. We *discretize* a simple co-dimension one foliation as follows. Let $h : M \to \mathbb{R}$ be the submersion of a simple co-dimension one foliation F. Given an interval $[a, b]$, a gridsize $\Delta = \frac{b-a}{N} > 0$ with $N \in \mathbb{Z}^+$, define the finite collection of points $W = \{a, a + \Delta, \ldots, b\}$. Then, $h^{-1}(W)$ is the discretization of F on $h^{-1}([a, b])$.

A bisimulation can be constructed using foliations by elaborating the following steps:

1. Find $(n - 1)$ simple co-dimension one tangential foliations on $K \subset M$, for each $f_l, l \in L$.
2. Construct either a local or global (on K) transversal foliation for each f_l.
3. Check the regularity condition for mutual transversality on K.
4. Discretize the foliations to obtain a cover C_l with mesh μ, for each $l \in L$.
5. Construct the approximate system A_μ by approximating the enabling and reset conditions, and the initial and final regions using C_l for each l.

Theorem 3 (Foliations). *Given hybrid automaton A, $\mu > 0$, and an open $U \subset M$ on which, $\forall l \in L$, $f_l \in \mathcal{X}(M)$ is non-vanishing, suppose there exists a*

*set of $n - 1$ simple, mutually transversal co-dimension one tangential foliations
on U. Then there exists $K \subset M$ homeomorphic to the closed, unit n-cube and a
collection of stable partitions on K such that A_μ has a finite bisimulation.*

Proof. Suppose that the collection of tangential foliations for each l is denoted
$\{F_i\}_{i=1,\ldots,n-1}^l$ and the associated submersions are $h_i^l, i = 1, \ldots, n - 1$. We can
find a closed set $K \subset U$ such that (1) $h_i(K) = [-1, 1]$ (by rescaling h_i, if
needed), and (2) there exists h_n^l independent of $h_i^l, i = 1, \ldots, n - 1$, for each
$l \in L$. Define the coordinates $y_1 = h_1, \ldots, y_n = h_n$. Fix $N \in \mathbb{Z}^+$ and define
$\Delta = \frac{1}{N} > 0$. Take the subcollection of submanifolds $y_1 = w_1, \ldots, y_n = w_n$, where
$w_i \in \{0, \pm\Delta, \pm 2\Delta, \ldots, \pm 1\}$. Call this collection of submanifolds $S = \{s_\alpha\}$ and
let $\overline{K} = K \setminus \cup_\alpha \{s_\alpha\}$. $\overline{K}$ is the union of $(2N)^n$ disjoint open sets $\{c_\beta\}$. Let
$\tilde{s}_\alpha = h^{-1}(s_\alpha)$ and $\tilde{c}_\beta = h^{-1}(c_\beta)$.

As in the Local Existence theorem, we can define the equivalence relation $\simeq$
on $L \times M$. For $p = (l, x)$ and $q = (l', x')$, we say $p \simeq q$ iff
1) $l = l'$,
2) $x \notin K$ iff $x' \notin K$,
3) if $x, x' \in K$, then $x \in \tilde{s}_\alpha$ iff $x' \in \tilde{s}_\alpha$ and $x \in \tilde{c}_\beta$ iff $x' \in \tilde{c}_\beta$, $\forall \alpha, \beta$.

$\simeq$ defines a stable partition on K with a finite number of equivalence classes,
so we can invoke the Stable Partitions Theorem to obtain the bisimulation of
A_μ.

Example [Timed automata] A timed automaton has dynamics, in Pfaffian form
(see section 5), given by $\{dx_1 - dt, \ldots, dx_n - dt\}$. There are $n - 1$ independent
tangential foliations defined by the submersions: $x_1 - x_2 = c_1, \ldots, x_{n-1} - x_n =
c_{n-1}$, where $c_i \in \mathbb{R}$. A transversal foliation is $x_n = d_n$ though the partition of [1]
uses more transversal foliations because of the nature of the enabling and reset
conditions: $x_1 = d_1, \ldots, x_n = d_n$. Each of the leaves of the transversal foliations
are transverse to every integral curve. The partition for timed automata is exact,
in the sense that it is not necessary to over-approximate regions.

Example [Brunovsky normal form] Consider the Brunovsky normal form for
linear systems in $\mathbb{R}^4$ given by

$$\dot{x}_1 = x_2$$
$$\dot{x}_2 = x_3$$
$$\dot{x}_3 = x_4$$
$$\dot{x}_4 = u.$$

The three tangential foliations are

$$x_1 - \frac{x_2 x_4}{u} + \frac{x_3 x_4^2}{2u^2} - \frac{x_4^4}{8u^3} = c_1$$

$$x_2 - \frac{x_3 x_4}{u} + \frac{x_4^3}{3u^2} = c_2$$

$$x_3 - \frac{1}{2u} x_4^2 = c_3.$$

A transversal foliation is $x_4 = c_4$. We confirm the regularity condition on the foliations by checking the rank of the matrix:

$$Dh = \begin{bmatrix} 1 & -\frac{x_4}{u} & \frac{x_4^2}{2u^2} & -\frac{x_2}{u} + \frac{x_3 x_4}{u^2} - \frac{x_4^3}{2u^3} \\ 0 & 1 & -\frac{x_4}{u} & -\frac{x_3}{u} + \frac{x_4^2}{u^2} \\ 0 & 0 & 1 & -\frac{x_4}{u} \\ 0 & 0 & 0 & 1 \end{bmatrix}.$$

This matrix is full rank for all $u \neq 0$; therefore, the partition is defined on all $\mathbb{R}^4$.

4.1 Topological conjugacy

Two vector fields f and g are *topologically conjugate* if there exists a homeomorphism $h : M \to N$, and h takes integral curves ϕ_t of f to integral curves ψ_t of g while preserving the parameter t. In particular, $h \circ \phi_t(x) = \psi_t(h(x))$ and $g = h_* f$. Suppose we have constructed a set of tangential and transversal foliations $\{F_1, \ldots, F_{n-1}, F_n = F_\perp)$ of $K \subset M$ for f.

Theorem 4. *Suppose f and g are topologically conjugate vector fields with homeomorphism $h : M \to N$ and the set of foliations $\{F_i\}$ defines a stable partition on $l \times K, K \subset M$ for f. Then there exists a stable partition on $l \times h(K), h(K) \subset N$ for g.*

Proof. Suppose each foliation F_i is constructed by submersion $\xi_i : M \to \mathbb{R}$. Define the set of foliations $\{G_i\}$ constructed by submersions $\eta_i = \xi_i \circ h^{-1} : N \to \mathbb{R}$. Then note that $L_g \eta_i = d(\xi_i \circ h^{-1})(h_* f) = d\xi_i \cdot f = L_f \xi_i$. Therefore, η_i form $(n-1)$ tangential foliations and one transverse foliation for g, and if ξ_i are independent, then so are η_i. Finally, the homeomorphism h maps fixed points of f to fixed points of g, so a stable partition defined on K for f non-vanishing on K, is well-defined for $h(K)$ and g is non-vanishing on $h(K)$.

5 Exterior differential systems

Tangential foliations of a vector field can be found using first integrals. A natural setting for finding first integrals is provided by exterior differential systems. The reader is referred to [10, 12] for background.

A set of independent one-forms $\omega^1, \ldots, \omega^q$ generates a Pfaffian system $I = \{\omega^1, \ldots, \omega^q\} = \{\sum f_k \omega^k | f_k \in C^\infty(M)\}$. The Frobenius theorem says that if I satisfies the Frobenius condition $d\omega^k \wedge \omega^1 \wedge \cdots \wedge \omega^q = 0$, for $k = 1, \ldots, q$, then it admits coordinates $h_1, \ldots, h_q$ such that $I = \{dh_1, \ldots, dh_q\}$. In this case the Pfaffian system is said to be *completely integrable* and the h_i are the first integrals of I. We adapt the proof of the Frobenius theorem to obtain our main result on existence of bisimulations.

Theorem 5 (First Integrals). *Given hybrid automaton A, $\mu > 0$, and an open $U \subset M$ on which, $\forall l \in L$, $f_l \in \mathcal{X}(M)$ is non-vanishing, there exists $K \subset M$ homeomorphic to the closed, unit n-cube and a collection of stable partitions such that A_μ has a finite bisimulation.*

Proof. The approach is to find a codistribution of one-forms $\{w^2, \ldots, w^n\}$ such that $w^i = dh_i = 0$. Then we will show that the $n-1$ independent functions $h_i : K \to \mathbb{R}$ are submersions and by construction first integrals. They will provide $n-1$ simple, co-dimension one tangential foliations, so we can invoke the Foliations theorem to show existence of a bisimulation.

Fix l, and let $f_1 = f_l$. On some open $V \subset U$ we can find $n-1$ smooth complementary vector fields $f_2, \ldots, f_n$ such that $span\{f_1, \ldots, f_n\} = \mathbb{R}^n$ at each $x \in V$ and $\{f_1, \ldots, f_n\}$ is clearly involutive on V. Let $\phi_t^i(x)$ be the flow of f_i. Fix $x^0 \in V$. There exists W, a neighborhood of 0 in $\mathbb{R}^n$ such that the map $G : W \to V$ given by

$$(a_1, \ldots, a_n) \mapsto \phi_{a_1}^1 \circ \cdots \circ \phi_{a_n}^n(x^0).$$

is well defined. Since the ϕ's commute, we can change the order of integration

$$\left(\frac{\partial G}{\partial a_i}\right)_0 = \frac{\partial}{\partial a_i} \phi_{a_i}^i \circ \phi_{a_1}^1 \circ \cdots \circ \phi_{a_{i-1}}^{i-1} \circ \phi_{a_{i+1}}^{i+1} \circ \cdots \circ \phi_{a_n}^n(x^0)$$
$$= f_i(x^0).$$

Since the f_i's are independent, $\frac{\partial G}{\partial a_i}$ is nonsingular, so G^{-1} exists locally on $V' \subset V$ by the Inverse Function Theorem. Let $[h_1(y), \ldots, h_n(y)]^T = G^{-1}(y), y \in V'$. By definition

$$\left[\frac{\partial G^{-1}}{\partial y}\right] \cdot \left[\frac{\partial G}{\partial a}\right] = I.$$

In particular,

$$\frac{\partial h_i}{\partial y} \cdot f_1 = 0$$

for $i = 2, \ldots, n$. So $h_2, \ldots, h_n$ are the desired functions. Since $G^{-1}(y)$ has rank n, the h_i are independent submersions.

Remark: The map G is nonsingular everywhere that $\{f_i\}$ are a complementary, involutive collection of vector fields, and V' is as large as the range of G.

5.1   Parallel composition

Bisimulation for hybrid systems is, in general, not closed under parallel composition of automata. Here we give a sufficient condition on the Pfaffian form of the continuous dynamics of each control location so that if two hybrid automata have a finite bisimulation, then so does their parallel composition. We refer the reader to [6] for the definition of composition of hybrid automata.

Theorem 6 (Parallel Composition). *Given hybrid automata* $A_1 = (L_1 \times M_1^n, \Sigma_1, D_1, Q_1^0, I_1, E_1, J_1, Q_1^f)$ *and* $A_2 = (L_2 \times M_2^m, \Sigma_2, D_2, Q_2^0, I_2, E_2, J_2, Q_2^f)$, *suppose there exist* $K_1 \subset M_1, K_2 \subset M_2$ *such that, via the First Integrals theorem, bisimulations for* $A_{1\mu}$ *and* $A_{2\mu}$ *exist. If for each pair* $(l, l'), l \in L_1, l' \in L_2$ *there exists a one-form of the Pfaffian system at* l

$$h(dx_1, \ldots, dx_n) - dt = 0,$$

and a one-form of the Pfaffian system at l'

$$h'(dx_{n+1}, \ldots, dx_{n+m}) - dt = 0,$$

such that the one-form

$$h(dx_1, \ldots, dx_n) - h'(dx_{n+1}, \ldots, dx_{n+m}) = d\alpha$$

is exact, and α *is independent of the first integrals on* K_1 *and* K_2 *of the vector fields at* l *and* l', *respectively, then a bisimulation of* $(A_1 \times A_2)_\mu$ *exists.*

Proof. ¿From the First Integrals theorem, we have $n - 1$ first integrals for each $f_l, l \in L_1$ and $m - 1$ first integrals for each $f_{l'}, l' \in L_2$, giving $n + m - 2$ first integrals for the vector field $f = [f_l \quad f_{l'}]^T$. But we require $n + m - 1$ first integrals to construct the bisimulation. The missing first integral is provided by the exact form α. Using the fact that $h(dx_1, \ldots, dx_n)$ has the form $\frac{dx_i}{f_i(x)}$ for some $i = 1, \ldots, n$, and similarly for h', it can be verified that α satisfies $L_f \alpha = 0$.

6 Applications

A domain of models that we wish to apply this theory to is kinematic models of rigid bodies. The symmetry in kinematics allows first integrals to be constructed. We demonstrate the ideas with several examples.

Example [Planar Aircraft] Consider the coordination problem of two aircraft A and B flying at a fixed altitude near an airport [11]. Each aircraft is modeled by a hybrid system in which an automaton location corresponds to an atomic maneuver performed with constant control inputs. The control inputs are changed instantaneously upon switching control locations. The state is $g \in SE(2)$ and X is an element of the Lie algebra $se(2)$. Assuming the aircraft does not exercise it's pitch control, the kinematic dynamics of aircraft A are given by $\dot{g} = gX$ where

$$g = \begin{bmatrix} \cos\phi & -\sin\phi & x \\ \sin\phi & \cos\phi & y \\ 0 & 0 & 1 \end{bmatrix}$$

and

$$X = \begin{bmatrix} 0 & -u_1 & u_2 \\ u_1 & 0 & 0 \\ 0 & 0 & 0 \end{bmatrix}.$$

ϕ is the yaw angle, and the inputs u_1, u_2 control the yaw and velocity, respectively, of the aircraft. There are two tangential foliations given by equations

$$u_1 x - u_2 \sin \phi = c_x$$
$$u_1 y + u_2 \cos \phi = c_y$$

and a transversal foliation given by $\phi = c_\phi$. Letting the state variables and inputs of aircraft B be ϕ_B, x_B, y_B, u_{1B}, and u_{2B}, analogous expressions for the tangential and transversal foliations are obtained for aircraft B. An additional tangential foliation is found for the parallel composition of the two systems given by

$$u_{1B}\phi_A - u_{1A}\phi_B = c_{AB}.$$

We check the regularity condition on the five tangential foliations and either of the two transversal foliations. Namely,

$$Dh = \begin{bmatrix} u_{1A} & 0 & -u_{2A}\cos\phi_A & 0 & 0 & 0 \\ 0 & u_{1A} & -u_{2A}\sin\phi_A & 0 & 0 & 0 \\ 0 & 0 & u_{1B} & 0 & 0 & -u_{1A} \\ 0 & 0 & 0 & u_{1B} & 0 & -u_{2B}\cos\phi_B \\ 0 & 0 & 0 & 0 & u_{1B} & -u_{2B}\sin\phi_B \\ 0 & 0 & 0 & 0 & 0 & 1 \end{bmatrix}.$$

This matrix has full rank so long as $u_{1A}, u_{1B} \neq 0$, so the partition is defined globally on $\mathbb{R}^4 \times \mathbf{T}^2$. If, in addition, $\frac{u_{1A}}{u_{1B}}$ is rational, a finite bisimulation on $K \times \mathbf{T}^2$, for compact $K \subset \mathbb{R}^4$, exists.

Example [Mobile robot]   Consider the coordination problem of two mobile robots A and B, operating in a closed workspace. The robots are modeled using hybrid automata, with each control location corresponding to an atomic maneuver, such as "move forward", or "change direction". Each location of the automaton has a kinematic model of the associated maneuver using constant control inputs. The control input changes instantaneously upon switching locations. The kinematic model for each robot, converted to chained form [8] is the following:

$$\dot{x}_1 = u_1$$
$$\dot{x}_2 = u_2$$
$$\dot{x}_3 = x_2 u_1$$
$$\dot{x}_4 = x_3 u_1.$$

There are three tangential foliations given by the equations

$$x_2 - \frac{u_2}{u_1} x_1 = c_2$$

$$x_3 - \frac{u_1}{2u_2} x_2^2 = c_3$$

$$x_4 + \frac{1}{3}\left(\frac{u_1}{u_2}\right)^2 x_2^3 - \frac{u_1}{u_2} x_2 x_3 = c_4.$$

and a transversal foliation given by: $x_1 = c_1$.

To show these foliations define a bisimulation for each robot, we must check the regularity condition:

$$Dh = \begin{bmatrix} 1 & 0 & 0 & 0 \\ -\frac{u_2}{u_1} & 1 & 0 & 0 \\ 0 & -\frac{u_1}{u_2}x_2 & 1 & 0 \\ 0 & -\frac{u_1}{u_2}x_3 + \left(\frac{u_1}{u_2}\right)^2 x_2^2 - \frac{u_1}{u_2}x_2 & 1 \end{bmatrix}.$$

This matrix has full rank so long as $u_1 \neq 0$ and $u_2 \neq 0$. Thus, the partition for each robot is defined globally on $\mathbb{R}^4$.

When we take their parallel composition, an extra tangential foliation is introduced:

$$u_{1B}x_{1A} - u_{1A}x_{1B} = c_{AB}.$$

A calculation similar to the previous example shows that a bisimulation for the parallel composition exists.

Example [Linear systems] Finally, we consider a hybrid automaton in which each location of the automaton contains an affine linear system. The dynamics of each location are given by:

$$\dot{x}_i = \lambda_i x_i + b_i, \quad i = 1, \ldots, n$$

where $\lambda_i, b_i \in \mathbb{R}$. We assume for each i that λ_i, b_i are not both zero. The tangential folations are

$$\frac{1}{\lambda_1} \ln |\lambda_1 x_1 + b_1| - \frac{1}{\lambda_2} \ln |\lambda_2 x_2 + b_2| = c_1$$

$$\vdots$$

$$\frac{1}{\lambda_{n-1}} \ln |\lambda_{n-1}x_{n-1} + b_{n-1}| - \frac{1}{\lambda_n} \ln |\lambda_n x_n + b_n| = c_{n-1}.$$

A transversal foliation is given by

$$\frac{1}{2\lambda_1}|\lambda_1 x_1 + b_1|^2 + \frac{1}{2\lambda_2}|\lambda_2 x_2 + b_2|^2 + \cdots + \frac{1}{2\lambda_n}|\lambda_n x_n + b_n|^2 = c_n.$$

We check the regularity condition as follows:

$$Dh = \begin{bmatrix} \frac{1}{|\lambda_1 x_1 + b_1|} & -\frac{1}{|\lambda_2 x_2 + b_2|} & 0 & \cdots & 0 \\ 0 & \frac{1}{|\lambda_2 x_2 + b_2|} & -\frac{1}{|\lambda_3 x_3 + b_3|} & \cdots & 0 \\ 0 & 0 & 0 & & 0 \\ \vdots & \vdots & & & \vdots \\ 0 & 0 & \frac{1}{|\lambda_{n-1}x_{n-1}+b_{n-1}|} & -\frac{1}{|\lambda_n x_n + b_n|} \\ |\lambda_1 x_1 + b_1| & |\lambda_2 x_2 + b_2| & \cdots & |\lambda_{n-1}x_{n-1} + b_{n-1}| & |\lambda_n x_n + b_n| \end{bmatrix}.$$

After some algebraic manipulation, we can show this matrix has full rank so long as $x_1 \neq -\frac{b_1}{\lambda_1}, \ldots, x_n \neq -\frac{b_n}{\lambda_n}$; that is, we avoid a set of hyperplanes. This divides $\mathbb{R}^n$ into quadrants where the bisimulation can be constructed.

7   Symbolic execution theory

In this section we consider the implementation of the theory of approximate verification in a symbolic model checking algorithm.

A theory $\mathcal{T}$ of A is a set of predicates that are assigned truth values by the states of A. We write $[p] \in Q$ for the set of states that satisfy predicate p. $\langle R \rangle$ denotes the set of formulas of $\mathcal{T}$ that define a region $R \subset Q$. A theory is *decidable* if it can be decided for each predicate p of $\mathcal{T}$ whether $[p]$ is empty. The theory $\mathcal{T}$ permits the symbolic analysis of A if (1) $\mathcal{T}$ is decidable, (2) $\mathcal{T}$ is closed under boolean operations and Pre and $Post$ operations, and (3) $\langle Q^f \rangle \in \mathcal{T}$, $\langle Q^0(l) \rangle \in \mathcal{T}, l \in L$.

Suppose the tangential and transversal foliations on K for each $l \in L$ are defined by submersions $h_i^l(x) = c_i$. Let $\mathcal{S}$ be the class of formulas

$$h_i^l(x) \ \% \ c_i$$

with $c_i \in \mathbb{R}$, $\% = \{\leq, <, =, >, \geq\}$, $l \in L$, $i = 1, \dots, n$, and all finite conjunctions and disjunctions of these expressions. A finite automaton with its symbolic execution theory is said to be *effectively presented* [5].

Theorem 7. A_μ *with the theory* $\mathcal{S}$ *is effectively presented.*

Proof. $\mathcal{S}$ is a symbolic execution theory of A_μ. For (1) the regions Q_μ^0, I_μ, J_μ, and Q_μ^f in $L \times K$ can be represented by formulas in $\mathcal{S}$, (2) $\langle Pre(R) \rangle \in \mathcal{S}$ and $\langle Post(R) \rangle \in \mathcal{S}$ for $\langle R \rangle \in \mathcal{S}$ by construction, and (3) $\mathcal{S}$ is decidable. Consider an atomic formula $\psi(x)$ for a closed region: $\exists x.(c_1 \leq h_1(x) \leq d_1) \wedge \cdots \wedge (c_n \leq h_n(x) \leq d_n)$. $\psi(x)$ is equivalent to the quantifier free expression $(c_1 \leq d_1) \wedge \cdots \wedge (c_n \leq d_n)$.

8   Critique and future work

This paper opens up avenues for applying model checking algorithms to the verification of safety problems for hybrid systems consisting of coordinating autonomous agents, and especially hybrid systems where the continuous level is a kinematic model. Model checking may provide a vast improvement in efficiency over simulation-based approaches for validating hybrid system performance, though potential gains may not be as great as those reported for model checking of circuit designs and protocols.

There are some limitations and obstacles to be overcome. First, it is likely that model checking will still be a computationally expensive tool. Initially, the number of autonomous agents will be small and the continuous dynamics will be low- dimensional, at least until further breakthroughs appear on this frontier. The approach becomes more interesting when more of the "burden of control" can be placed at the logic level. Some work that remains to be done is obtaining the approximate automaton automatically, given the analytical representation of its bisimulation.

The paper suggests some areas for future investigation. First, the paper develops a local geometric theory of bisimulation. A global theory is needed. The most promising approach is to use symmetry to obtain global first integrals. Also, a theory of robustness of hybrid systems is needed in light of the approximations that are introduced to complete the verification. We plan to report on these directions in future papers.

Acknowledgments The author is grateful to André de Carvalho, Tom Henzinger, and Charles Pugh, for their insights and many valuable discussions, and to Peter Caines, whose talks at Berkeley initiated this investigation.

References

1. R. Alur and D. L. Dill. Automata for modeling real-time systems. In *"Proc. 17th ICALP: Automata, Languages and Programming*, LNCS 443, Springer-Verlag, 1990.
2. P. Caines and Y. Wei. The hierarchical lattices of a finite machine. *Systems and Control Letters*, vol. 25, no. 4, pp. 257-263, July, 1995.
3. P. Caines and Y. Wei. On dynamically consistent hybrid systems. In P. Antsaklis, W. Kohn, A. Nerode, eds., *Hybrid Systems II*, pp. 86-105, Springer-Verlag, 1995.
4. L.O. Chua, M. Komuro, and T. Matsumoto. The double scroll family - part I: rigorous proof of chaos. *IEEE Transactions on Circuits and systems* vol. 33, no. 11, pp. 1072-1097, November, 1986.
5. T. Henzinger. Hybrid automata with finite bisimulations. In *"Proc. 22nd ICALP: Automata, Languages and Programming*, LNCS 944, pp. 324-335, Springer-Verlag, 1995.
6. T. Henzinger. The theory of hybrid automata. In *Proc. 11th IEEE Symposium on Logic in Computer Science*, pp. 278-292, New Brunswick, NJ, 1996.
7. H. B. Lawson. The Quantitative theory of foliations. *Regional Conference Series in Mathematics*, no. 27. American Mathematical Society, Providence, 1977.
8. R. Murray and S. Sastry. Nonholonomic motion planning: steering using sinusoids. *IEEE Transactions on Automatic Control*, vol.38, no.5, pp. 700-16, May, 1993.
9. J. Palis and W. de Melo. *Geometric Theory of Dynamical Systems: an Introduction*. Springer-Verlag, New York, 1982.
10. W. Sluis. *Absolute Equivalence and its Applications to Control Theory*. Ph.D. thesis, University of Waterloo, 1992.
11. C. Tomlin, G. Pappas, J. Lygeros, D. Godbole, and S. Sastry. Hybrid Control Models of Next Generation Air Traffic Management. In P. Antsaklis, W. Kohn, A. Nerode, and S. Sastry, eds., *Hybrid Systems IV*, LNCS 1273, pp. 378-404, Springer-Verlag, 1997.
12. F. Warner. *Foundations of Differential Manifolds and Lie Groups*. Springer-Verlag, New York, 1983.

Verification of Polyhedral-Invariant Hybrid Automata Using Polygonal Flow Pipe Approximations

Alongkrit Chutinan and Bruce H. Krogh

Department of Electrical and Computer Engineering
Carnegie Mellon University
Pittsburgh, PA 15213-8390
ac4c@andrew.cmu.edu/krogh@ece.cmu.edu

Abstract. This paper presents a computational technique for verifying properties of hybrid systems with arbitrary continuous dynamics. The approach is based on the computation of approximating automata, which are finite-state approximations to the (possibly infinite-state) discrete-trace transition system for the hybrid system. The fundamental computation in the generation of approximating automata is the mapping of sets of continuous states to the boundaries of the location invariants. This mapping is computed by intersecting flow pipes, the sets of reachable states for continuous systems, with the invariant boundaries. Flow pipes are approximated by sequences of overlapping convex polygons. The paper presents an application of the computational procedure to a benchmark hybrid system, a batch evaporator.

1 Introduction

Hybrid system behaviors can be described by an infinite-state transition system [7]. A standard approach to verifying properties of a hybrid system is to find an equivalent transition system called a *bisimulation* with a finite number of states [10,11]. There are two principle difficulties with this approach. First, a finite-state bisimulation exists only for certain classes of hybrid systems [6,10,11]. Since a finite-state bisimulation may not exist in general, one cannot guarantee a procedure for computing a bisimulation will terminate. The second problem is that any procedure for computing a bisimulation for a hybrid system requires the computation of *flow pipes*, that is, the collection of continuous-time trajectories emanating from a set of initial states [12]. Although flow pipes can be computed exactly in certain cases [1,3,10,11], sets of continuous trajectories can only be approximated numerically for most continuous systems. Moreover, the errors in these approximations typically grow with simulation time, making them too conservative to obtain meaningful results.

This paper concerns an approach to verification that addresses both of these problems. The first problem is dealt with by verifying finite-state transition systems that are conservative approximations (i.e. simulations), rather than bisimulations, of the infinite-state transition system. Previous papers have presented algorithms for computing finite-state transition systems for hybrid systems, called

F.W. Vaandrager and J.H. van Schuppen (Eds.): HSCC'99, LNCS 1569, pp. 76–90, 1999.
© Springer-Verlag Berlin Heidelberg 1999

approximating automata [2–4]. In this paper we observe that the algorithm for constructing and refining approximating automata can be viewed as a modification of the iterative procedure for computing a bisimulation. Although it cannot be guaranteed the verification question can be resolved using approximating automata, there are cases where verification can be accomplished with this approach even when a finite-state bisimulation does not exist [2–4].

The second problem is addressed by computing polygonal approximations to flow pipes for continuous dynamic systems. The procedure for computing flow pipe approximations from [4] is summarized briefly in this paper. An attractive feature of this procedure is that the approximation error does not grow with the simulation time. We show how the flow pipe approximations are used in the algorithm for computing approximating automata. The complete procedure is illustrated for the benchmark hybrid system problem from [8].

2   Transition Systems and Bisimulations

This section introduces the formal definitions and procedures used for verifying properties of hybrid systems. The fundamental abstract structure for representing the system dynamics is the transition system defined as follows.

Definition 1. A *transition system* T is a 3-tuple $T = (Q, \rightarrow, Q_0)$ where Q is the set *states*, $\rightarrow \subseteq Q \times Q$ is the set of *transitions*, and $Q_0 \subseteq Q$ is the set of *initial states*.

Given q and $q' \in Q$, the notation $q \rightarrow q'$ indicates that $(q, q') \in \rightarrow$. Sets of valid trajectories for a transition system are defined in the obvious way for sequences of states. Trajectories can be of finite length or infinite length.

Definition 2. (Pre/Postcondition Sets) Given a transition system $T = (Q, \rightarrow, Q_0)$, and a set $P \subseteq Q$, the *precondition* of P, denoted $Pre(P)$, is defined as $Pre(P) = \{q \in Q \mid \exists p \in P, q \rightarrow p\}$; the *postcondition* of P, denoted $Post(P)$, is defined as $Post(P) = \{q \in Q \mid \exists p \in P, p \rightarrow q\}$.

Definition 3. (Simulation) Let $T_1 = (Q_1, \rightarrow_1, Q_{01})$ and $T_2 = (Q_2, \rightarrow_2, Q_{02})$ be transition systems. A *simulation relation* of T_1 by T_2 is a binary relation $\preceq \subseteq Q_1 \times Q_2$ such that:

i. If $q_1 \preceq q_2$ and $q_1 \rightarrow_1 q_1'$, then there exists q_2' such that $q_2 \rightarrow_2 q_2'$ and $q_1' \preceq q_2'$;
ii. For each $q_1 \in Q_{01}$, there exists $q_2 \in Q_{02}$ such that $q_1 \preceq q_2$.

We say T_2 simulates T_1, denoted $T_1 \preceq T_2$, if there exists a simulation relation of T_1 by T_2. T_2 is also called a simulation of T_1.

Definition 4. (Bisimulation) Let $T_1 = (Q_1, \rightarrow_1, Q_{01})$ and $T_2 = (Q_2, \rightarrow_2, Q_{02})$ be transition systems. A *bisimulation relation* between T_1 and T_2 is a binary relation $\equiv \subseteq Q_1 \times Q_2$ such that $\equiv$ is a simulation relation of T_1 by T_2 and $\equiv^{-1} \subseteq Q_2 \times Q_1$ is a simulation relation of T_2 by T_1.

We say the transition systems T_1 and T_2 bisimulate each other, denoted $T_1 \equiv T_2$, if there exists a bisimulation relation between T_1 and T_2. An approach to finding a bisimulation of a transition system uses *quotient transition systems*, defined as follows.

Definition 5. (Quotient Transition System) [6,10,11] Given a transition system $T = (Q, \rightarrow, Q_0)$ and a partition $\mathcal{P}$ of Q, the *quotient transition system* of T is defined as $T/\mathcal{P} = (\mathcal{P}, \rightarrow_\mathcal{P}, Q_0/\mathcal{P})$, where for all $P, P' \in \mathcal{P}$, $P \rightarrow_\mathcal{P} P'$ iff there exist $q \in P$ and $q' \in P'$ such that $q \rightarrow q'$, or equivalently $Post(P) \cap P' \neq \emptyset$, and $Q_0/\mathcal{P} = \{P \in \mathcal{P} \mid P \cap Q_0 \neq \emptyset\}$.

It is easy to see that the relation $\preceq = \{(q, P) \in Q \times \mathcal{P} \mid q \in P\}$ is a simulation relation of T by $T/\mathcal{P}$. Therefore, $T \preceq T/\mathcal{P}$. The quotient system $T/\mathcal{P}$ is a bisimulation of T with the relation $\equiv = \{(q, P) \in Q \times \mathcal{P} \mid q \in P\}$ if and only if the partition $\mathcal{P}$ satisfies

$$\forall P, P' \in \mathcal{P}, \text{ either } P \cap Pre(P') = \emptyset \text{ or } P \cap Pre(P') = P. \tag{1}$$

In words, (1) states that all states in each $P \in \mathcal{P}$ behaves uniformly; given another $P' \in \mathcal{P}$, either all states or no state in P can reach some state in P' in one transition. Condition (1) leads to the following general procedure for computing a bisimulation of a transition system using quotient systems.

Bisimulation Procedure (BP) [6, 10, 11]:
set $\mathcal{P} = \mathcal{P}_0$
% check termination condition
while $\exists P, P' \in \mathcal{P}$ such that $\emptyset \neq P \cap Pre(P') \neq P$ {
    % partition refinement
    split P into $P_1 = P \cap Pre(P')$ and $P_2 = P \setminus Pre(P')$
    set $\mathcal{P} = (\mathcal{P} \setminus \{P\}) \cup \{P_1, P_2\}$
}

Note that (1) is precisely the termination condition of BP. In each iteration of BP, the partition refinement scheme uses the information obtained from $Pre(P')$ to split P into the part that can reach P' and the part that cannot. Also note that the $Pre(\cdot)$ operation in BP is with respect to the transition relation $\rightarrow$ of T, not the transition relation $\rightarrow_\mathcal{P}$ of the quotient system $T/\mathcal{P}$.

The termination condition serves to guarantee that the quotient system is a bisimulation of T when BP terminates. However, when transition systems are used to represent hybrid systems there is no guarantee the procedure will terminate. In the sequel we show that it may be possible to perform verification successfully using the quotient transition system obtained from a partition before the termination condition is satisfied. One can also use an approximation to the quotient transition system, rather than the exact quotient system, provided it is based on a conservative approximation to the *Post* operator. Moreover, we note that the partition refinement can be performed using heuristics if it is not convenient to compute the *Pre* operator as required in BP. This may affect the rate at which a useful quotient transition system is obtained, but it may be adequate since it is not necessary to converge to a bisimulation. These concepts are developed and illustrated in this paper. Further theoretical analysis of the use of simulations and approximations for verification and control is given in [9].

3 Polyhedral-Invariant Hybrid Automata

We consider a class of hybrid system called *polyhedral-invariant hybrid automata* (PIHA) defined as follows using the formalism from [10, 11] (with some restrictions).

Definition 6. A *polyhedral-invariant hybrid automaton* is a tuple $H = (X, X_0, F, E, I, G, R)$ where

- $X = X_C \times X_D$ where $X_C \subseteq R^n$ is the continuous state space and X_D is a finite set of discrete locations.
- F : a function that assigns to each discrete location $u \in X_D$ a vector field $f_u(\cdot)$ on X_C.
- $I : X_D \to 2^{X_C}$ assigns to $u \in X_D$ an *invariant* set of the form $I(u) \subseteq X_C$ where $I(u)$ is a nondegenerate convex polyhedron.
- $E \subseteq X_D \times X_D$ is a set of discrete transitions.
- $G : E \to 2^{X_C}$ assigns to $e = (u, u') \in E$ a guard set that is a union of faces of $I(u)$.
- $X_0 \subseteq X$ is the set of initial states of the form $X_0 = \bigcup_i (P_i, u_i)$ where each $P_i \subseteq I(u_i)$ is a polytope and $u_i \in U$. Here, the notation (P, u) means the set $\{(x, u) \in X \mid x \in P\}$.
- I,G, and E must satisfy the following *coverage* assumption.

$$\forall e = (u, u') \in E, G(e) \subseteq I(u').$$

Definition 7. Given an initial hybrid system state (x_0, u), the continuous trajectory in location u, denoted by $\zeta_{(x_0,u)}(\cdot)$, evolves according to the following conditions:

i. $\zeta_{(x_0,u)}(0) = x_0$; and
ii. $\dot{\zeta}_{(x_0,u)}(t) = f_u(\zeta_{(x_0,u)}(t)), \forall t \geq 0$ (until a discrete transition occurs).

PIHA are equivalent to switched continuous dynamic systems in which there are no jumps in the continuous state trajectory. The set of states through which a location is entered is called an *entry set* which is defined as follows.

Definition 8. Given a PIHA H, the set of *entry states* for the locations is defined as

$$X_{entry} = \{ (x, u') \in X \mid \text{for some } (x, u) \in X \text{ and } (u, u') \in E, x \in G((u, u'))\}$$

We assume that at each $(x, u) \in X_{entry}$, $f_u(x)$ points into the interior of $I(u)$ (denoted $int(I(u))$). This means the continuous state trajectory always initially enters the interior of the invariant when the system makes a transition to the new location.

   Transition systems provide an effective formalism for defining the semantics of a hybrid automaton. Henzinger [7] described the complete untimed behavior of a hybrid automaton using the *time-abstract transition system*. In this paper, we are interested in the behaviors of the PIHA H only at the times when discrete transitions occur. To abstract away the continuous dynamics and obtain the projection of the hybrid system behaviors onto the instants of discrete transitions, we define the *discrete-trace transition system* for H as follows.

Definition 9. Given a PIHA H, its *discrete-trace transition system* is given by $T_H = \{Q_H, \rightarrow_H, X_0\}$ where $Q_H = X_0 \cup X_{entry} \bigcup_{u \in X_D} \{q_u^\perp\}$ and the transition relation $\rightarrow_H$ is defined as

i. **Discrete Transitions.** $(x, u) \rightarrow_H (x', u')$ iff $u' \neq u$ and there exist $e = (u, u') \in E$ and $t_1 > 0$ such that $\zeta_{(x,u)}(t_1) = x'$, $x' \in G(e)$, and $\zeta_{(x,u)}(t) \in int(I(u))$ for all $t \in (0, t_1)$.

ii. **Null Transitions.** $(x, u) \rightarrow_H q_u^\perp$ iff $\zeta_{(x,u)}(t) \in I(u)$ for all $t \geq 0$.

The discrete transition comprises all the continuous-state trajectories in the hybrid system between location transitions. The null transition comprises all the continuous-state trajectories that remain in a location indefinitely. Although null transitions can occur in general, we will assume that all continuous-state trajectories in a PIHA eventually lead to a location transition. This is similar to the liveness assumption in [10].

4   Verification Using Approximations

In this section we introduce the verification problem and describe how it can be solved using an alternative to BP.

4.1   CTL Specifications

Temporal logic is a well-established formulation for specification of finite-state transition systems [5]. The specification problem is formulated by attaching a finite set of *atomic propositions* to a transition system whose individual values are either true or false for each state in the transition system. The transition graph of the transition system is unfolded into an infinite *computation tree*. A temporal logic called *computation tree logic* (CTL) can then be used to specify system evolutions in terms of the atomic propositions along some or all paths of the computation tree. It has been shown in [5] that a CTL formula for a transition system corresponds to a region in the state space of the transition system for which the CTL expression is true. Furthermore, the region corresponding to a CTL formula can be computed using fixed-point iterations. For a finite transition system, such fixed-point computations are guaranteed to terminate. The transition system satisfies the specification if all initial states are included in the region corresponding to the CTL specification.

The same concept for specification of finite-state transition systems can be extended to infinite-state transition systems (e.g. T_H). Atomic propositions can be assigned to each state of T_H and the computation tree for T_H (with an uncountable number of nodes) can be obtained by unfolding its transition graph. CTL specification can be interpreted as before in the case of finite-state transition systems. The problem here is that the fixed-point computation for the CTL formula will not terminate because the state space of the transition system is uncountable.

4.2 Verification Using Simulations (Approximating Automata)

In the bisimulation approach to hybrid system verification, BP is applied to find a partition of the state space that will give a finite-state bisimulation of the infinite-state transition system. Verification of properties for the bisimulation is then equivalent to verification of properties of the original transition system. The problem with this approach is that a finite bisimulation may not exist, meaning BP will not converge and the verification step cannot be performed. Even when a finite-state bisimulation exists, it may take a very long time for BP to converge to the solution.

We propose using finite-state simulations, called *approximating automata*, rather than bisimulations to verify properties of the original hybrid system. Since quotient systems are, in general, simulations of the underlying transition system, one could attempt to perform verification on the quotient system in any iteration of BP. If the property is verified, there would be no need to refine the quotient system further. If not, another refinement iteration can be executed and verification can be attempted on the new quotient transition system. Continuing this process, it is sometimes possible to conclude whether or not the hybrid system satisfies the desired property before a bisimulation is achieved [2–4]. The restriction here is that since simulations are conservative approximations, only *universal* properties can be verified. In the context of CTL, a universal property is a property that holds along all paths in the computational tree.

To slow down the state explosion resulting from the partition refinement, we refine only the states in the quotient system that are relevant to the CTL specification. In the process of a CTL verification, one obtains the set of initial states in the current quotient system that satisfy the CTL specification. Since the specification is universal, the verification result cannot be improved by refining these states and their descendants. Thus, one should only refine the initial states that do not satisfy the CTL specification and their descendants. Integrating this additional refinement criterion into the bisimulation procedure, we obtain the *Approximating Automata Procedure (AAP)* shown in figure 1. In AAP, TBR is the set of states "to be refined" in each iteration. It consists of the set $REFINE$, the states that should be refined from the bisimulation requirement, and the set $reach((\mathcal{P} \setminus SPEC) \cap X_0/\mathcal{P}_N)$, the states that should be refined from the CTL specification requirement. Choices of *bisimulation_termination_condition* and *refinement_method* are left unspecified since there are various alternatives for these steps, as discussed in section 2.

5 Approximating the PIHA Quotient System

This section discusses a computational method for approximating a quotient system of the discrete-trace transition system T_H where H is a PIHA, the first step in AAP. Given a partition $\mathcal{P}$, we use the flow pipe approximation [4] to conservatively approximate postcondition sets which are essential in defining the transition relation $\rightarrow_{\mathcal{P}}$ of $T_H/\mathcal{P}$ (see definition 5). We assume that the partition of the state space for T_H is of the form

$$\mathcal{P} = \{(\pi, u) \in X_0 \cup X_{entry} \mid \pi \text{ is a convex polytope and } u \in X_D\}.$$

Approximating Automata Procedure (AAP):
initialize $N = 0$ and $\mathcal{P}_N = \mathcal{P}_0$
repeat for the partition $\mathcal{P}_N$
    compute (or approximate) $T_H / \mathcal{P}_N$ (see section 5)
    compute $SPEC = \{P \in \mathcal{P}_N \mid P$ satisfies CTL specification for $T_H / \mathcal{P}_N\}$
    if $X_0 / \mathcal{P}_N \subseteq SPEC$
        $stop = 1$ % specification is satisfied
    else
        compute $REFINE = \{P \in \mathcal{P}_N \mid \exists P' \in \mathcal{P}_N$
            violating *bisimulation_termination_condition*$\}$
        if $REFINE == \emptyset$
            $stop = 1$ % bisimulation obtained and specification is false
        else
            compute $TBR = reach((\mathcal{P}_N \setminus SPEC) \cap X_0 / \mathcal{P}_N) \cap REFINE$
            if $TBR == \emptyset$
                $stop = 1$ % no state worth refining and specification is false
            else
                $stop = 0$ % refine partition
                $\mathcal{P}_{N+1} = \mathcal{P}_N$
                for each $P \in TBR$
                    split P using *refinement_method* into P_1, P_2 such that
                        $P_1 \cup P_2 = P$ and $P_1 \cap P_2 = \emptyset$
                    set $\mathcal{P}_{N+1} = (\mathcal{P}_{N+1} \setminus \{P\}) \cup \{P_1, P_2\}$
                endfor
                $N = N + 1$
            endif
        endif
    endif
until ($stop == 1$)

Fig. 1. Verification procedure using simulations

To approximate the postcondition set for each $(\pi, u) \in \mathcal{P}$, we start by computing an outer approximation to the set on the boundary of $I(u)$ that can be first reached from π under the flow equation $\dot{x} = f_u(x)$ starting from continuous states in π. We will refer to such set as the *forward mapping* of π for location u, denoted $FMAP_u(\pi)$. The procedure for computing $FMAP_u(\pi)$ is given in figure 2. The forward mapping procedure computes outer approximations to the flow pipe segments from π under the differential equation $\dot{x} = f_u(x)$, denoted $\hat{\mathcal{R}}^u_{[t,t+\Delta t]}(\pi)$, and takes the intersection of the flow pipe segment approximation with the boundary of $I(u)$ in each time step until $\hat{\mathcal{R}}^u_{[t,t+\Delta t]}(\pi)$ lies completely outside of $I(u)$. (Note that $\hat{\mathcal{R}}^u_{[t,t+\Delta t]}(\pi)$ is guaranteed to eventually leave the location under the assumption there are no null events for PIHA.)

The approximation to the flow pipe segment for each time step $[t, t + \Delta t]$ is computed as follows. First, a set of *enclosing* normal vectors, c_i, is chosen. By enclosing, we mean that the half-space intersection $\bigcap_i \{x \mid c_i^T x \leq d_i\}$ forms a closed polyhedron for some constants d_i. The flow pipe segment is computed by

Forward Mapping Procedure:
initialize $t = 0, FMAP_u(\pi) = \emptyset$, and $stop = 0$
repeat
    compute flow pipe segment $\hat{\mathcal{R}}^u_{[t,t+\Delta t]}(\pi)$
    if $\left(\hat{\mathcal{R}}^u_{[t,t+\Delta t]}(\pi) \cap I(u) == \emptyset\right)$
        $stop = 1$
    else
        $FMAP_u(\pi) = FMAP_u(\pi) \cup \left(\hat{\mathcal{R}}^u_{[t,t+\Delta t]}(\pi) \cap \delta I(u)\right)$
        $t = t + \Delta t$
    endif
until $(stop == 1)$

Fig. 2. Procedure for computing forward mapping set given (π, u)

solving, for each normal vector c_i, the optimization problem

$$\max_{x_0,t} c_i^T \zeta_{(x_0,u)}(t)$$

$$\text{s.t. } x_0 \in \pi, t \in [t, t+\Delta t] \tag{2}$$

Let d_i^* be the solution to (2) for c_i. The flow pipe segment is given by $\hat{\mathcal{R}}^u_{[t,t+\Delta t]}(\pi)$ $= \bigcap_i \{x | c_i^T x \le d_i^*\}$.

To find the set of normal vectors for the optimization problems, we use the heuristics depicted in figure 3a. The trajectories starting from the vertices of π are simulated using an ODE solver to the two time points t and $t + \Delta t$. The normal vectors are then taken from the normal vectors on the faces of the convex hull of these points. After obtaining the normal vectors, we solve (2) for each c_i to obtain the flow pipe segment as depicted in figure 3b. For more details on the flow pipe approximation, see [4].

$V_t(\pi)$ $V_t(\pi)$

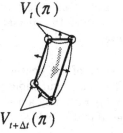   

$V_{t+\Delta t}(\pi)$ $V_{t+\Delta t}(\pi)$

(a) Initial directions from convex hull   (b) Solving optimization problems
                                                          for flow pipe segment

Fig. 3.

Using the forward mapping procedure, the overall procedure for approximating the quotient system $T_H/\mathcal{P}$ is given in figure 4. The procedure constructs the transition relation $\rightarrow_\mathcal{P}$ directly from the forward mapping sets. Alternatively, the postcondition sets can be used to define the quotient system as in definition 5. The postcondition sets can be obtained from the mapping sets by

$$Post((\pi, u)) = \bigcup_{u' s.t. (u,u') \in E} (FMAP_u(\pi) \cap I(u'), u').$$

Quotient System Approximation Procedure:
```
% compute forward mapping sets
for each (π, u) ∈ P
   compute FMAPᵤ(π)
endfor
% define transition relation
for each (π, u) ∈ P
   for each (π', u') ∈ P, (π', u') ≠ (π, u)
      define (π, u) →ₚ (π', u') if FMAPᵤ(π) ∩ π' ≠ ∅ and (u, u') ∈ E
   endfor
endfor
```

Fig. 4. Procedure for approximating the quotient system $T_H/\mathcal{P}$

6   Application: Verification of a Batch Evaporator

The example hybrid system considered in this paper is a batch evaporator taken from Kowalewski and Stursberg [8]. The evaporation process diagram is shown in figure 5. The process follows the following production sequence. First, tank T_1 is filled with a solution which is evaporated until a desired concentration is reached. Tank T_1 is then drained as soon as tank T_2 is emptied from the previous batch. For safety reasons, the heating is shut off when the alarm temperature, T_{alarm}, is reached. When the temperature in tank T_1 falls below a certain temperature, T_{crys}, crystallization will occur and spoil the batch. Our objective is to verify that the alarm temperature is chosen appropriately such that from a given set of initial conditions the temperature in tank T_1 never falls below the crystallization temperature before T_1 is completely drained.

The control inputs to the system are the status of the heater (on/off) and the valve positions V_{15} and V_{18} (open/closed). Three configurations of the inputs are currently employed. We specify an input configuration by a discrete variable u. The three configurations u_1, u_2, and u_3, are tabulated below.

Configuration	Heating	V_{15}	V_{18}	Description
u_1	on	closed	open	heating T_1
u_2	off	closed	open	cooling T_1/drain T_2
u_3	off	open	closed	cooling/drain T_1

The continuous state variables are the liquid level in T_1, the liquid level in T_2, and the temperature in T_1, denoted by H_1, H_2, and T, respectively. The

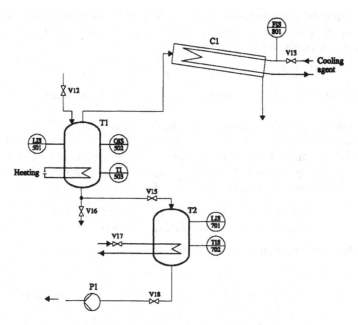

Fig. 5. Batch evaporation diagram

continuous dynamics depends on the input configuration. For configuration u_1, the state equations are

$$\dot{H}_1 = 0$$
$$\dot{H}_2 = -3.333 \cdot 10^{-4} \sqrt{19.62 H_2}$$
$$\dot{T} = \frac{5000 - 24(T - 283)}{1.23 \cdot 10^5 H_1 - 1.327 \cdot 10^9 T^{-2} + 2.819 \cdot 10^6 T^{-1} + 6.433 \cdot 10^3 - 10.513 T}.$$

For u_2, the state equations are

$$\dot{H}_1 = 0 \qquad \dot{H}_2 = -3.333 \cdot 10^{-4} \sqrt{19.62 H_2}$$
$$\dot{T} = \frac{-24(T - 283)}{1.23 \cdot 10^5 H_1 + step(T - 373) R(T)}$$
$$R(T) = \left(-1.327 \cdot 10^9 T^{-2} + 2.819 \cdot 10^6 T^{-1} + 6.433 \cdot 10^3 - 10.513 T\right),$$

where $step(\cdot)$ denotes the standard step function. Finally, the state equations for u_3 are

$$\dot{H}_1 = -6.667 \cdot 10^{-4} \sqrt{19.62 H_1}$$
$$\dot{H}_2 = 3.333 \cdot 10^{-4} \sqrt{19.62 H_1}$$
$$\dot{T} = \frac{-0.036(T - 283)(0.03 + 0.628 H_1)}{29.1 H_1}.$$

The production sequence described previously can be translated into the PIHA shown in figure 6. Locations $u_1, u_2,$ and u_3 correspond directly to the input configurations u_1, u_2, u_3. Locations u_4 and u_5 are empty locations with trivial continuous dynamics. They are used to indicate the *failure* and *success* of the production sequence, respectively. Locations u_1, u_2, and u_3 have the same invariant given by the rectangular box,

$$\{H_{1min} \leq H_1 \leq H_{1max} \wedge H_{2min} \leq H_2 \leq H_{2max} \wedge T_{crys} \leq T \leq T_{alarm}\},$$

which represents the physical limits and thresholds on the continuous state variables.

The system starts with location u_1 and initial conditions $H_1 \in [0.2, 0.22]$ m, $H_2 \in [0.28, 0.3]$ m, and $T = 373K$. Since the liquid level in each tank in the ODE model can only reach zero asymptotically, we approximate the event that a tank is empty by a small threshold $H_{imin}, i = 1, 2$. The numerical values for the limits on the rectangular invariant box for all locations are

$$H_{imin} = 0.04 \text{ m}, \qquad H_{imax} = 0.4 \text{ m}, i = 1, 2,$$
$$T_{crys} = 338 \text{ K}, \text{ and } T_{alarm} = 395 \text{ K}.$$

Our verification problem can be restated as

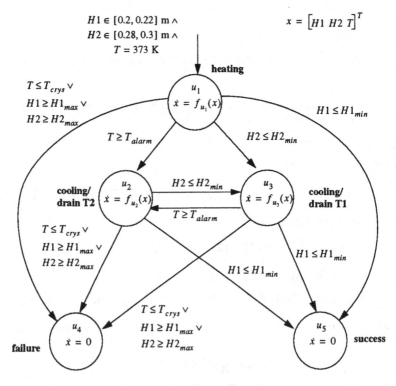

Fig. 6. PIHA for the batch evaporator

Verify that all trajectories from the initial continuous state set $X_0^C = \{0.2 \leq H_1 \leq 0.22, 0.28 \leq H_2 \leq 0.3, T = 373\}$ in location u_1 eventually reach location u_5.

which translates directly into the CTL expression $AF(u = u_5)$.

The verification was performed using AAP with the quotient system $T_H/\mathcal{P}$ approximated as described in section 5. The continuous part of the entry set for each location is the face(s) of the invariant box that is(are) the guard(s) on all its incoming transitions as shown in figure 6. The heuristics based on the vector field variation that is used to obtain the initial partition for each face is described briefly as follows.

Taking the whole invariant face as the starting point for the partition, we estimate the vector field variation on the face by taking the samples of the quantity $c^T f_u(x_i)$, where c is the normal vector to the face and u is the parent location, from points x_i on the face. We take the difference between the maximum and minimum samples as the variation estimate. If the variation is greater than the tolerance, the polytope is divided into two subsets by a hyperplane that pass roughly through the middle of the polytope. Each polytope is divided recursively until all polytopes in the partition satisfy the vector field variation tolerance.

The initial entry set partition $\mathcal{P}_0$ is shown in figure 8. Since each face of the invariant box can be identified with a unique location, the location associated with each face is not shown in the figure. Each continuous subset in the entry set partition will be referred to as a patch. The initial continuous state set X_0^C is not partitioned further because it is already a small set.

Using the forward mappings computed for the partition $\mathcal{P}_0$, we compute the transition relation for the quotient system $T_H/\mathcal{P}_0$ as described in section 5. The forward mapping is illustrated in Fig. 7.

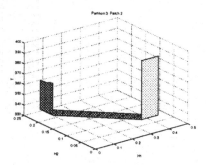

Fig. 7. Forward mapping approximation: the light-shaded patch is the initial set and its forward mapping to the boundary of the invariant box is composed of dark-shaded patches.

The transition relation for the initial quotient system $T_H/\mathcal{P}_0$ is shown in figure 8. By our convention, the initial state is always the last state in the transition table. For example, patch 21 is the initial state representing (X_0^C, u_1) in figure 8. Some patches are identified as "indeterminate." These are patches containing singularity points for the vector field and the mapping cannot be computed for them. However, as we shall see, these patches finally become unreachable from the initial set as the refinement proceeds with each iteration of the verification procedure. Thus, they do not affect our verification result.

The TBR set for the first iteration of the verification procedure is also shown in figure 8. Although patch 21 should be in the TBR set, we choose not to refine the initial state at all in this example to keep the presentation simple as there will only one initial state in each quotient system. Patches in the TBR set are simply split in half to form the next partition $\mathcal{P}_1$. We do not use the refinement method in BP because it is costly and difficult to implement with flow pipe approximations and convex polyhedral representation of sets. The quotient system $T_H/\mathcal{P}_1$ is computed with the new partition and the process continues.

After three iterations, we find that the TBR set is empty. This is because the initial state (patch 40 in figure 9) already satisfies the specification. Since $T_H/\mathcal{P}_3$ is a simulation of T_H, every trajectory of T_H is contained in $T_H/\mathcal{P}_3$ and we conclude that T_H for the batch evaporator also satisfies the specification.

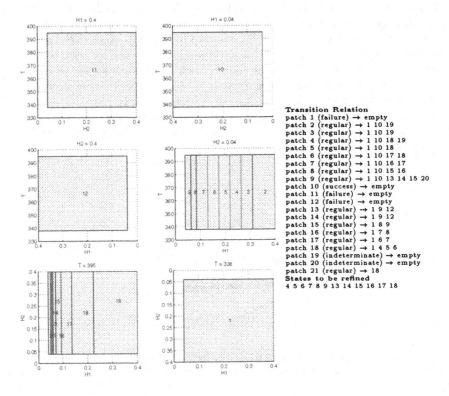

Fig. 8. Quotient System $T/\mathcal{P}_0$

7   Discussion

This paper presents a computational method for verifying properties of polyhedral-invariant hybrid automata (PIHA) with arbitrary continuous dynamics in the locations. The verification procedure is based on approximating automata

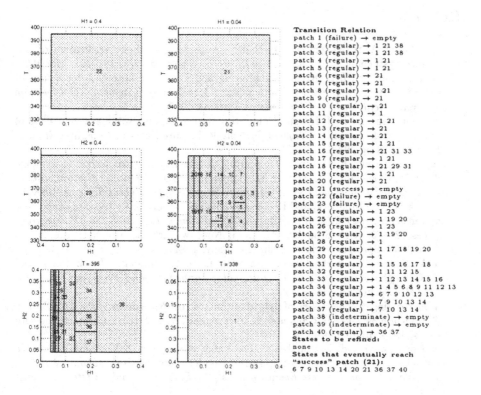

Fig. 9. Quotient System $T/\mathcal{P}_3$

computed as the quotient transition systems from partitions of the infinite state space for the discrete-trace transition system. The continuous-state flow pipes are approximated using sequences of convex polyhedra. Important research issues to be explored include: extending the flow pipe approximation technique to handle systems with uncertain dynamics, such as differential inclusions; numerical methods to guarantee outer approximations using floating-point arithmetic; and incorporating methods for identifying null transitions, that is, identifying when there are trajectories that do not leave invariants for a location. The computational procedures described in this paper are currently being incorporated into a Matlab/Simulink tool for specifying and analyzing PIHA.

References

1. R. Alur, T.A. Henzinger, and P.-H. Ho. Automatic symbolic verification of embedded systems. *IEEE Trans. on Software Engineering*, 22(3):181–201, Mar 1996.
2. A. Chutinan and B.H. Krogh. Computing approximating automata for a class of hybrid systems. *submitted to Mathematical Modeling of Systems: Special Issue on Discrete Event Models of Continuous Systems*, 1997.
3. A. Chutinan and B.H. Krogh. Computing approximating automata for a class of linear hybrid systems. In *Hybrid Systems V*, Lecture Notes in Computer Science. Springer-Verlag, 1998.
4. A. Chutinan and B.H. Krogh. Computing polyhedral approximations to dynamic flow pipes. *submitted to 37th IEEE Conference on Decision and Control: Invited Session on Synthesis and Verification of Controllers for Hybrid Systems*, 1998.
5. E. Clarke, O. Grumberg, and D. Long. Verification tools for finite-state concurrent systems. In *Proceedings of A Decade of Concurrency: Reflections and Perspectives*, pages 124–75, REX School/Symposium, Noordwijkerhout, The Netherlands, 1-4 June 1993. Springer-Verlag, Berlin, Germany, 1994.
6. T.A. Henzinger. Hybrid automata with finite bimulations. In Z. Fülöp and F. Gécseg, editors, *ICALP 95: Automata, Languages, and Programming*, pages 324–335. Springer-Verlag, 1995.
7. T.A. Henzinger. The theory of hybrid automata. In *Proceedings of the 11th Annual Symposium on Logic in Computer Science*, pages 278–292. IEEE Computer Society Press, 1996. Invited tutorial.
8. S. Kowalewski and O. Stursberg. The batch evaporator: A benchmark example for safety analysis of processing systems under logic control. *submitted to: 4th Int. Workshop on Discrete Event Systems (WODES '98), Cagliari (Italy)*, August 1998.
9. B.H. Krogh and A. Chutinan. Hybrid systems: Modeling and control. In P.M. Frank, editor, *Advances in Control*. Springer-Verlag, 1999. To appear.
10. G. Lafferriere, G. J. Pappas, and S. Sastry. Hybrid systems with finite bisimulations. Technical Report UCB/ERL M98/15, University of California at Berkeley, April 1998.
11. G. Lafferriere, G. J. Pappas, and S. Yovine. Decidable hybrid systems. Technical Report UCB/ERL M98/39, University of California at Berkeley, June 1998.
12. Feng Zhao. *Automatic Analysis and Synthesis of Controllers for Dynamical Systems Based on Phase-Space Knowledge*. PhD thesis, MIT Artificial Intelligence Laboratory, 1992.

Path Planning and Flight Controller Scheduling for an Autonomous Helicopter

M. Egerstedt[1], T.J. Koo[2], F. Hoffmann[2], and S. Sastry[2]

[1] Optimization and Systems Theory, Royal Inst. of Technology
SE-100 44 Stockholm, Sweden
`magnuse@math.kth.se`
[2] Department of EECS, University of California
Berkeley, CA 94720, USA
`{koo, fhoffman, sastry}@eecs.berkeley.edu`

Abstract. In this article we investigate how to generate flight trajectories for an autonomous helicopter. The planning strategy that we propose reflects the controller architecture. It is reasonable to identify different flight modes such as take-off, cruise, turn and landing, which can be used to compose an entire flight path. Given a set of nominal waypoints we generate trajectories that interpolate close to these points. This path generation is done for two different cases, corresponding to two controllers that either govern position or velocity of the helicopter. Based on a given cost functional, the planner selects the optimal one among these multiple paths. This approach thus provide a systematic way for generating not only the flight path, but also a suitable switching strategy, i.e. when to switch between the different controllers.

1 Introduction

For autonomous mobile robots in general and for aerial based ones in particular, the need to function in a dynamic, changing environment is a crucial and important feature in a successful design. If a robot detects an obstacle, or a helicopter flies too close to the ground, an immediate, appropriate action is required. In a purely reactive control system, this problem can be addressed by introducing an obstacle avoidance behavior based on, for instance, potential field methods [1, 2]. However, a control system that is commanded to track reference trajectories, which is the case in this article, has to replan the trajectories on-line. Therefore any solution to the planning problem that relies too heavily on time consuming optimization techniques is likely to run into problems.

In this article we propose a solution to the trajectory planning problem that is based directly on the controller architecture itself. It turns out that it is reasonable to identify different flight modes such as take-off, cruise, turn and landing, as well as different controllers, such as velocity and position controllers, tracking different reference signals [7]. Given a set of waypoints defined as states of a linear system (position, velocity and acceleration), the planning task becomes to generate a trajectory that interpolates among these points, subject to additional

F.W. Vaandrager and J.H. van Schuppen (Eds.): HSCC'99, LNCS 1569, pp. 91–102, 1999.
© Springer-Verlag Berlin Heidelberg 1999

smoothness constraints on the path. The waypoints are used as soft constraints in the construction of the path, allowing for a trade off between accuracy and smoothness.

This path generation is done for two different linear systems, each corresponding to a different controller. A cost functional determines which of the two paths is optimal with respect to smoothness and interpolation accuracy. We thus provide a systematic way for generating not only a feasible flight path, but also a suitable switching strategy, i.e. when to switch between the different controllers.

Our solution is based on techniques from linear optimal control theory. The main idea is to compose the flight path from motion primitives, normally referred to as flight modes, such as take-off and landing. The planner finds an optimal path and decides what controller provides the optimal solution for that path segment.

Our approach offers the major advantage that it does not only propose reference trajectories, but also advises the control system when to switch between the different controllers.

Fig. 1. The helicopter tracks a reference path through given way points. Different controllers are active at different parts of the route.

The article is structured as follows: In Section 2 we introduce the helicopter model followed by, in Section 3, a discussion of the switched path planning algorithm. We then conclude with some simulation results, showing the numerical feasibility of our proposed method.

2   The Helicopter Model

In [4], an approximate model of the helicopter dynamics is derived, based on the assumption that some of the cross-coupling terms can be neglected. Under this assumption, the model becomes

$$\ddot{P} = R \begin{pmatrix} 0 \\ 0 \\ -T_M \end{pmatrix} + \begin{pmatrix} 0 \\ 0 \\ 1 \end{pmatrix}$$

$$\dot{\Theta} = \Psi \omega$$

$$\dot{\omega} = J^{-1}(\tau - \omega \times J\omega),$$

(1)

where $P = [x, y, z]$ are the scaled Cartesian coordinates of the helicopter's center of mass and $R \in SO(3)$. Furthermore, Θ are the Euler angles (roll (ϕ), pitch

(θ) and yaw (ψ)), ω the helicopter's angular velocity and τ describes the external torque that is applied to the helicopter's center of gravity. This nonlinear, dynamic system is driven by the inputs $[T_M, T_T, a_{1s}, b_{1s}]$, where T_M and T_T are the normalized main and tail rotor thrusts respectively while a_{1s} and b_{1s} are the longitudinal and lateral tilt of the main rotor.

Notice that the model (1) is highly nonlinear. Therefore control theoretical issues such as stability and tracking become much more complicated than for other mobile robots, for which the dynamics are fairly simple. In the helicopter case, we can not, for instance, let the robot stop and "think" while reasoning about the next action. Since we can not rely on computationally expensive optimization algorithms it seems questionable whether optimal planning for the full dynamic helicopter model is a realistic option. This is something that we need to take into account when our planner is designed.

The complex dynamics give rise to the need for additional considerations. In order to use a *behavior based* approach where, traditionally [1], desired motion directions are proposed, we in the helicopter case need to generate trajectories that correspond to the desired outputs of the active behavior.

In this article we only investigate point-to-point motions to describe the behavioral spectrum of the robot. From a behavior based view-point, this path generation can be regarded as an integral part of a general motion behavior.

2.1 Controller Design

In [7], different control designs are investigated, such as linear robust control, fuzzy control, and nonlinear feedback linearization. Due to the fact that the number of states is greater than the number of inputs, different controllers are designed to govern specific outputs, such as position or velocity tracking[1]. Therefore the overall behavior of the helicopter mainly depends on the currently active set of controllers. Different flight modes, such as take-off, cruising or hovering, employ their own collection of controllers, and in this article, we show how to select these controllers in a systematic way.

There are a number of reasons for introducing these different flight modes. First of all, flight modes provide building blocks for composing more complex behaviors such as searching or investigating objects on the ground. In addition, behaviors can be decomposed and analyzed in terms of these motion primitives[8], also used by human pilots. Furthermore, the flight mode approach allows us, at least partially, to decouple the state variables and to guarantee the stability of individual controllers for certain flight envelopes.

From the perspective of this article, on the other hand, we simplify the planning task by partitioning the overall path into smaller segments. This divide and conquer approach to the planning problem significantly reduces the computational complexity of the trajectory generation.

---

[1] This is due to practical considerations and not a theoretical consequence of the system dynamics.

2.2 Differential Flatness

Before we describe the planning task, some comments about differential flatness need to be made. The model (1) is differentially flat, as shown in [4], since it can be feedback linearized. In other words, there exists a diffeomorphism from the states of the nominal trajectory to the states in the helicopter model. If we assume that we fly in the so called *coordinated flight* mode, where we actively keep the side slip angle zero, then the heading of the helicopter can be reconstructed directly from the nominal trajectory. From this, all of the remaining states in the model (1) can be calculated, as shown in [4].

Hence we can recover the states of the helicopter from the flight trajectory and its derivatives, and this is a desired property for two reasons. First of all we want to be able to, given a desired trajectory, use this for controlling all of the states, and thus the flatness property makes it possible for us to obtain the desired state trajectory as well as the nominal inputs of the helicopter. Secondly, we could use this property for imposing constraints on the trajectory, constraints that come from the fact that we only can apply limited inputs in order to avoid saturation.

So, what the flatness property can help us with is to transform the nonlinear, coupled system into a system with decoupled x, y, z-states. This makes it possible for us to view these states as separate when planning the paths, at the same time as we still design output trajectories that are compatible with the nonlinear helicopter dynamics [6] (under the assumption that we do not saturate the actuators.)

The way this can be viewed is that the nominal trajectory provides the reference inputs for the actual tracking controller, designed on the full helicopter model. This controller can also be augmented by error feedback if necessary [4].

3 Planning for the Switched Control System

Based on the discussion in the previous section, the coarse behavior of the helicopter dynamics can be simplified into a linear system governed by either the position or the velocity controller. Although this simplification can not be used for the controller design itself, it serves as a valid abstraction of the dynamics for planning purposes. The trajectories generated by the planner are then tracked using the controllers, designed with respect to the full, nonlinear helicopter model.

We want the trajectories to have at least continuous second derivatives. Thus the paths fed into the position or velocity controller are produced by control systems on the form

$$x^{(3)} = k_p(u - x) \quad \text{position control}$$
$$x^{(3)} = k_v(u - \dot{x}) \quad \text{velocity control.} \tag{2}$$

The reason for this construction is that if we minimize the L_2-norm of the control input (which will be the case in the next subsection) we get smooth signals. This means that in the position control case, x will stay close to the smooth controlled input. This results in small variations in x as well, which is a desired feature when the position controller is used. The same argument can then be applied in the velocity controller case.

Notice that these linear systems are used for trajectory planning only and not for control. We can thus neglect the stability issues since our control law is designed in a way that drives the linear system from waypoint to waypoint [3].

By setting $\bar{x} = (x, \dot{x}, \ddot{x})^T$ and using a similar notation in the y- and z-direction, and letting

$$A_p = \begin{pmatrix} 0 & 1 & 0 \\ 0 & 0 & 1 \\ -k_p & 0 & 0 \end{pmatrix}, b_p = \begin{pmatrix} 0 \\ 0 \\ k_p \end{pmatrix}$$
$$A_v = \begin{pmatrix} 0 & 1 & 0 \\ 0 & 0 & 1 \\ 0 & -k_v & 0 \end{pmatrix}, b_v = \begin{pmatrix} 0 \\ 0 \\ k_v \end{pmatrix} \quad (3)$$

the overall system becomes

$$\begin{pmatrix} \dot{\bar{x}} \\ \dot{\bar{y}} \\ \dot{\bar{z}} \end{pmatrix} = \begin{pmatrix} A\bar{x} + bu_x \\ A\bar{y} + bu_y \\ A\bar{z} + bu_z \end{pmatrix} \quad \text{where} \quad \begin{pmatrix} A \\ b \end{pmatrix} = \begin{cases} \begin{pmatrix} A_p \\ b_p \end{pmatrix} & \text{position control} \\ \begin{pmatrix} A_v \\ b_v \end{pmatrix} & \text{velocity control} \end{cases} \quad (4)$$

This system decouples into three subsystems, which evolve independently in $\mathbb{R}^3$. Even though we are not obliged to switch between position and velocity control simultaneously in all the three subsystems, we prefer to consider them as one system, since the output (either position or velocity) is what the helicopter is asked to track.

3.1 Trajectory Planning

In this subsection, a general framework for generating trajectories for linear, single-input, multiple outputs control systems is presented based on linear optimal control theory. The main idea is to use soft constraints on the position, velocity and acceleration of the system at given, discrete times, $t_i, i = 1, \ldots, m$. The planning task becomes to find the control, u, which drives the individual subsystems of (4) close to the prespecified points in state space, $(\bar{x}_k, \bar{y}_k, \bar{z}_k, t_k)$. Notice that a general interpolation point can be formed by an arbitrary combination of constraints on position, velocity and acceleration along each dimension.

For example, the first subsystem of (4) becomes

$$\dot{\bar{x}} = A\bar{x} + bu, \quad \bar{x} \in \mathbb{R}^3, u \in \mathbb{R}. \quad (5)$$

In addition to the soft constraints, we want to minimize the L_2-norm of the control signal

$$\int_0^T \frac{1}{2} u^2(s) ds \quad (6)$$

because it smoothes the generated path. The flatness property, discussed earlier, implies that the actual variations in the internal states of the nonlinear helicopter model become smooth as well. There is much to gain in terms of performance and smoothness by using soft interpolation constraints rather than demanding exact interpolation.

Given a set of basis functions

$$g_i(t) = \begin{cases} e^{A(t_i - t)}b & t \le t_i \\ 0 & t > t_i, \end{cases} \tag{7}$$

we obtain the (translated) states, $\hat{x}$, of (5) to be

$$\hat{x}(t_i) = \bar{x}(t_i) - e^{At_i}\bar{x}_0 = \int_0^{t_i} e^{A(t_i - s)}bu(s)ds$$

$$= \int_0^T g_i(s)u(s)ds, \quad i = 1, \ldots, m. \tag{8}$$

The convex cost functional that we want to minimize, with respect to u, becomes

$$J(u) = \int_0^T \frac{1}{2}\rho u(s)^2 ds + \sum_{k=1}^m \frac{1}{2}(\hat{x}(t_k) - \alpha_k)^T \tau_k(\hat{x}(t_k) - \alpha_k), \tag{9}$$

where α_k is the waypoint that we want the system to drive closely to at time t_k, and

$$\tau_k = \begin{pmatrix} \tau_{k_1} & 0 & 0 \\ 0 & \tau_{k_2} & 0 \\ 0 & 0 & \tau_{k_3} \end{pmatrix}. \tag{10}$$

Here, τ_{k_j} indicates how important it is that $\hat{x}$'s jth component is close to the desired point at the time t_k. If no constraint on that component is imposed at this time, we simply let $\tau_{k_j} = 0$.

Taking the Fréchet derivative of this functional [5] with respect to u and setting it equal to 0, based on the guess that

$$u(t) = \xi^T g(t), \tag{11}$$

where

$$g(t) = \left(g_1(t)^T, \ldots, g_m(t)^T\right)^T, \tag{12}$$

gives us

$$\xi = (\rho \mathcal{I} + \mathcal{T}\mathcal{G})^{-1}\mathcal{T}\alpha. \tag{13}$$

In (13), the terms $\mathcal{T}$ and $\mathcal{G}$ are defined as

$$\mathcal{T} = \begin{pmatrix} \tau_1 & & 0 \\ & \ddots & \\ 0 & & \tau_m \end{pmatrix} \tag{14}$$

and

$$\mathcal{G} = \int_0^T g(s)g(s)^T \, ds. \tag{15}$$

It should be pointed out that in [3] it was proved that this problem is a convex optimization problem. Thus our necessary optimality conditions are in fact sufficient ones as well.

One advantage of this approach is that even though we only control a single input, we are able to impose constraints on all of the states as shown in Figure 2.

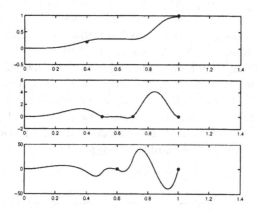

Fig. 2. Interpolation through points specified simultaneously for the position (upper graph), velocity (middle graph) and acceleration (lower graph). The reason why the trajectory seems to interpolate through and not just close to the desired points is due to the fact that we, in this case, chose to let the τ_k:s be large, giving a higher priority to interpolation rather than smoothing.

3.2 Motion Primitives

The planning task becomes to generate a feasible trajectory assuming a specific motion primitive for the current segment. In the *cruising* case, we ask the helicopter to maintain a given altitude, h, while moving horizontally at a constant velocity, v_c. Assuming, without loss of generality, that we want to fly in

x-direction, the constraints become

$$
\begin{pmatrix} x \\ \dot{x} \\ \ddot{x} \\ y \\ \dot{y} \\ \ddot{y} \\ z \\ \dot{z} \\ \ddot{z} \end{pmatrix}(t_0) = \begin{pmatrix} \star \\ v_c \\ \star \\ \star \\ 0 \\ \star \\ h \\ 0 \\ \star \end{pmatrix} \rightarrow \begin{pmatrix} x \\ \dot{x} \\ \ddot{x} \\ y \\ \dot{y} \\ \ddot{y} \\ z \\ \dot{z} \\ \ddot{z} \end{pmatrix}(t_1) = \begin{pmatrix} \star \\ v_c \\ \star \\ \star \\ 0 \\ \star \\ h \\ 0 \\ \star \end{pmatrix}, \tag{16}
$$

where the $\star$ indicates that no constraint is imposed on that state at that time.

The same type of way points can be identified for other motion primitives. The planning for each of the segments can be done separately, using the final state configuration from one motion as the initial value for the next. Although this divide and conquer approach does not guarantee a global optimum, it provides a good solution requiring a minimal computational effort. For the linear optimal control problem in itself, this may not be such a big benefit, but we already in the introduction talked about the switched planning. This refers to the case when we switch between different A:s and b:s and, as we will see in the next subsection, this leads to a combinatorial optimization problem which complexity increases with the number of waypoints.

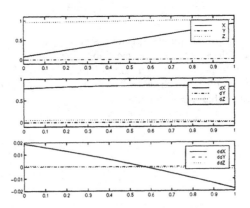

Fig. 3. Here a cruise motion in the x-direction is planned and the upper graph shows the position, the middle the velocity while the lower shows the acceleration of the planned path. In the figures, x is solid, y is dotted, and z is dash-dotted. The reason why $\ddot{x} \neq 0$ is that in this case our initial condition was not $\ddot{x}(t_0) = 0$.

3.3 Controller Scheduling

In order to make the optimization problem numerically tractable, we assume that switchings are only allowed to occur at the waypoints. In the future we plan to employ reinforcement learning to learn an optimal control switching policy. Whenever the reinforcement learner inserts new switching points into the coarse flight path, the planner replans the trajectory ON-line using these new waypoints.

For a motion primitive composed of M waypoints, the number of possible switching policies is $3 \cdot 2^M$. Since we keep the number of waypoints small within each motion primitive, this does not lead to a too large number of possible policies from a computational point of view.

We define a functional for evaluating the performance for each of these different $3 \cdot 2^M$ solutions, as

$$
\begin{aligned}
J_j(u) = \frac{1}{2}\rho \int_0^T & (u_{jx}(t)^2 + u_{jy}(t)^2 + u_{jz}(t)^2)dt \\
+ \frac{1}{2}\sum_{k=1}^m & \left\{ (\hat{x}_j(t_k) - \alpha_{xk})^T \tau_{xk}(\hat{x}_j(t_k) - \alpha_{xk}) \right. \\
+ & (\hat{y}_j(t_k) - \alpha_{yk})^T \tau_{yk}(\hat{y}_j(t_k) - \alpha_{yk}) \\
+ & \left. (\hat{z}_j(t_k) - \alpha_{zk})^T \tau_{zk}(\hat{z}_j(t_k) - \alpha_{zk}) \right\},
\end{aligned}
\tag{17}
$$

where u_{jx} is the control signal corresponding to the j:th switching strategy for the x-subsystem. In the same way, $\hat{x}_j$ are the states for this subsystem driven by u_{jx}.

It should be noted that we do not need to calculate the entire functional for each j since old results can be reused in order to reduce the numerical complexity.

The planner generates the final trajectory using the path number $j^\star \in [1, 3 \cdot 2^M]$ with the lowest value on the functional. This corresponds to the path that is optimal with respect to a weighted sum of a waypoint-fitting and a smoothness criterion.

Figure (5) shows an entire flight path.

4 Conclusions

This article investigates the problem of generating optimal flight trajectories for an autonomous helicopter. We propose a planning strategy that partitions the optimization problem into isolated segments. The planner can employ different controllers in order to generate a smooth trajectory that minimizes the deviation from the given, constraining waypoints.

Our approach constitutes a systematic way for not only generating the flight path, but also provides a suitable switching strategy, i.e. when to switch between the different controllers.

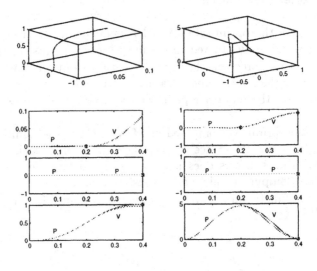

(a) Take-off motion.

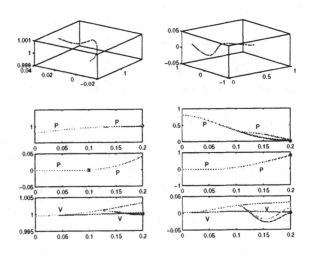

(b) Turn motion.

Fig. 4. Simulation of different motion primitives.The upper figures shows 3D-plots of the position and the velocity respectively in the motion primitive. The lower left plots show x, y and z while the right ones show $\dot{x}, \dot{y}$ and $\dot{z}$. The P:s and the V:s indicate what controller is active at each part of the path.

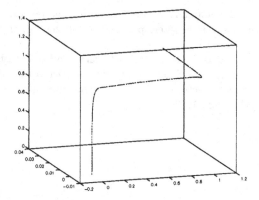

Fig. 5. Here, a simulation of a planned route, built up from different motion primitives (take-off, cruise and turn) and controllers is displayed.

Our approach also offers an analytical solution to the planning problem that does not require any computationally expensive numerical optimization. Therefore the planner is able to generate trajectories ON-line under the real time constraints given by the operation of the helicopter.

5 Acknowledgment

This work was supported by the Army Research Office under grant DAAH 04-96-1-0341, the ONR under grant N00014-97-1-0946, the Deutsche Forschungsgemeinschaft under grant Ho 1790/2-1 and the Swedish Foundation for Strategic Research through its Centre for Autonomous Systems at KTH, Sweden.

References

1. R.C. Arkin: *Behavior-Based Robotics*, The MIT Press, Cambridge, Massachusetts, 1998.
2. R. Brooks: A Robust Layered Control System for a Mobile Robot, *IEEE Journal of Robotics and Automation*, Vol. RA-2, No. 1, pp. 14-23, 1986.
3. M. Egerstedt and C.F. Martin: Trajectory Planning for Linear Control Systems with Generalized Splines, proceedings of the *Mathematical Theory of Networks and Systems* in Padova, Italy, 1998.
4. T.J. Koo and S. Sastry: Output Tracking Control Design of a Helicopter Model Based on Approximate Linearization, proceedings of the *37th IEEE CDC*, Tampa, Florida, 1998.
5. D.G. Luenberger: *Optimization by Vector Space Methods*, John Wiley and Sons, Inc., New York, 1969.
6. M.J. van Nieuwstadt and R.M. Murray: Real Time Trajectory Generation for Differentially Flat Systems, to appear in *International Journal of Robust and Nonlinear Control*.

7. H. Shim, T.J. Koo, F. Hoffman and S.S. Sastry: A Comprehensive Study on Control Design of Autonomous Helicopter, proceedings of the *37th IEEE CDC*, Tampa, Florida, 1998.
8. C.J. Tomlin, G.J. Papas, J. Košecká, J. Lygeros and S.S. Sastry: Advanced Air Traffic Automation: A Case Study in Distributed Decentralized Control, *Control Problems in Robotics*, Lecture Notes in Control and Information Sciences 230, Springer-Verlag, London, 1998.

Reachability Analysis Using Polygonal Projections

Mark R. Greenstreet[1] and Ian Mitchell[2]

[1] Department of Computer Science
University of British Columbia
Vancouver, BC V6T 1Z4, Canada
mrg@cs.ubc.ca
[2] Scientific Computing and Computational Mathematics
Stanford University
Stanford, CA 94305-9025, USA
mitchell@sccm.stanford.edu

Abstract. This paper presents Coho, a reachability analysis tool for systems modeled by non-linear, ordinary differential equations. Coho represents high-dimensional objects using projections onto planes corresponding to pairs of variables. This representation is compact and allows efficient algorithms from computational geometry to be exploited while also capturing dependencies in the behaviour of related variables. Reachability is performed by integration where methods from linear programming and linear systems theory are used to bound trajectories emanating from each face of the object. This paper has two contributions: first, we describe the implementation of Coho and, second, we present analysis results obtained by using Coho on several simple models.

1 Overview

Reachability analysis is the basis for many verification tasks. This paper addresses reachability for systems modeled by ordinary differential equations (ODEs). In this context, the state of the system is a point in $\mathcal{R}^d$, where d is the dimension (i.e. number of variables) of the system. Given two regions, $A \subseteq B \subseteq \mathcal{R}^d$, the reachability problem is to show that all trajectories that start in A remain in B, either during some time interval, $[0, t_{\text{end}}]$, or for all time.

To verify a safety property, one must show that all trajectories are contained in a region satisfying the property. Examples of safety properties include: aircraft are adequately separated [TPS97], an arbiter circuit never asserts grants to both of its clients simultaneously [MG96], and the level of water in a tank is in a specified interval[ACH+95]. To verify these properties, one can determine a region that contains all possible trajectories of the system. If this region is contained in the region that satisfies the desired property, then the safety property holds for the system. This paper presents a method for constructing a region that contains all possible trajectories of a system.

F.W. Vaandrager and J.H. van Schuppen (Eds.): HSCC'99, LNCS 1569, pp. 103–116, 1999.

In this paper we describe a technique for reachability analysis of systems modeled by ODEs. Such systems present two challenges. First, closed form solutions exist only for special cases (e.g. linear models and a few others). Mathematicians have proven enough negative results for closed form solutions that it is clear that little progress can be made by strictly analytical means. Thus, we must use approximation techniques (such as numerical integration) to analyze real systems. With care, these techniques can be designed in a way that ensures the approximations always lead to an over estimation of the reachable space. Thus, our verification is sound—incorrect designs will never be falsely verified, but we may fail to verify a correct system because of our approximations.

The second challenge is that we are interested in systems with moderately high dimensionality. The circuit models that motivate our work typically have five to twenty variables. Algorithms to represent and manipulate general d-dimensional polyhedra typically have time and space complexities with exponents of d or $d/2$ [PS85]. Thus, we will consider a restricted class of high-dimensional objects that can be efficiently represented and manipulated.

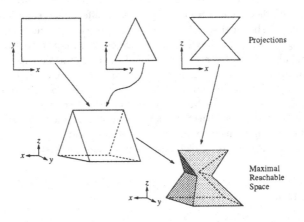

Fig. 1. A three dimensional "projectahedron"

In [GM98] we presented the theory for an approach to reachability analysis where high dimensional objects are represented by their projections onto two dimensional subspaces with each projection corresponding to a pair of variables. For example, figure 1 shows how a three-dimensional object (the "anvil") can be represented by its projection onto the xy, yz, and xz planes. The high dimensional object is the largest set of points that satisfies the constraints of each projection. We call objects that are represented by this technique "projectahedra."

Projectahedra offer several advantages. Ignoring degeneracies, faces of the object represented by a projectahedron correspond to edges of its projection polygons. Our reachability analysis requires flows from each face to be considered, and the projectahedron representation allows many operations on faces to be carried out as simple operations on polygon edges. The projection polygons are two-dimensional and can be manipulated efficiently using well-known algorithms from computational geometry [PS85].

We do not require the projection polygons to be convex; thus, non-convex, high-dimensional objects can be represented by projectahedra. As is shown in section 4, the reachable regions arising from ODE models are often highly non-convex. However, even non-convex projections cannot represent all possible high-dimensional polyhedra. For example, projectahedra cannot represent objects with indentations on their faces (i.e. a cube where some or all of the faces have hemispherical concavities). Instead, such objects will be mapped to projectahedra where their indentations are filled-in. Since we are only verifying safety properties, the resulting over approximation of the reachable space does not compromise the soundness of our analysis.

Given an object represented by a projectahedron, we must determine how this object evolves according to the ODE model for the system. As this problem cannot be solved analytically for general models, we pursue a numerical approach. We require the ODE model to have bounded derivatives. This means that trajectories are continuous, and it is sufficient to consider the set of points reachable from points on the boundary of the projectahedron. Our algorithm approximates the non-linear model with a linear model and an error bound. By constructing a separate approximation for each face at each time step, the error bounds can be fairly tight. The face is then transformed by the linear approximation and moved outward by the worst-case error. This approach allows general models to be used. Currently, linearization is done manually, and model generation requires significant effort.

This paper makes two contributions to the verification of systems with ODE models. Section 3 describes Coho, our implementation of the techniques mentioned above. For brevity, we focus on the top-level structure of the implementation and describe a few of the "surprises" we encountered. Then section 4 presents several examples where we have used Coho. We start with the linear two- and three-dimensional models presented in [DM98]. We then analyze two systems with non-linear models: a Van der Pol oscillator, and a three-dimensional "play-doh^{TM}" example, where the region is compressed in one dimension while being stretched and folded along another.

2   Computing Reachability

Coho is based on the techniques for reachability presented in [GM98]. This section presents a brief summary of this approach. Consider a general, ODE model:

$$\dot{x} = f(x) \tag{1}$$

where $x \in \mathbb{R}^d$. Given $A_0 \subseteq \mathbb{R}^d$ such that $x_0 \in A_0$, we are interested in answering reachability questions such as: Given $t \in \mathbb{R}^+$, find $A_t \subseteq \mathbb{R}^d$ such that $x(t) \in A_t$. For general ODE models, closed form solutions are not possible. Therefore, we do not hope to compute the smallest reachable set. Instead, we want to compute a conservative projectahedron for A_t. Our reachability computation is an iterative, integration algorithm. We describe a single time-step of this algorithm below.

Each edge of a projection polygon corresponds to a $d - 1$ dimensional face of the projectahedron. At each step, we compute a conservative estimate of the convex hull of this face, and then determine a new convex region that contains any point reachable from this hull at the end of the time step. We then project this hull back onto the basis for the projection polygon corresponding to the face. This computation is performed for each edge of the projection polygon to compute a bounding projection polygon at the end of the time step. By updating each projection polygon in this manner, we obtain a projectahedron that contains all points reachable at the end of the time step.

To compute a conservative approximation of the convex hull of a face, we intersect the constraints for the polygon edge with the constraints for the convex hulls of each of the projection polygons. To compute the set of points that are reachable from this hull, we approximate the model from equation 1 with the differential inclusion:

$$x \in H \Rightarrow \dot{x} \in Ax + b + U \qquad (2)$$

where H is the conservative approximation of the convex hull for the face, $A \in \mathbb{R}^d \times \mathbb{R}^d$ is a matrix, $b \in \mathbb{R}^d$ is a vector, and $U \in (\mathbb{R} \times \mathbb{R})^d$ is a hypercube (i.e., the Cartesian product of d intervals). For the examples presented in this paper, we compute A, b, and U by performing a power-series expansion about a point near the center of H. Because H is convex, it can be represented by a linear program, and we can use linear optimization techniques to obtain fairly accurate approximations for f.

We now consider the inhomogeneous linear system

$$\dot{x} = Ax + b + u(t) \qquad (3)$$

where u is any function such that $u(t) \in U$. By chosing a worst-case u, we obtain a conservative approximation of the reachable region. In our implementation, we approximate u with a linear bounds. This produces a linear program for the approximation of the points reachable from the face, and we obtain a projection of the face back to the coordinates of the projection polygon from this linear program.

3   Implementation

At the top-level, Coho is divided into one component that performs numeric computations and another that performs geometric operations. As shown in figure 2, at each time step the numeric component inputs a projectahedron, updates each face, and outputs a new projectahedron. In this process, each edge of each projection polygon is transformed to a polygon that contains the projection of the corresponding face at the end of the time step. The geometric component merges

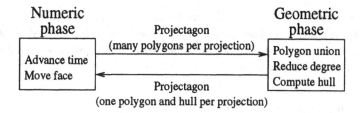

Fig. 2. Top-level of Coho

the polygons associated with a single projection, and computes an approximation with fewer vertices. The geometric component outputs the resulting polygon
and its convex hull. This completes a single step of the integration. Details of
the algorithm are presented in [GM98].

3.1   Matlab and Java

Currently, Coho implements the numeric component in Matlab [The92] and the
geometric component in Java [AG96]. This approach builds on the strengths of
both environments. Matlab provides comprehensive, optimized and well-tested
implementations of linear programming, matrix exponentiation, and other matrix operations—greatly simplifying the implementation of the integrator. Furthermore, the interactive interface and plotting capabilities facilitated the analysis and visualization of the systems described in this paper.

Although Matlab currently provides a few simple geometric operations, its
capabilities are not sufficient for our purposes. Thus, we implemented a computational geometry package in Java. With its type safety, garbage collection,
and object oriented abstractions, Java is much better suited for implementing
non-numeric algorithms than Matlab's scripting language. These features also
gave Java a clear advantage over C or C++: the geometry package was developed without the use of or need for a debugger; our experience suggests that
this would not have been possible in C or C++.

The geometric operations were encapsulated as a filter, taking one type of
projectahedron and returning a different format of a similar projectahedron (see
figure 2). This filter is a Matlab script that writes its parameter projectahedron
to a file, invokes the Java program as a shell command, and reads the result
from a second file.

We were initially concerned that the time to start the Java Virtual Machine
(JVM) would be prohibitive. In practice, starting the JVM each time step takes
about as long as the numeric computations for the time step. The time spent
performing geometric operations is small by comparison. We believe that the
cost of the numeric operations is dominated by the time spent solving the large
number of linear optimization problems that occur in our formulation. An implementation optimized for our application would almost certainly result in much
improved performance. Likewise, with some programming effort we could change

the interface between Matlab and Java so that the JVM would be started only once. For the examples described in this paper, the total elapsed time was a few seconds per time step running on a 250 MHz UltraSparc 2 workstation with Matlab 5.1 and JDK 1.2-beta4.

3.2   Numerical computation

The numeric phase of a time step begins by loading a polygon and its convex hull for each projection of the system. The convex hulls are then bloated outward slightly for safety. Each projection's bloated convex hull can be translated into a set of linear inequalities in the projection's two coordinates. The combination of all the projections' linear inequalities describes a convex region containing the projectahedron.

At this point, the movement of each edge of each projection's polygon can be computed independently. Each edge represents a face of the projectahedron, and the objective is to compute the furthest outward that points on the face could move during a time step. For each face, the following computations occur.

Restriction: The convex region computed from the convex hulls is further restricted to a box around the edge in the coordinates of the edge by four more linear inequalities. In the full dimensional space this is equivalent to constructing a slab around the face being examined.

Linearize Model: A user supplied function computes a linearization of the system derivatives which is valid over the slab. This model includes linear and constant terms, and must give bounds on the error introduced by the linearization within the slab. The user function has access to the slab's description in terms of the collection of linear inequalities computed in the previous step. Typically, linear programs are run to find bounds on each variable within the slab, from which the linearization and errors computed.

Advance Time: The linear model is used to move the slab forward in time—currently by matrix exponential, although future versions may use better integration routines.

Map Extent: The slab's end of step shape will still be described by a collection of linear inequalities after time is advanced. Building a polygon from these inequalities requires mapping out the region they contain. The mapping is accomplished by running a series of linear programs on the time advanced set of inequalities. Note that the slab may rotate during the time step, so its projection may not be a simple rectangle.

Add Errors: So far, the slab's movement is entirely controlled by the linearized model. To treat the error, we add a constant derivative offset within the error bounds throughout the time step, in such a way as to bloat the slab's projection outward as much as possible.

Each edge of each projection's polygon therefore produces an "edge polygon" at the end of the time step; this polygon contains the projection of all points that could be reached from the corresponding face within the time step. The union of all such polygons, and the region contained within that union, is the projection of an over approximation of the projectahedron at the end of the time step. Since the next time step must start with a single polygon for each projection, the geometric filter is called at this point to simplify the projectahedron 's description.

3.3 Geometric computation

The input to the geometric phase is a list of edge polygons for each projection: the union of these polygons contains the boundary of the new projection. The resulting polygon may have many more edges than the polygon at the beginning of the time step. To avoid unbounded growth in the number of polygon edges, we conservatively reduce the vertex count. Finally, as projectahedra evolve, degeneracies may occur in the projection polygons: edges may become very short, or vertex angles may become highly acute or highly obtuse. In fact, these degenaracies occur frequently when the projectahedron becomes very narrow along one or more axes, which is typical when approaching an attractor or similar phase space feature.

The five steps of the geometric phase are described below. The constant ϵ is used to test for potential numerical degeneracies—the current implementation uses $\epsilon = 10^{-12}$.

Short edge removal: If the length of an edge is less than ϵ times the distance from the origin of the endpoint furthest from the origin, then one of the vertices is deleted. If only one vertex remains at the end of this step, it is replaced by a square with edges of length 2ϵ. If exactly two vertices remain, the segment is replaced with a bounding rectangle whose major axis is parallel to the segment and that encloses the segment by ϵ.

Special case for highly acute vertices: Each edge polygons is convex. If a vertex of one of these polygons is highly acute (angle less than ϵ radians), then the edge polygon must be very thin. Such polygons are replaced by a bounding rectangle whose major axis is parallel to the bisector of the angle and that encloses the original edge polygon by ϵ.

Polygon merge: For each projection, the edge polygons are merged to produce the boundary of the new projection polygon.

Topological simplification: Having computed the boundary in the previous step, we now find the left most vertex, which must be on the outer boundary. An edge tour starting at this vertex gives the outer boundary of the projection's new polygon. This operation "fills-in" the interior of the projection polygon.

Vertex count reduction: Typically, the polygon produced by the preceding steps of the geometric phase will have many more vertices than the projection polygon had at the beginning of the time step. To prevent an explosion in the number of vertices, we must compute a conservative approximation of the polygon that has a reasonable vertex count.

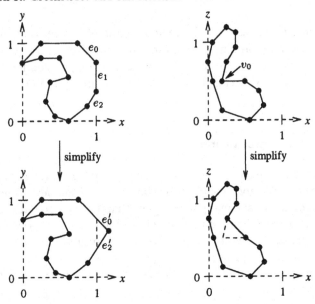

Fig. 3. Vertex count reduction

We say that a vertex is convex (resp. concave) if the polygon is locally convex (resp. concave) at that vertex. Consider a pair of adjacent convex vertices (such as the two rightmost vertices of the xy projection in the top half of figure 3). Let e_1 be the edge joining these two vertices and e_0 and e_2 be the other two edges incident on these vertices. If e_0 and e_2 intersect on the outside of e_1, then the two vertices can be replaced by this intersection. The resulting polygon contains all points in the original polygon; thus, this simplification is safe. Likewise, any concave vertex can be safely eliminated (as in deletion of v_0 from the xz polygon in figure 3).

The selection of vertices to remove is done in a greedy manner. All operations increase the size of the polygon, and each has a cost which is a weighted sum of the increase in the area of the projection polygon and the increase in area of its hull. Currently, the two weights are equal. Vertices are deleted until the total cost reaches a preset fraction of the original polygon area. In the examples below, we used a threshold of 2%—the resulting polygons typically had less than fifteen vertices.

The removal of concave vertices can create short edges or highly obtuse vertices (angles within ϵ of π). Such degeneracies are eliminated when they occur by deleting appropriate vertices.

3.4   Surprises

Of course, not all went as expected when we first used Coho. Originally, we only used the area of the polygon in computing the cost of deleting a vertex. However, projection polygons are approximated by their convex hulls in many places in our algorithm; thus, an approximation that enlarges the convex hull is in some sense more costly than one that does not. We found that by including the area of the convex hull in our cost function, we obtained tighter bounds with our reachability analysis.

A second surprise was the difficulty caused by infeasible vertices. Recall that at the beginning of each time step, each edge of each projection polygon corresponds to a face of the projectahedron. The numerical phase of the algorithm computes a convex bound for this face, moves it forward in time, and project it back to the basis plane to produce an edge polygon. The edge polygons for adjacent edges should overlap, and this is guaranteed if the vertex where the edges met was feasible at the beginning of the time step. We discovered that the over approximations used in our algorithm can produce infeasible vertices. However, the extent of the over estimate is not necessarily the same for all projections. This can produce sets of edge polygons that fail to form a closed boundary.

For example, consider the projectahedron as depicted in figure 3. Assume that both polygons have an extent of $[0, 1]$ in x before the vertex reduction step. The vertex elimination operation in the geometric component replaces the two rightmost vertices of the xy projection with a single vertex. This gives the xy projection polygon an x extent of $[0, 1.1]$. Vertex elimination for the xz polygon eliminates a single vertex along the concave section of the boundary, leaving the extent of the polygon unchanged. After vertex elimination, the rightmost vertex of the xy polygon is infeasible (because no point with that x value lies in the xz polygon).

Infeasible vertices led to incomplete boundaries in some of our earlier trials. Our solution is to detect infeasible vertices (using linear programming) and to treat a sequence of adjacent edges as a single piece of the boundary, extending the sequence until both endpoints are feasible. This procedure guarantees that the numeric component will produce a complete boundary that contains the true boundary at each time step. In the example above, edges e_0' and e_2' would be treated as a single triangle instead of two separate edges in the next numeric phase of the analysis.

4   Examples

This section presents our initial experience applying Coho to examples from the literature as well as some of our own design.

4.1   Dang and Maler's linear examples

In [DM98], Dang and Maler analyzed five systems with linear models and two with non-linear models. Here, we use Coho to analyze their five linear systems. Four of these are two-dimensional models from [HS74]. Figure 4 show the models and Coho's analysis for two of those examples, the node and sink. We also ran Coho on their center and saddle examples with similar results. When our

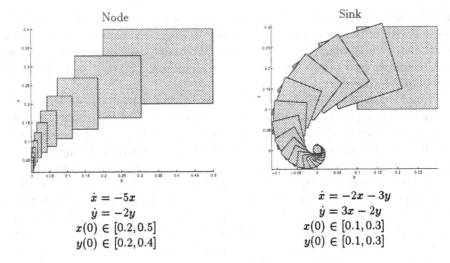

$$\dot{x} = -5x$$
$$\dot{y} = -2y$$
$$x(0) \in [0.2, 0.5]$$
$$y(0) \in [0.2, 0.4]$$

$$\dot{x} = -2x - 3y$$
$$\dot{y} = 3x - 2y$$
$$x(0) \in [0.1, 0.3]$$
$$y(0) \in [0.1, 0.3]$$

Fig. 4. Two-dimensional, linear models

analysis is compared with the "face-lifting" technique of Dang and Maler, Coho appears to be much more accurate. This can be seen in the node example where the boundaries of our polygons approach the origin without touching the axes, as should be the case for that model. With face-lifting, a much larger area is computed for the node, and it has extensive contact with the axes. Our sink analysis is likewise more accurate, clearly showing distinct cycles of the spiral where face-lifting merges them together.

The greater accuracy of Coho arises from several factors. First, Coho's "approximate" linearizations of the models are exact (to within the accuracy of double precision floating point arithmetic) for these linear examples; the error bound for these models is zero. Second, face-lifting quantizes the reachable region on a relatively coarse fixed grid, while Coho can place polygon vertices at any location representable in double precision. On the other hand, the fixed quantization of face-lifting may make it more amenable for use with symbolic techniques such as timed automata. This does not appear practical for Coho's polygonal projections.

Figure 5 shows Dang and Maler's 3-dimensional example. We have not yet implemented reconstruction of 3-dimensional objects from their projections, so the figure just shows the two projections. The figure in [DM98] does not provide enough detail to support a comparison of accuracy. However, our figure clearly shows how Coho automatically increases the number of vertices in the projection polygons to maintain the requested accuracy. The evolution from rectangle to elliptical "blobs" shows how Coho adjusts the orientation of edges according to its linearization of the ODE model.

Two non-linear models were presented in [DM98]. At the present time, we must manually derive code to linearize a non-linear model and compute the error

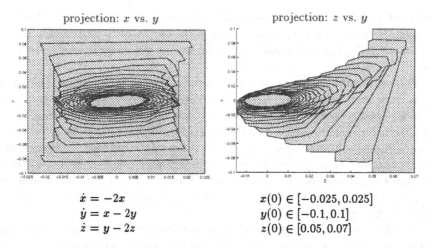

projection: x vs. y — projection: z vs. y

$$\dot{x} = -2x$$
$$\dot{y} = x - 2y$$
$$\dot{z} = y - 2z$$

$$x(0) \in [-0.025, 0.025]$$
$$y(0) \in [-0.1, 0.1]$$
$$z(0) \in [0.05, 0.07]$$

Fig. 5. Dang and Maler's 3-dimensional Model

bounds. This derivation is the most tedious and error-prone aspect of using Coho, and we are looking into ways to automate it. However, this, and the very recent completion of Coho, are the reasons that we have not yet analyzed Dang and Maler's non-linear examples. Instead, we present two examples of our own design that exhibit behaviors that are qualitatively different than those of linear models.

4.2 Van der Pol's oscillator

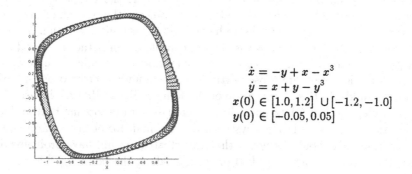

$$\dot{x} = -y + x - x^3$$
$$\dot{y} = x + y - y^3$$
$$x(0) \in [1.0, 1.2] \cup [-1.2, -1.0]$$
$$y(0) \in [-0.05, 0.05]$$

Fig. 6. Van der Pol's oscillator

Our first non-linear example is a Van der Pol oscillator adopted from [HS74]. Our model is symmetric in x and y; the equations and a cycle of the oscillator are shown in figure 6. This system also shows how invariants can be verified using reachability analysis: to establish an invariant set, it is sufficient to choose a region Q_0 and integrate for one period of the oscillator to produce Q_1. If Q_1 is contained in Q_0, then the region traced out during the integration is an invariant set.

During our first attempts at this example, the reachable region quickly became very long and skinny, stretching along the trajectory of the oscillation. This stretching occurs because the non-linear terms in the ODE stabilize the amplitude but not the phase of trajectories. Recall that Coho uses error bounds from a model's linear approximation to bloat edges outward—a conservative strategy to maintain the soundness of our analysis. While over estimation of the reachable region's amplitude is damped by the non-linear terms of the oscillator, over estimation of the region's phase tends to accumulate, and the region gets longer and longer.

On the straight segments of the oscillator's trajectory, this stretching causes little harm. However, as the oscillator makes a sharp turn at the corners, the over estimated phase can spill into large over estimates of amplitude. If the amplitude isn't damped before the next corner, the region grows without bound.

To prevent this explosion in region size, we divide a complete oscillation into two portions. The starting region lies on the positive x-axis, and we track this region through half a cycle until it has completely crossed the negative x-axis. We manually identify the segment of the negative x-axis where any portion of the region crossed, and restart from this segment. This second region is tracked around to its crossing of the positive x-axis. Figure 6 clearly shows that all trajectories starting in the original region cross through the second starting region, and all trajectories from the second starting region cross through the original starting region. Thus, we have identified an invariant set.

This technique can be extended to allow verification of a hybrid system whose dynamics depend on a current discrete "mode". As the region being tracked crosses a boundary between two modes, we can record the portion of the boundary touched by the region. Once we have finished tracking the region in the old mode—most likely because it has moved completely into the new mode—the analysis is restarted using the new mode's continuous dynamics. The initial conditions for this restart are those portions of the boundary crossed by the region in the old mode. In the computation of the Van der Pol oscillator's invariant set, for example, different ODEs could have been used for the top and bottom halves of the state space, simulating a system with one mode for positive y and another for negative y. It should be noted that this strategy would have problems dealing with a region which straddled or jittered along the boundary between two modes.

4.3   Squishing "Play-Doh"

Our example of a three dimensional non-linear system corresponds to squishing a lump of modeling clay. Consider a region shaped roughly like an octagonal hockey puck (or tuna can, for those unfamiliar with the Canadian obsession). Orient the puck so that its projection in the x-y plane is the octagon, and in the

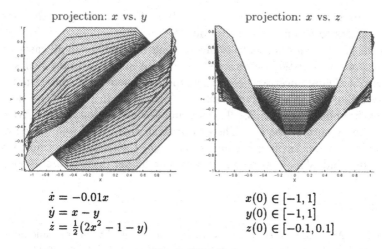

$$\dot{x} = -0.01x$$
$$\dot{y} = x - y$$
$$\dot{z} = \tfrac{1}{2}(2x^2 - 1 - y)$$

$$x(0) \in [-1, 1]$$
$$y(0) \in [-1, 1]$$
$$z(0) \in [-0.1, 0.1]$$

Fig. 7. Play-Doh

x-z plane a rectangle. The playdoh system takes this puck and squishes (in the y direction) toward the $y = x$ line, while bending the puck in its narrow dimension (the z direction) into a "V" shape. A small shrinking factor is applied in the x direction to offset Coho's over approximations. The resulting region is a thick, bent V lying at a 45 degree angle. While the squishing process is linear, bending is accomplished by a quadratic non-linear derivative function.

Although encountering many infeasible vertices and requiring many vertices introduced during numeric phases to be removed, Coho manages to track the two projections without an explosion in vertex count. Note, however, that the regions do not stay symmetric; a result caused by degeneracy handling code in the geometric phase.

5   Conclusions

This paper has presented Coho, a reachability analysis tool for systems modeled by ordinary differential equations. Coho uses projectahedra, an efficient method for representing high-dimensional objects as their projections onto two-dimensional subspaces. Non-linear models are handled by creating local linearizations for each face of the projectahedron. Each linear approximation includes an error bound which ensures the soundness of the analysis. We described the implementation of Coho, and have presented several example analyses, including linear and non-linear systems in two and three dimensions.

Implementation of Coho was completed recently; clearly, more examples will be needed to thoroughly validate our approach. The initial results presented in this paper are encouraging. The reachable state space estimates computed by Coho are more accurate than published results by other methods. The increased accuracy can be attributed to Coho's use of exact methods for analyzing linear systems combined with a flexible representation of reachable space that does not require vertices to lie on fixed grid-points.

Our set of examples has at least one obvious limitation—in all cases, system derivatives are completely determined by current state. In many real systems, input and modeling uncertainties lead to models where only constraints on the derivatives, but not exact values, can be determined. In [Gre96], we described several such models based on Brockett's annulus construction [Bro89]. We intend to try Coho on similar models. The examples in this paper were based on two and three dimensional systems. Coho was designed with models of up to twenty variables in mind. We are eager to try Coho on higher dimensional models.

References

[ACH+95] R. Alur, C. Courcoubetis, N. Halbwachs, et al. The algorithmic analysis of hybrid systems. *Theoretical Computer Science*, 138:3–34, 1995.

[AG96] Kenneth Arnold and James Gosling. *The Java Programming Language*. Addison-Wesley, 1996.

[Bro89] R. W. Brockett. Smooth dynamical systems which realize arithmetical and logical operations. In Hendrik Nijmeijer and Johannes M. Schumacher, editors, *Three Decades of Mathematical Systems Theory: A Collection of Surveys at the Occasion of the 50th Birthday of J. C. Willems*, volume 135 of *Lecture Notes in Control and Information Sciences*, pages 19–30. Springer, 1989.

[DM98] Thao Dang and Oded Maler. Reachability analysis via face lifting. In Thomas A. Henzinger and Shankar Sastry, editors, *Proceding of the First International Workshop on Hybrid Systems: Computation and Control*, pages 96–109, Berkeley, California, April 1998.

[GM98] Mark R. Greenstreet and Ian Mitchell. Integrating projections. In Thomas A. Henzinger and Shankar Sastry, editors, *Proceding of the First International Workshop on Hybrid Systems: Computation and Control*, pages 159–174, Berkeley, California, April 1998.

[Gre96] Mark R. Greenstreet. Verifying safety properties of differential equations. In *Proceedings of the 1996 Conference on Computer Aided Verification*, pages 277–287, New Brunswick, NJ, July 1996.

[HS74] Morris W. Hirsch and Stephen Smale. *Differential Equations, Dynamical Systems, and Linear Algebra*. Academic Press, San Diego, CA, 1974.

[MG96] Ian Mitchell and Mark Greenstreet. Proving Newtonian arbiters correct, almost surely. In *Proceedings of the Third Workshop on Designing Correct Circuits*, Båstad, Sweden, September 1996.

[PS85] Franco P. Preparata and Michael I. Shamos. *Computational Geometry: An Introduction*. Texts and Monographs in Computer Science. Springer, 1985.

[The92] The Mathworks Inc., Natick, Mass. *Matlab: High-Performance Numeric Computation and Visualization Software*, 1992. http://www.matlab.com.

[TPS97] Claire Tomlin, George Pappas, and Shankar Sastry. Conflict resolution for air traffic management: A case study in multi-agent hybrid systems. Technical Report UCB/ERL M97/33, Electronics Research Laboratory, University of California, Berkeley, 1997. to appear in IEEE Transactions on Automatic Control.

Scale-Independent Hysteresis Switching[*]

João P. Hespanha[1] and A. Stephen Morse[2]

[1] Dept. Electrical Eng. & Comp. Science, University of California at Berkeley
275M Cory Hall #1770, Berkeley, CA 94720-1770
hespanha@robotics.eecs.berkeley.edu
[2] Dept. Electrical Engineering, Yale University
P.O. Box 208267, New Haven, CT 06520-8267
morse@sysc.eng.yale.edu

Abstract. This paper introduces a new switching logic inspired by the hysteresis switching logic considered in [7, 11]. The new logic also uses hysteresis to prevent chatter, but unlike its predecessor in [7, 11], it is "scale-independent" as well. The logic is shown to have the requisite properties for adaptive control applications.

1 Introduction

"Scale-independence" is a property of certain switching algorithms used in an adaptive context which is key to proving an algorithm's correctness when operating in the face of noise and disturbance inputs [9]. The concept of *dwell-time switching* – exploited in [9] and elsewhere – has the advantage of being scale-independent. However, the existence of a prescribed dwell-time makes it impossible to rule out the possibility of finite escape in applications of dwell-time switching to the adaptive control of nonlinear systems [6]. On the other hand, the popular idea of *hysteresis switching* [7, 11] does not have this shortcoming. Unfortunately, hysteresis switching is not a scale-independent algorithm. The objective of this paper is to introduce a new form of chatter-free switching that does not employ a prescribed dwell-time and which is scale independent. We call this logic "scale-independent hysteresis switching" and we prove its correctness for applications to adaptive control [6].

Consider the hybrid dynamical system

$$\dot{x} = f_\sigma(x, t), \qquad\qquad x(0) = x_0 \qquad (1)$$

where $\{f_p : p \in \mathcal{P}\}$ is an indexed family of locally Lipschitz functions taking values on a finite dimensional space $\mathcal{X}$ and defined on $\mathcal{X} \times [0, \infty)$, and σ is a piecewise constant *switching signal* taking values in $\mathcal{P}$. The switching signal σ is chosen so as to cause the *performance signals*

$$\pi_p \stackrel{\triangle}{=} \Pi(p, x, t), \qquad\qquad p \in \mathcal{P} \qquad (2)$$

---

[*] This research was supported by the Air Force Office of Scientific Research, the Army Research Office, and the National Science Foundation.

F.W. Vaandrager and J.H. van Schuppen (Eds.): HSCC'99, LNCS 1569, pp. 117–122, 1999.
© Springer-Verlag Berlin Heidelberg 1999

to have certain desired properties. Here, Π is a *performance function* from $\mathcal{P} \times \mathcal{X} \times [0, \infty)$ to $\mathbb{R}$ that is continuous with respect to the second and third arguments for frozen values of the first.

The algorithm used to generate σ considered in this paper is called a *scale-independent hysteresis switching logic* and can be regarded as a hybrid dynamical system $\mathbb{S}_{\mathbb{H}}$ whose input is x and whose state and output are both σ. To specify $\mathbb{S}_{\mathbb{H}}$ it is necessary to first pick a positive number $h > 0$ called a *hysteresis constant*. $\mathbb{S}_{\mathbb{H}}$'s internal logic is then defined by the computer diagram shown in Figure 1 where the π_p are defined by (2) and, at each time t, $q \overset{\triangle}{=} \arg\min_{p \in \mathcal{P}} \Pi(p, x, t)$. In interpreting this diagram it is to be understood that σ's value at each of its

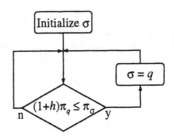

Fig. 1. Computer Diagram of $\mathbb{S}_{\mathbb{H}}$.

switching times $\bar{t}$ is equal to its limit from the right as $t \downarrow \bar{t}$. Thus if $\bar{t}_i$ and $\bar{t}_{i+1}$ are two consecutive switching times, then σ is constant on $[\bar{t}_i, \bar{t}_{i+1})$. The functioning of $\mathbb{S}_{\mathbb{H}}$ is roughly as follows. Suppose that at some time t_0, $\mathbb{S}_{\mathbb{H}}$ has just changed the value of σ to p. σ is then held fixed at this value unless and until there is a time $t_1 > t_0$ at which $(1 + h)\pi_q \leq \pi_p$ for some $q \in \mathcal{P}$. If this occurs, σ is set equal to q and so on. This type of logic has numerous applications in adaptive and supervisory control [7, 11, 8, 1, 6, 2, 3, 5].

The main result of this paper is the scale-independent hysteresis switching theorem. This theorem states that under appropriate "open-loop" assumptions, switching will stop at some finite time. Being able to establish that switching stops in finite time is crucial to the stability analysis of adaptive and supervisory control algorithms using hysteresis switching.

The switching logic described above is new. Its main advantage over the hysteresis switching logic considered in [7, 11] is that it is "scale-independent" in that its output σ remains unchanged if its performance function/input signal pair $\{\Pi, x\}$ is replaced by another pair $\{\bar{\Pi}, \bar{x}\}$ satisfying

$$\bar{\Pi}(p, \bar{x}, t) = \vartheta \Pi(p, x, t), \qquad \forall p \in \mathcal{P}, \, t \geq 0$$

where ϑ is a positive time function. This is because, for any fixed time t, (i) the value of p that minimizes $\Pi(p, x, t)$ is the same value of p that minimizes

$\bar{\Pi}(p, \bar{x}, t)$ and (ii) $(1 + h)\Pi(q, x, t) \leq \Pi(p, x, t)$ is exactly equivalent to $(1 + h)\bar{\Pi}(q, \bar{x}, t) \leq \bar{\Pi}(p, \bar{x}, t)$ for every $p, q \in \mathcal{P}$. Scale-independence often simplifies considerably the analysis of estimator-based supervisory control algorithms [9, 10, 6, 2, 3, 5].

This paper is organized as follows. Section 2 contains the statement and proof of the scale-independent hysteresis switching theorem. This theorem states that under appropriate "open-loop" assumptions, switching will stop at some finite time. In Sect. 3 it is illustrated how the scale-independence property can be used to apply the scale-independent hysteresis switching theorem to some systems for which the "open-loop" assumptions stated in Sect. 2 might be violated. Section 4 contains some concluding remarks.

2   Scale-Independent Hysteresis Switching Theorem

Let $\mathcal{X}_0$ denote a given subset of $\mathcal{X}$, and $\mathcal{S}$ the class of all piecewise-constant functions $s : [0, \infty) \to \mathcal{P}$. In what follows, for each pair $\{x_0, s\} \in \mathcal{X}_0 \times \mathcal{S}$, $T_{\{x_0, s\}}$ is the length of the maximal interval of existence of solution to the equations

$$\dot{x} = f_{s(t)}(x, t), \qquad\qquad x(0) = x_0$$

and $x_{\{x_0, s\}}$ is the corresponding solution. The following "open-loop" assumptions are made:

Assumption 1 (Open-Loop). *For each pair $\{x_0, s\} \in \mathcal{X}_0 \times \mathcal{S}$ the following is true:*

1. *There exists a positive constant ϵ such that for each $p \in \mathcal{P}$, the performance signal $\pi_p \overset{\triangle}{=} \Pi(p, x_{\{x_0, s\}}, t)$ is bounded below on $[0, T_{\{x_0, s\}})$ by ϵ.*
2. *For each $p \in \mathcal{P}$, the performance signal $\pi_p \overset{\triangle}{=} \Pi(p, x_{\{x_0, s\}}, t)$ has a limit[1] {which may be infinite} as $t \to T_{\{x_0, s\}}$.*
3. *There exists at least one $p^* \in \mathcal{P}$ such that the performance signal $\pi_{p^*} \overset{\triangle}{=} \Pi(p^*, x_{\{x_0, s\}}, t)$ is bounded on $[0, T_{\{x_0, s\}})$.*

First note that, because of the definition of $\mathbb{S}_{\mathbb{H}}$, $\sigma(0)$ must be such that $\pi_{\sigma(0)}(0)$ is strictly smaller than $(1 + h)\pi_p(0)$ for every $p \in \mathcal{P}$. Thus, because of the Lipschitz continuity of the f_p and the continuity of Π, there must exist an interval $[0, t_1)$ of maximal length on which $\pi_{\sigma(0)}$ remains strictly smaller than $(1 + h)\pi_p$ for every $p \in \mathcal{P}$. This means that σ is constant on $[0, t_1)$. Either this interval is the maximal interval of existence for x or it is not, in which case x is bounded on $[0, t_1)$ {cf. [4]}. If the latter is true, a switch must occur at t_1 and therefore $\pi_{\sigma(t_1)}(t_1) \leq \pi_p(t_1)$ for every $p \in \mathcal{P}$. Again, because of the Lipschitz continuity of the f_p, the continuity of Π, and the assumed boundedness below of each $h\pi_p$ by a positive constant, there must exist an interval $[t_1, t_2)$ of

---

[1] This assumption holds, for example, if the performance signals are monotone functions of time.

maximal length on which $\pi_{\sigma(t_1)}$ remains strictly smaller than $(1+h)\pi_p$ for every $p \in \mathcal{P}$. σ will then be constant on such an interval. Continuing this reasoning one concludes that there must be an interval $[0, T)$ of maximal length on which there is a unique pair $\{x, \sigma\}$ with x continuous and σ piecewise constant, which satisfies (1) with σ generated by $\mathbb{S}_\mathbb{H}$. Moreover, on each strictly proper subinterval $[0, \tau) \subset [0, T)$, σ can switch at most a finite number of times.

To establish existence of solution to (1) with σ generated by $\mathbb{S}_\mathbb{H}$, only the first Open-Loop Assumption was used. The remaining assumptions enable us to draw conclusions regarding the limiting behavior of σ as $t \to T$. The following is the main result of this chapter.

Theorem 1 (Scale-Independent Hysteresis Switching). *Let $\mathcal{P}$ be a finite set, assume that the Open-Loop Assumptions 1 hold and, for fixed initial state $\{x_0, \sigma_0\} \in \mathcal{X}_0 \times \mathcal{P}$, let $\{x, \sigma\}$ denote the unique solution to (1) with σ generated by $\mathbb{S}_\mathbb{H}$ with input x. If $[0, T)$ is the largest interval on which this solution is defined, there is a time $T^* < T$ beyond which σ is constant and $\pi_{\sigma(T^*)} \overset{\triangle}{=} \Pi(\sigma(T^*), x, t)$ is bounded on $[0, T)$.*

Proof. Let $\{x, \sigma\}$ denote the unique solution to (1) with σ generated by $\mathbb{S}_\mathbb{H}$ and suppose that $[0, T)$ is the largest interval on which this solution is defined. Also, let ϵ be a positive constant that bounds below each performance signal π_p, $p \in \mathcal{P}$. For each $t \in [0, T)$, the switching logic guarantees that

$$\pi_{\sigma(t)}(t) < (1+h)\pi_q(t) \tag{3}$$

with $\pi_q(t) \leq \pi_p(t)$ for every $p \in \mathcal{P}$. This is because if at some time t (3) were violated, σ would have to switch {at that precise instant of time} to that value q in $\mathcal{P}$ that minimizes $\pi_q(t)$ and therefore (3) would still be valid.

Because of the third Open-Loop Assumption 1, there is some $p^* \in \mathcal{P}$ such that $\pi_{p^*}(t)$ is bounded on $[0, T)$ by some positive constant K. From this and (3) one must have $\pi_{\sigma(t)}(t) \leq (1+h)K$ on $[0, T)$. Because of this and the monotonicity of the π_p, if a performance signal π_p becomes larger than $(1+h)K$, σ will not switch to p ever again that. Thus, after some finite time $\bar{T} < T$, $\sigma(t)$ must remain inside the following nonempty set

$$\mathcal{P}^* \overset{\triangle}{=} \left\{ p \in \mathcal{P} : |\pi_p(\tau)| \leq (1+h)K, \ \tau \in [0, T) \right\}$$

To conclude the proof it is then enough to shown that there is a time $T^* \in [\bar{T}, T)$ beyond which σ actually becomes constant. Because of the second Open-Loop Assumption 1, for each $p \in \mathcal{P}^*$, $\pi_p(t)$ converges to a finite limit as $t \to \infty$. Thus, for each $p \in \mathcal{P}^*$, there must exist a time $T_p \in [\bar{T}, T)$ such that

$$|\pi_p(t) - \pi_p(\tau)| < \frac{h\epsilon}{2}, \qquad \forall t, \tau \in [T_p, T) \tag{4}$$

Because $\mathcal{P}$ is a finite set, $\mathcal{P}^*$ is also a finite set and therefore the set $\{T_p : p \in \mathcal{P}^*\}$ has a maximum element. Let $T_1 \in [\bar{T}, T)$ be such element. If there is no switching in $[T_1, T)$ then one can take $T^* \triangleq T_1$ and the proof is finished. Otherwise, let $T_2 \in (T_1, T)$ be the time instant at which the next switching occurs. For a given $t \in [T_2, T)$, let q be that element in $\mathcal{P}$ that minimizes $\pi_q(t)$. Since $\sigma(T_2), q \in \mathcal{P}^*$, $T_1 \geq T_{\sigma(T_2)}$, and $T_1 \geq T_q$, from (4) one concludes that

$$\pi_{\sigma(T_2)}(T_2) > \pi_{\sigma(T_2)}(t) - \frac{h\epsilon}{2} \qquad \text{and} \qquad \pi_q(T_2) < \pi_q(t) + \frac{h\epsilon}{2} \qquad (5)$$

But $\sigma(T_2)$ is the element $p \in \mathcal{P}$ that minimizes $\pi_p(T_2)$, thus $\pi_{\sigma(T_2)}(T_2) \leq \pi_q(T_2)$. From this and (5) one concludes that

$$\pi_{\sigma(T_2)}(t) - \frac{h\epsilon}{2} < \pi_{\sigma(T_2)}(T_2) \leq \pi_q(T_2) < \pi_q(t) + \frac{h\epsilon}{2}$$

and therefore that

$$\pi_{\sigma(T_2)}(t) < \pi_q(t) + h\epsilon \qquad (6)$$

Moreover, because of the first Open-Loop Assumption 1, $\pi_q(t) > \epsilon$. From this and (6) one concludes that

$$\pi_{\sigma(T_2)}(t) < (1 + h)\pi_q(t)$$

Thus there can be no more switching at any time $t \in [T_2, T)$ and one can take $T^* \triangleq T_2$. $\qquad\qquad\qquad\qquad\qquad\qquad\qquad\qquad\qquad\qquad\qquad\qquad\qquad\qquad\qquad\square$

3   Relaxing the Open-Loop Assumptions

The scale-independence property can be use to somewhat relax the "open-loop" assumptions stated in the previous section. Suppose that the following assumptions hold.

Assumption 2 (Relaxed Open-Loop). *There exists a positive time-function ϑ such that for each pair $\{x_0, s\} \in \mathcal{X}_0 \times \mathcal{S}$ the following is true:*

1. *There exists a positive constant ϵ such that for each $p \in \mathcal{P}$, the scaled performance signal $\bar{\pi}_p \triangleq \vartheta \Pi(p, x_{\{x_0,s\}}, t)$ is bounded below on $[0, T_{\{x_0,s\}})$ by ϵ.*
2. *For each $p \in \mathcal{P}$, the scaled performance signal $\bar{\pi}_p \triangleq \vartheta \Pi(p, x_{\{x_0,s\}}, t)$ has a limit {which may be infinite} as $t \to T_{\{x_0,s\}}$.*
3. *There exists at least one $p^* \in \mathcal{P}$ such that the scaled performance signal $\bar{\pi}_{p^*} \triangleq \vartheta \Pi(p^*, x_{\{x_0,s\}}, t)$ is bounded on $[0, T_{\{x_0,s\}})$.*

In light of the scale independence property, $\mathbb{S}_{\mathbb{H}}$'s output σ remains unchanged if its performance function/input signal pair $\{\Pi, x\}$ is replaced by another pair $\{\bar{\Pi}, \bar{x}\}$ satisfying

$$\bar{\Pi}(p, \bar{x}, t) = \vartheta \Pi(p, x, t), \qquad\qquad \forall p \in \mathcal{P}$$

Since Assumptions 2 guarantee that the Open-Loop Assumptions 1 hold for the pair $\{\bar{\Pi}, \bar{x}\}$, we obtain the following corollary of Theorem 1.

Corollary 1 (Scale-Independent Hysteresis Switching). *Let $\mathcal{P}$ be a finite set, assume that the Relaxed Open-Loop Assumptions 2 hold and, for fixed initial state $\{x_0, \sigma_0\} \in \mathcal{X}_0 \times \mathcal{P}$, let $\{x, \sigma\}$ denote the unique solution to (1) with σ generated by $\mathbb{S}_\mathbb{H}$ with input x. If $[0, T)$ is the largest interval on which this solution is defined, there is a time $T^* < T$ beyond which σ is constant and $\bar{\pi}_{\sigma(T^*)} \overset{\triangle}{=} \vartheta \Pi(\sigma(T^*), x, t)$ is bounded on $[0, T)$.*

It should be emphasized that in supervisory control it is often the case that Assumptions 1 might be violated, whereas Assumptions 2 can be shown to hold [6, 2, 3, 5].

4   Conclusions

A new type of switching logic called scale-independent hysteresis switching was introduced. This logic is inspired on the one in [7, 11], but has the advantage that it is scale-independent in that its output remains unchanged if the performance signals are scaled by a positive time function. The reader is referred to [7, 11, 8, 1, 6, 2, 3, 5] for the use of hysteresis switching logics in adaptive and supervisory control.

References

1. J. Balakrishnan. *Control Systems Design Using Multiple Models, Switching, and Tuning.* PhD thesis, Yale University, 1995.
2. G. Chang, J. P. Hespanha, A. S. Morse, M. Netto, and R. Ortega. Supervisory field-oriented control of induction motors with uncertain rotor resistance. In *Proc. of the 1998 IFAC Workshop on Adaptive Control and Signal Processing, Glasgow, Scotland,* Aug. 1998.
3. S. Fujii, J. P. Hespanha, and A. S. Morse. Supervisory control of families of noise suppressing controllers. To be presented at the 37th Conf. on Decision and Control, Dec. 1998.
4. J. K. Hale. *Ordinary Differential Equations.* Wiley-Interscience, 1969.
5. J. P. Hespanha, D. Liberzon, and A. S. Morse. Towards the supervisory control of uncertain nonholonomic systems. Submitted to the 1999 American Contr. Conf., July 1998.
6. J. P. Hespanha and A. S. Morse. Certainty equivalence implies detectability. To appear in Systems & Control Letters.
7. R. H. Middleton, G. C. Goodwin, D. J. Hill, and D. Q. Mayne. Design issues in adaptive control. *IEEE Trans. Automat. Contr.,* 33(1):50–58, Jan. 1988.
8. A. S. Morse. Control using logic-based switching. In A. Isidori, editor, *Trends in Control: A European Perspective,* pages 69–113. Springer-Verlag, London, 1995.
9. A. S. Morse. Supervisory control of families of linear set-point controllers—part 1: exact matching. *IEEE Trans. Automat. Contr.,* 41(10):1413–1431, Oct. 1996.
10. A. S. Morse. Supervisory control of families of linear set-point controllers—part 2: robustness. *IEEE Trans. Automat. Contr.,* 42(11):1500–1515, Nov. 1997.
11. A. S. Morse, D. Q. Mayne, and G. C. Goodwin. Applications of hysteresis switching in parameter adaptive control. *IEEE Trans. Automat. Contr.,* 37(9):1343–1354, Sept. 1992.

Well-Posedness of a Class of Piecewise Linear Systems with No Jumps [*]

Jun-ichi Imura[1] and Arjan van der Schaft[2]

[1] Division of Machine Design Engineering, Hiroshima University
Higashi-Hiroshima 739, Japan
imura@mec.hiroshima-u.ac.jp, or imura@math.utwente.nl
[2] Faculty of Mathematical Sciences, University of Twente
P.O. Box 217, 7500 AE Enschede, the Netherlands, and
CWI, P.O. Box 94079, 1090 GB Amsterdam, the Netherlands
a.j.vanderschaft@math.utwente.nl

Abstract. In this paper, a well-posedness (existence and uniqueness of solutions) problem of bimodal systems given by two linear systems is addressed, where the definition of solutions of Carathéodory is used. This problem is a basic problem in the study of well-posedness for discontinuous dynamical systems. We give here a complete answer to this problem. The obtained result shows that the well-posedness of bimodal systems can be characterized by two properties: the preservation property of the lexicographic inequality relation between the two regions specifying the two modes, and the smooth continuation property.

1 Introduction

Various kinds of models describing hybrid systems with a corresponding theoretical framework have been developed in the literatures (see e.g., [8, 11]). However, when we consider general hybrid systems, it is not easy to analyze properties of their solutions and to obtain systematic tools for deriving controllers as in the case of linear dynamical systems or smooth nonlinear dynamical systems. As a result, several recent papers treat special classes of hybrid systems to make the analysis easier and to obtain general theoretical tools ([13, 9, 15, 10, 16, 6] etc.). Especially, in [15, 10, 16, 6], a special class of hybrid systems called complementarity systems has been formulated. This system is a generalization of physical systems with jump phenomena which occurs between unconstrained motion and constrained motion, such as the collision of a mass to a hard wall, and can describe various hybrid systems including relay type systems. In addition, several algebraic and checkable conditions for the well-posedness, that is, the existence and uniqueness of the solutions of such systems, have been derived.

On the other hand, for physical phenomena such as the collision to an elastic wall, the system has a discontinuous vector field and does not exhibit jumps and

---

[*] This research has been performed while the first author being a research fellow of Canon foundation at Faculty of Mathematicak Sciences in University of Twente.

F.W. Vaandrager and J.H. van Schuppen (Eds.): HSCC'99, LNCS 1569, pp. 123–136, 1999.
© Springer-Verlag Berlin Heidelberg 1999

sliding modes. As far as we know, necessary and sufficient conditions for the well-posedness of such systems have not been obtained within a general framework, although we can find some results relating to well-posedness of such systems in [5, 7, 14, 4, 1, 17]. When we consider solutions without jumps, there are, roughly speaking, two kinds of definitions of solutions, namely, Carathéodory's definition and Filippov's definition [4]. The latter yields the concept of a sliding mode. In the case of physical systems such as the collision to an elastic wall, the solution belongs to the former, although we need to extend Carathéodory's definition, in a straightforward manner, to the case of discontinuous vector fields.

Thus in this paper, we discuss a well-posedness problem in the sense of Carathéodory of a bimodal system where two linear dynamics are interconnected by the rule that some criterion value is negative or positive. The resulting dynamics belongs to the class of so-called piecewise linear systems. This problem is a most fundamental one of well-posedness problems for discontinuous dynamical systems, because we do not treat jump phenomena and we do not use the concept of a sliding mode. Therefore, as a first step to various developments of hybrid systems, it will be very meaningful to clarify to what extent this fundamental problem can be analyzed. Furthermore, it is desirable from practical points of view that there exist no sliding modes in feedback control systems.

After the description of the bimodal systems to be studied here, we give some mathematical preliminaries on the lexicographic inequality and the smooth continuation of the solution, which will play a significant role in the answer to the well-posedness problem. Next, we derive a necessary and sufficient condition for the bimodal system to be well-posed under Carathéodory's definition of solutions. The obtained condition is easily checkable, and several extensions will be possible. Finally, as an application to switching control systems, in the case that two state feedback gains are switched according to a criterion depending on the state, we give a characterization of all feedback gains for which the closed loop system remains well-posed.

In the sequel, we will use the following notation for lexicographic inequalities: for $x \in \mathcal{R}^n$, if for some i, $x_j = 0$ $(j = 1, 2, \cdots, i - 1)$, while $x_i > (<)0$, we denote it by $x \succ (\prec)0$. In addition, if $x = 0$ or $x \succ (\prec)0$, we denote it $x \succeq (\preceq)0$. We use the notation $*$ representing any fixed but unspecified number or matrix. Finally, I_n denotes the $n \times n$ identity matrix.

2   Bimodal Systems

2.1   Description of Bimodal Systems

Consider the piecewise linear system given by

$$\Sigma_O \begin{cases} \text{mode 1}: \dot{x} = Ax, \text{ if } y = Cx \geq 0 \\ \text{mode 2}: \dot{x} = Bx, \text{ if } y = Cx \leq 0 \end{cases} \tag{1}$$

where $x \in \mathcal{R}^n$, $y \in \mathcal{R}$, and A and B are $n \times n$ matrices (in general different).

For simplicity of notation, we use here $\dot{x}(t)$ in (1), although there may be a set (of measure 0) of points of time where the solution is not differentiable. Formally, the system Σ_O is given by its integral form which is called the Carathéodory equation:

$$x(t) = x(t_0) + \int_{t_0}^{t} f(x(\tau))d\tau \qquad (2)$$

where $f(x)$ is a discontinuous vector field given by the right hand side of (1). We call the $x(t)$ given by this equation the solution in the sense of Carathéodory.

Then the well-posedness for the system Σ_O is defined as follows.

Definition 1. *The system Σ_O is said to be well-posed at x_0 if there exists a unique solution locally in the sense of Carathéodory at the initial state x_0 in $\mathcal{R}^n$. In addition, the system Σ_O is said to be well-posed if it is well-posed at every initial state $x_0 \in \mathcal{R}^n$.*

The following result shows that we only have to prove local existence and uniqueness of solutions at every initial state in order to show the well-posedness of the system Σ_O.

Lemma 1. *If there exists an $\varepsilon > 0$ such that a unique solution $x(t)$ of Σ_O exists on $[0, \varepsilon)$ in the sense of Carathéodory from every initial state $x_0 \in \mathcal{R}^n$, then the system Σ_O is well-posed and the solution is absolutely continuous on any interval of $\mathcal{R}$.*

Proof. Since there exists a local unique solution from every initial state, we can make a successively connected solution. Then the solution $x(t)$ in (2) is given by $x(t) = e^{S_i(t-t_i)}e^{S_{i-1}(t_i-t_{i-1})} \cdots e^{S_0 t_1}x(0)$ for all $t \in [t_i, t_i + \varepsilon)$, where $i \in \{0, 1, 2, \cdots\}$ is the switching number, t_j is a switching time ($t_0 = 0$), and $S_j = A$ or B ($j = 0, 1, 2, \cdots, i$). Since there exists a positive real number a such that $\max\{\| e^{At} \|, \| e^{Bt} \|\} \le e^{at}$ for all $t \ge 0$, it follows that $\| x(t) \| \le e^{at} \| x(0) \|$ for all $t \in [t_i, t_i + \varepsilon)$ and all $i \in \{0, 1, 2, \cdots\}$. Noting that there exists a unique solution for all $t \ge t_\infty$ even when $t_\infty < \infty$ (i.e., a finite accumulation point of switching times exists), we have $x \in \mathcal{L}_{\infty e}$ (extended $\mathcal{L}_\infty$ space). Thus there exists a unique solution $x(t)$ on $[0, \infty)$. In addition, since $f(x) \in \mathcal{L}_{1e}$ (with $f(x)$ defined by (2)) holds from $x \in \mathcal{L}_{\infty e}$, it follows from Lebesgue integral theory that the solution given by (2) is absolutely continuous on any interval of $\mathcal{R}$. $\square$

It is well-known that a sufficient condition for a system given by a first-order differential equation to be well-posed is that it satisfies a global Lipschitz condition. When we apply this to the system Σ_O, it follows that a sufficient condition for well-posedness is that there exists a K such that $B = A + KC$. Note that in this case the vector field is necessarily continuous in the state x.

Now, how about the case of discontinuous vector fields? Let us consider the following example shown in Fig. 1. The equations of motion of this system are given by

$$\begin{cases} \text{mode 1} : \begin{bmatrix} \dot{x}_1 \\ \dot{x}_2 \end{bmatrix} = \begin{bmatrix} 0 & 1 \\ 0 & 0 \end{bmatrix} \begin{bmatrix} x_1 \\ x_2 \end{bmatrix}, & \text{if } y = \begin{bmatrix} 1 & 0 \end{bmatrix} \begin{bmatrix} x_1 \\ x_2 \end{bmatrix} \ge 0 \\ \text{mode 2} : \begin{bmatrix} \dot{x}_1 \\ \dot{x}_2 \end{bmatrix} = \begin{bmatrix} 0 & 1 \\ -k & -d \end{bmatrix} \begin{bmatrix} x_1 \\ x_2 \end{bmatrix}, & \text{if } y = \begin{bmatrix} 1 & 0 \end{bmatrix} \begin{bmatrix} x_1 \\ x_2 \end{bmatrix} \le 0 \end{cases} \qquad (3)$$

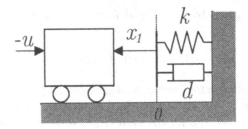

Fig. 1. Collision to elastic wall

From simple calculations, we see that this system is well-posed (without jumps and sliding modes), although the vector field is discontinuous in x when $d \neq 0$.

On the other hand, we can easily find an example which is not well-posed, as shown below.

$$
\begin{cases}
\text{mode } 1: \begin{bmatrix} \dot{x}_1 \\ \dot{x}_2 \end{bmatrix} = \begin{bmatrix} 0 & 1 \\ 0 & 0 \end{bmatrix} \begin{bmatrix} x_1 \\ x_2 \end{bmatrix}, & \text{if } y = \begin{bmatrix} 1 & 0 \end{bmatrix} \begin{bmatrix} x_1 \\ x_2 \end{bmatrix} \geq 0 \\[2ex]
\text{mode } 2: \begin{bmatrix} \dot{x}_1 \\ \dot{x}_2 \end{bmatrix} = \begin{bmatrix} 0 & -1 \\ 0 & 0 \end{bmatrix} \begin{bmatrix} x_1 \\ x_2 \end{bmatrix}, & \text{if } y = \begin{bmatrix} 1 & 0 \end{bmatrix} \begin{bmatrix} x_1 \\ x_2 \end{bmatrix} \leq 0
\end{cases}
$$

In fact, if the initial state $x(0)$ satisfies $x_1(0) = 0$ and $x_2(0) = 1$, then the solution $x(t)$ in mode 1 belongs to the region $x_1 > 0$, and the solution $x(t)$ in mode 2 belongs to the region $x_1 < 0$. Thus there exist two solutions for this initial state.

Within the type of physical systems as given by (3), there will exist many systems with discontinuous vector fields, but which are well-posed. What algebraic structure exists in the vector field of such a system? Can we extend this property to the general case from the mathematical viewpoint? The purpose of this paper is to address these questions.

Remark 1. Let us consider the system given by

$$
\begin{cases}
\text{mode } 1: \begin{bmatrix} \dot{x}_1 \\ \dot{x}_2 \end{bmatrix} = \begin{bmatrix} 0 & -1 \\ 0 & 0 \end{bmatrix} \begin{bmatrix} x_1 \\ x_2 \end{bmatrix}, & \text{if } y = \begin{bmatrix} 1 & 0 \end{bmatrix} \begin{bmatrix} x_1 \\ x_2 \end{bmatrix} \geq 0 \\[2ex]
\text{mode } 2: \begin{bmatrix} \dot{x}_1 \\ \dot{x}_2 \end{bmatrix} = \begin{bmatrix} 0 & 1 \\ 0 & 0 \end{bmatrix} \begin{bmatrix} x_1 \\ x_2 \end{bmatrix}, & \text{if } y = \begin{bmatrix} 1 & 0 \end{bmatrix} \begin{bmatrix} x_1 \\ x_2 \end{bmatrix} \leq 0
\end{cases}
$$

The system is not well-posed at $(x_1, x_2) = (0, 1)$ in the sense of Definition 1. However, if we use Filippov's definition, there exists a unique solution from the initial state $(x_1(0), x_2(0)) = (0, 1)$. In fact, the system Σ_O can be rewritten by $\dot{x} = \frac{1}{2}(1 + u)Ax + \frac{1}{2}(1 - u)Bx$, using a relay-type input of $u = \text{sgn}(y)$. Thus when $(x_1(0), x_2(0)) = (0, 1)$, there exists a unique solution given by the equivalent control input $u = 0$. Certainly, Filippov's definition is very important from the practical viewpoint as well as the mathematical viewpoint. However, in this paper, as a first step, we concentrate on the well-posedness in Definition 1.

Remark 2. When we consider the case of $d \to \infty$ in the example (3), a jump of the solution will occur. Such a system can be treated within the framework of complementarity systems. Thus we conjecture that there is some relation between complementarity systems and the system given by (1). In other words, there may be some possibility to approximate the complementarity system, i.e., the discontinuous dynamical system with jumps, by a system without jumps given by (1). Some researchers have already studied the relation between two solutions for a simple physical system as in Fig. 1 (see e.g., [2]), and we plan to return to this issue in a future paper.

2.2 Equivalent Representation of Bimodal System Σ_O

For the system Σ_O, define the following row-full rank matrices.

$$
T_A \triangleq \begin{bmatrix} C \\ CA \\ \vdots \\ CA^{h-1} \end{bmatrix}, \quad T_B \triangleq \begin{bmatrix} C \\ CB \\ \vdots \\ CB^{k-1} \end{bmatrix} \tag{4}
$$

where h and k are the observability indexes of the pairs (C, A) and (C, B) (cf. [3]), respectively. In addition, let $\mathcal{S}_A^+$, $\mathcal{S}_A^-$, $\mathcal{S}_B^+$, and $\mathcal{S}_B^-$ be sets defined by

$$
\mathcal{S}_N^+ \triangleq \{x \in \mathcal{R}^n \mid T_N x \succeq 0\}, \quad \mathcal{S}_N^- \triangleq \{x \in \mathcal{R}^n \mid T_N x \preceq 0\} \tag{5}
$$

for $N = A, B$. Then noting that $T_A x = [y, \dot{y}, \cdots, y^{(h-1)}]^T$ for the system $\dot{x} = Ax$ and $T_B x = [y, \dot{y}, \cdots, y^{(k-1)}]^T$ for the system $\dot{x} = Bx$, we consider the system given by

$$
\Sigma_{AB} \begin{cases} \text{mode } 1 : \dot{x} = Ax, & \text{if } x \in \mathcal{S}_A^+ \\ \text{mode } 2 : \dot{x} = Bx, & \text{if } x \in \mathcal{S}_B^- \end{cases} . \tag{6}
$$

We call T_A and T_B the rule (or observability) matrices of the system Σ_{AB}. The well-posedness for the system Σ_{AB} is defined similar to Definition 1. The following result shows that the system Σ_O is well-posed if and only if the system Σ_{AB} is well-posed.

Lemma 2. *The system Σ_{AB} is equivalent to the original system Σ_O, i.e., both systems have the same solutions.*

Proof. If $y(t) = Cx(t) \geq 0$ for $\dot{x} = Ax$, then $T_A x(t) \succeq 0$. Conversely, if $T_A x(t) \succ 0$ for $\dot{x} = Ax$, then $y(t) = Cx(t) \geq 0$ is obvious. When $T_A x(t) = 0$, the definition of the observability index implies that $y(t) \equiv 0$. The case of $\dot{x} = Bx$ is similar. Thus modes 1 and 2 of Σ_{AB} are equivalent to those of Σ_O, respectively, which implies that they have the same solutions. □

Thus, we will discuss the well-posedness of the system Σ_{AB} in the next sections. Note that the claim in Lemma 1 is still true for Σ_{AB}.

3 Preliminaries on Lexicographic Inequalities and Smooth Continuation

3.1 Lemmas on Lexicographic Inequalities

First, we give some lemmas on lexicographic inequalities. All the proofs are omitted due to limited space [12]. Throughout this subsection, x will be a vector in $\mathcal{R}^n$.

Lemma 3. *Let T be a $m \times n$ real matrix with $m \leq n$ and rank $T =$ rank $T_1 = r$, where $T = [T_1^T \; T_2^T]^T$ and $T_1 \in \mathcal{R}^{r \times n}$. Then $Tx \succeq (\preceq)0$ if and only if $T_1 x \succeq (\preceq)0$.*

This lemma shows that the row full-rank submatrix T_1 of the matrix T is enough for representing the relation by the lexicographic inequality. Thus the following result is obtained: let T be a $m \times n$ matrix and t_i^T be the ith row vector of T. Let also $T_i \triangleq [t_1 \; t_2 \; \cdots t_i]^T$. Suppose that rank $T_i = $ rank $T_{i+1} = i$. Then from Lemma 3, we can use, in place of T, $\tilde{T} = [t_1 \; \cdots \; t_i \; t_{i+2} \; \cdots \; t_m]$, which is made by removing the $i + 1$th column t_{i+1} from T. Hence we can assume that T is row-full rank, whenever we consider $Tx \succeq (\preceq)0$.

Definition 2. *Let $\mathcal{L}^n$ be the set of $n \times n$ lower-triangular matrices. In addition, let $\mathcal{L}^n_+$ be the set of elements in $\mathcal{L}^n$ with all diagonal elements positive.*

The following lemma shows that the set $\mathcal{L}^n_+$ characterizes the coordinate transformations conserving the lexicographic inequalities property.

Lemma 4. *Let T be an $n \times n$ real matrix. Then $x \succeq (\preceq)0 \leftrightarrow Tx \succeq (\preceq)0$ if and only if $T \in \mathcal{L}^n_+$.*

From the definition of the lexicographic inequality, if follows that for any nonsingular $n \times n$ matrix T we have the property

$$\{x \in \mathcal{R}^n \mid Tx \succeq 0\} \bigcup \{x \in \mathcal{R}^n \mid Tx \preceq 0\} = \mathcal{R}^n,$$
$$\{x \in \mathcal{R}^n \mid Tx \succeq 0\} \bigcap \{x \in \mathcal{R}^n \mid Tx \preceq 0\} = \{0\}.$$

The following lemma generalizes this property to the singular matrix case.

Lemma 5. *Let T and S be $l \times n$ and $m \times n$ real matrices with rank $T = l$, rank $S = m$, and $l \geq m$. Then the following statements are equivalent.*
(i) $\{x \in \mathcal{R}^n \mid Tx \succeq 0\} \bigcup \{x \in \mathcal{R}^n \mid Sx \preceq 0\} = \mathcal{R}^n$.
(ii) $S = [M \; 0]T$ for some $M \in \mathcal{L}^m_+$.

3.2 Characterization of Smooth Continuation Property

If all the solutions of the n-dimensional linear system $\dot{x} = Ax$ locally preserve the lexicographic inequality relation, that is, for each initial state $x(0)$ satisfying $x(0) \succ (\prec)0$, there exists an $\varepsilon > 0$ such that $x(t) \succ (\prec)0$ for all $t \in [0, \varepsilon]$, then we say that the system has the smooth continuation property, or smooth continuation in the system is possible [15]. In this subsection, we derive a necessary and sufficient condition for this property.

Definition 3. *Let $\mathcal{G}_0^n$ be the set defined by*

$$
\mathcal{G}_0^n \triangleq \left\{ \Gamma \in \mathcal{R}^{n \times n} \,\middle|\, \Gamma = \begin{bmatrix} * & \gamma_{12} & 0 & \cdots & 0 \\ \vdots & \ddots & \ddots & \ddots & \vdots \\ \vdots & & \ddots & \ddots & 0 \\ \vdots & & & \ddots & \gamma_{n-1,n} \\ * & \cdots\cdots\cdots & & & * \end{bmatrix}, \gamma_{i,i+1} \geq 0, i = 1, 2, \cdots, n-1 \right\}
$$

where γ_{ij} is the (i,j) element of the matrix Γ. In addition, if $\gamma_{i,i+1} > 0$ for all $i \in \{1, 2, \cdots, n-1\}$, we denote it by $\mathcal{G}_+^n$.

The set $\mathcal{G}_0$ will characterize the smooth continuation property of linear systems as follows.

Lemma 6. *Consider the system $\dot{x} = Ax$, where $x \in \mathcal{R}^n$. Then the following conditions are equivalent.*
(i) The system has the smooth continuation property.
(ii) $A \in \mathcal{G}_0^n$
(iii) There exists a matrix $T \in \mathcal{L}_+^n$ such that

$$
TAT^{-1} = \begin{bmatrix} \tilde{A}_{11} & 0 & \cdots & 0 \\ \tilde{A}_{21} & \tilde{A}_{22} & \ddots & \vdots \\ \vdots & \ddots & \ddots & 0 \\ \tilde{A}_{p1} & \cdots & \tilde{A}_{p,p-1} & \tilde{A}_{pp} \end{bmatrix} \tag{7}
$$

where

$$
\tilde{A}_{ii} = \begin{bmatrix} 0 & 1 & 0 & \cdots & 0 \\ 0 & 0 & 1 & \ddots & \vdots \\ \vdots & \vdots & \ddots & \ddots & \vdots \\ 0 & 0 & \cdots & 0 & 1 \\ * & \cdots\cdots\cdots & & & * \end{bmatrix} \in \mathcal{R}^{n_i \times n_i}, \quad \tilde{A}_{ij} = \begin{bmatrix} * & \cdots & * \\ \vdots & & \vdots \\ * & \cdots & * \end{bmatrix} \in \mathcal{R}^{n_i \times n_j}, \quad \text{for } i > j,
$$

and $n = n_1 + n_2 + \cdots + n_p$ ($p \in \{1, 2, \cdots, n\}$).

4   Characterization of Well-posedness of Bimodal Systems

In this section, we discuss the well-posedness of Σ_O, or equivalently Σ_{AB}. First, we give a result in the case where both pairs (C, A) and (C, B) are observable. This will clarify a fundamental issue in the algebraic structure for the well-posed bimodal system. Next, the unobservable case is treated, as a generalization of the observable case.

4.1   Observable Case

In this subsection, we assume that the pairs (C, A) and (C, B) are observable, namely, T_A and T_B are nonsingular, where

$$T_A \triangleq \begin{bmatrix} C \\ CA \\ \vdots \\ CA^{n-1} \end{bmatrix}, \quad T_B \triangleq \begin{bmatrix} C \\ CB \\ \vdots \\ CB^{n-1} \end{bmatrix}. \tag{8}$$

In addition, we consider the following two systems:

$$\Sigma_A \begin{cases} \text{mode 1}: \dot{x} = Ax, \text{ if } x \in \mathcal{S}_A^+ \\ \text{mode 2}: \dot{x} = Bx, \text{ if } x \in \mathcal{S}_A^- \end{cases}, \tag{9}$$

$$\Sigma_B \begin{cases} \text{mode 1}: \dot{x} = Ax, \text{ if } x \in \mathcal{S}_B^+ \\ \text{mode 2}: \dot{x} = Bx, \text{ if } x \in \mathcal{S}_B^- \end{cases} \tag{10}$$

where $\mathcal{S}_N^+$ and $\mathcal{S}_N^-$ $(N = A, B)$ are given by (5). Utilizing the fact that $\mathcal{S}_A^+ \bigcup \mathcal{S}_A^- = \mathcal{R}^n$, the system Σ_A is given by a rule matrix T_A only. The system Σ_B is also defined by a rule matrix T_B, in a similar way. Then we get the following result.

Theorem 1. *Suppose that both pairs (C, A) and (C, B) are observable. Then the following statements are equivalent.*
(i) Σ_{AB} is well-posed.
(ii) Σ_A is well-posed.
(iii) Σ_B is well-posed.
(iv) $\mathcal{S}_A^+ \bigcup \mathcal{S}_B^- = \mathcal{R}^n$ and $\mathcal{S}_A^+ \bigcap \mathcal{S}_B^- = \{0\}$.
(v) $T_B T_A^{-1} \in \mathcal{L}_+^n$.
(vi) $T_A B T_A^{-1} \in \mathcal{G}_+^n$.
(vii) $T_B A T_B^{-1} \in \mathcal{G}_+^n$.

Proof. First, we prove (i)→(v)→ (iv) → (i).

(i)→(v). $\mathcal{S}_A^+ \bigcup \mathcal{S}_B^- = \mathcal{R}^n$ is obviously necessary for the well-posedness. From Lemma 5, there exists a $M \in \mathcal{L}_+^n$ such that $T_B = M T_A$. (v)→ (iv) follows from Lemma 5. (iv) → (i). Note that $T_A A T_A^{-1} \in \mathcal{G}_0$ and $T_B B T_B^{-1} \in \mathcal{G}_0$, because T_A and T_B are the observability matrices. From Lemma 6, this guarantees the smooth continuation property for each mode. Hence, (iv) implies that the system Σ_{AB} has a unique solution at every initial state.

Next, we prove (v)→(ii)→(vi)→ (v).

(v)→(ii). Since (v) implies from Lemma 4 that $\mathcal{S}_A^- = \mathcal{S}_B^-$, Σ_{AB} is equivalent to Σ_A. Since Σ_{AB} is well-posed from (v), Σ_A is also well-posed. (ii)→(vi). Transforming the system Σ_A into the new coordinates $z = [z_1 \ z_2 \cdots z_n]^T = T_A x$, we get

$$\tilde{\Sigma}_A \begin{cases} \text{mode 1}: \dot{z} = T_A A T_A^{-1} z, \text{ if } z \succeq 0 \\ \text{mode 2}: \dot{z} = T_A B T_A^{-1} z, \text{ if } z \preceq 0 \end{cases}.$$

Then (ii) implies that smooth continuation occurs for each mode of Σ_A. Thus from Lemma 6, (ii) implies $T_A B T_A^{-1} \in \mathcal{G}_0$. Letting γ_{ij} be the element of $\Gamma \overset{\Delta}{=} T_A B T_A^{-1}$, and noting $C T_A^{-1} = [1\ 0\ \cdots\ 0]$, we get

$$
\begin{aligned}
CB &= C T_A^{-1} \Gamma T_A &&= [*\ \gamma_{12}\ 0\ \cdots\cdots\ 0] T_A, \\
CB^2 &= C T_A^{-1} \Gamma^2 T_A &&= [*\ *\ \gamma_{12}\gamma_{23}\ 0\ \cdots\ 0] T_A, \\
&\ \vdots \\
CB^{n-1} &= C T_A^{-1} \Gamma^{n-1} T_A &&= [*\ \cdots\cdots\ *\ \Pi_{i=1}^{n-1}\gamma_{i,i+1}] T_A.
\end{aligned}
\tag{11}
$$

Thus we obtain

$$
T_B = L T_A \tag{12}
$$

where

$$
L \overset{\Delta}{=}
\begin{bmatrix}
1 & 0 & \cdots & \cdots & & 0 \\
* & \gamma_{12} & \ddots & & & \vdots \\
\vdots & \ddots & \gamma_{12}\gamma_{23} & \ddots & & \vdots \\
\vdots & & & \ddots & \ddots & 0 \\
* & \cdots & \cdots & & * & \Pi_{i=1}^{n-1}\gamma_{i,i+1}
\end{bmatrix}.
\tag{13}
$$

This implies that all $\gamma_{i,i+1}$ are positive, since T_A and T_B are nonsingular. Hence $T_A B T_A^{-1} \in \mathcal{G}_+$. (vi)→(v). In a similar way to (11), we get (12) from (vi). Since $L \in \mathcal{L}_+$, we get (v).

The proof of (v)→(iii)→(vii)→ (v) is similar. $\qquad\qquad\qquad\qquad\qquad$ □

Remark 3. From Theorem 1, it turns out that the well-posedness property of the bimodal system Σ_{AB} with both (C, A) and (C, B) observable is characterized by either one of the following two properties: (i) the preservation property of the lexicographic inequality relation between two rule matrices T_A and T_B, which is characterized by the set $\mathcal{L}_+^n$, and (ii) the smooth continuation property which is characterized by the set $\mathcal{G}_+^n$ (or $\mathcal{G}_0^n$). The former corresponds to the condition (iv) or (v) in Theorem 1, and the latter to (vi) or (vii). Note also that the well-posedness property of Σ_{AB} can be given by the equivalence between Σ_{AB}, Σ_A, and Σ_B. From (vi), it follows that a parameterization of all matrices B for which Σ_{AB} is well-posed is given by the form $B = T_A^{-1} \Gamma T_A$ for any $\Gamma \in \mathcal{G}_+^n$.

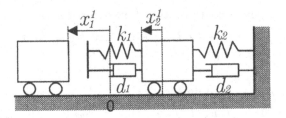

Fig. 2. Elastic collision between 2 objects

Example 1. Consider the physical system in Fig. 2. The equations of motion of this system are given by

$$\text{mode 1}: \begin{cases} \dot{x}^1 = \begin{bmatrix} 0 & 1 \\ 0 & 0 \end{bmatrix} x^1 \\ \dot{x}^2 = \begin{bmatrix} 0 & 1 \\ -k_2 & -d_2 \end{bmatrix} x^2 \\ y = [1 \ \ 0 \ -1 \ \ 0]x \geq 0, \end{cases}$$

$$\text{mode 2}: \begin{cases} \begin{bmatrix} \dot{x}^1 \\ \dot{x}^2 \end{bmatrix} = \begin{bmatrix} 0 & 1 & 0 & 0 \\ -k_1 & -d_1 & k_1 & d_1 \\ 0 & 0 & 0 & 1 \\ k_1 & d_1 & -k_1-k_2 & -d_1-d_2 \end{bmatrix} \begin{bmatrix} x^1 \\ x^2 \end{bmatrix} \\ y = [1 \ \ 0 \ -1 \ \ 0]x \leq 0 \end{cases}$$

where $x = [(x^1)^{\mathrm{T}} \ (x^2)^{\mathrm{T}}]^{\mathrm{T}} = [x_1^1 \ x_2^1 \ x_1^2 \ x_2^2]^{\mathrm{T}}$. These provide

$$A = \begin{bmatrix} 0 & 1 & 0 & 0 \\ 0 & 0 & 0 & 0 \\ 0 & 0 & 0 & 1 \\ 0 & 0 & -k_2 & -d_2 \end{bmatrix}, \quad B = \begin{bmatrix} 0 & 1 & 0 & 0 \\ -k_1 & -d_1 & k_1 & d_1 \\ 0 & 0 & 0 & 1 \\ k_1 & d_1 & -k_1-k_2 & -d_1-d_2 \end{bmatrix},$$

$$C = [1 \ \ 0 \ -1 \ \ 0].$$

Simple calculations show that the pair (C, A) is observable if and only if $k_2 \neq 0$, and also the pair (C, B) is observable if and only if $k_2 \neq 0$. Thus we here assume $k_2 \neq 0$.

From

$$T_A = \begin{bmatrix} 1 & 0 & -1 & 0 \\ 0 & 1 & 0 & -1 \\ 0 & 0 & k_2 & d_2 \\ 0 & 0 & -k_2 d_2 & k_2 - d_2^2 \end{bmatrix}, \quad T_B = \begin{bmatrix} 1 & 0 & -1 & 0 \\ 0 & 1 & 0 & -1 \\ -2k_1 & -2d_1 & 2k_1+k_2 & 2d_1+d_2 \\ \alpha_1 & \alpha_2 & \alpha_3 & \alpha_4 \end{bmatrix}$$

where

$$\alpha_1 = (4d_1 + d_2)k_1,$$
$$\alpha_2 = -2k_1 + (4d_1 + d_2)d_1,$$
$$\alpha_3 = -(4d_1 + d_2)k_1 - (2d_1 + d_2)k_2,$$
$$\alpha_4 = (2k_1 + k_2) - 4d_1^2 - 3d_1 d_2 - d_2^2,$$

it follows that

$$T_B T_A^{-1} = \begin{bmatrix} 1 & 0 & 0 & 0 \\ 0 & 1 & 0 & 0 \\ * & * & 1 & 0 \\ * & * & * & 1 \end{bmatrix}$$

which belongs to the set $\mathcal{L}_+$. Hence the system is well-posed. We also have

$$T_A B T_A^{-1} = \begin{bmatrix} 0 & 1 & 0 & 0 \\ * & * & 1 & 0 \\ * & * & * & 1 \\ * & * & * & * \end{bmatrix}$$

which belongs to the set $\mathcal{G}_+$.

4.2   Unobservable Case

Next, we give a result in the case that both pairs are unobservable.

Theorem 2. *Suppose that the observability indexes of the pairs (C, A) and (C, B) are m_A and m_B, respectively, and $m_A \geq m_B$. Then the following statements are equivalent.*
(i) Σ_{AB} is well-posed.
(ii) The following conditions are satisfied.
(a) $m_A = m_B$.
(b) $T_B = M T_A$ for some $M \in \mathcal{L}_+^{m_A}$.
(c) $(A - B)x = 0$ for all $x \in Ker T_A$.
(iii) The following conditions are satisfied.
(a) $m_A = m_B$.
(b) $T_A B = \Gamma T_A$ for some $\Gamma \in \mathcal{G}_+^{m_A}$.
(c) $(A - B)x = 0$ for all $x \in Ker T_A$.

The proof is omitted due to limits of space [12].

Remark 4. If $m_A = m_B = n$, (ii) and (iii) in Theorem 2 generalize (v) and (vi) in Theorem 1, respectively. Note also that the condition $T_B = M T_A$, which is a necessary and sufficient condition for the well-posedness in the observable case, is not sufficient for the well-posedness in the unobservable case, even if $m_A = m_B$. In other words, it is required that the solutions in both modes in $Ker T_A = Ker T_B$ are the same. This allows us to conclude that whenever the pair (C, A) is observable and the pair (C, B) is unobservable, the system Σ_{AB} is not well-posed.

Remark 5. The conditions in Theorem 2 are checked as follows. First, check the condition (iii)(a). If it is not satisfied, we conclude that the system is not well-posed. Otherwise, check (b) and (c) in (iii). So pick any matrix $\tilde{T}_A$ such that $T = [T_A^T \ \tilde{T}_A^T]^T$ is nonsingular. Then note that (b) and (c) are equivalent to

$$[I_{m_A} \ 0] T B T^{-1} \begin{bmatrix} I_{m_A} \\ 0 \end{bmatrix} \in \mathcal{G}_+^{m_A} \tag{14}$$

and

$$[0 \quad I_{n-m_A}]T(A - B)T^{-1} \begin{bmatrix} 0 \\ I_{n-m_A} \end{bmatrix} = 0, \tag{15}$$

respectively. Thus if both of these two conditions are satisfied, we conclude that the system is well-posed. Otherwise, we conclude that the system is not well-posed. Note here that we only have to check the condition for some $\tilde{T}_A$.

Example 2. Consider the system in Example 1 again. Assume that $k_2 = 0$ and $d_2 \neq 0$. Then since

$$T_A = \begin{bmatrix} C \\ CA \\ CA^2 \end{bmatrix} = \begin{bmatrix} 1 & 0 & -1 & 0 \\ 0 & 1 & 0 & -1 \\ 0 & 0 & 0 & d_2 \end{bmatrix}, \quad T_B = \begin{bmatrix} C \\ CB \\ CB^2 \end{bmatrix} = \begin{bmatrix} 1 & 0 & -1 & 0 \\ 0 & 1 & 0 & -1 \\ -2k_1 & -2d_1 & 2k_1 & 2d_1 + d_2 \end{bmatrix},$$

we have $m_A = 3$ and $m_B = 3$. Thus (iii)(a) in Theorem 2 is satisfied. Letting $\tilde{T}_A \triangleq [0 \ 0 \ 1 \ 0]$, we have

$$TAT^{-1} = \begin{bmatrix} 0 & 1 & 0 & 0 \\ 0 & 0 & 1 & 0 \\ 0 & 0 & -d_2 & 0 \\ 0 & 0 & 1/d_2 & 0 \end{bmatrix}, \quad TBT^{-1} = \begin{bmatrix} 0 & 1 & 0 & 0 \\ -2k_1 & -2d_1 & 1 & 0 \\ k_1 d_2 & d_1 d_2 & -d_2 & 0 \\ 0 & 0 & 1/d_2 & 0 \end{bmatrix}.$$

Using (14) and (15) in Remark 5, we can show that (b) and (c) in (iii) are satisfied. Therefore, the system is well-posed.

5   Application to Well-posedness Problem in Control Switching

The well-posedness conditions as obtained in the previous sections can be applied to several issues in hybrid systems theory. Especially, by combining a stability condition of piecewise linear systems by Johansson and Rantzer [9] with our result, we can judge stability of those systems where the existence of a unique solution without sliding modes is guaranteed.

As the other application, in this section, we discuss a well-posedness problem of switching control systems where state feedback gains are switched according to a criterion depending on the state.

Consider the following problem: let the control system be given by

$$\dot{x} = Ax + Bu \tag{16}$$

where $x \in \mathcal{R}^n$ and $u \in \mathcal{R}^m$. For this system, let us consider a state feedback controller with two modes given by

$$u = \begin{cases} K_1 x & \text{if } Cx \geq 0 \\ K_2 x & \text{if } Cx \leq 0 \end{cases} \tag{17}$$

where $C \in \mathcal{R}^{1 \times n}$, and K_1 and K_2 are given according to some design specification of the closed loop system. Then the problem is to check whether this closed loop system is well-posed or not.

As a simple example, consider the system given by

$$\begin{bmatrix} \dot{x}_1 \\ \dot{x}_2 \end{bmatrix} = \begin{bmatrix} 0 & 1 \\ 0 & 0 \end{bmatrix} \begin{bmatrix} x_1 \\ x_2 \end{bmatrix} + \begin{bmatrix} 0 \\ 1 \end{bmatrix} u$$

and $K_1 = [k_1 \ k_2]$, $K_2 = [\bar{k}_1 \ \bar{k}_2]$, and $C = [c_1 \ c_2]$. Then letting T_{A+BK_1} and T_{A+BK_2} be the rule matrices (i.e., the observability matrices) for the pairs $(C, A+BK_1)$ and $(C, A + BK_2)$, and assuming that these matrices are nonsingular, we obtain

$$T_{A+BK_2} T_{A+BK_1}^{-1} = \begin{bmatrix} 1 & 0 \\ * & a \end{bmatrix}$$

where $a \triangleq \frac{c_1(c_1 + c_2 \bar{k}_2) - c_2^2 \bar{k}_1}{c_1(c_1 + c_2 k_2) - c_2^2 k_1}$. Thus from Theorem 2, we conclude that the closed loop system is well-posed if and only if $a > 0$.

This example shows the following significant fact: even if each controller stabilizes each system in the usual sense, the total system is not necessarily well-posed. For example, consider the case of $c_1 = 1$, $c_2 = 1$, $k_1 = -1$, $k_2 = -3$, $\bar{k}_1 = -1$ and $\bar{k}_2 = -1$. Then $A + BK_1$ and $A + BK_2$ are stable, but $a < 0$. Note that such a case is not rare and the stability in the usual sense for each mode does not automatically provide the well-posedness of the closed loop system.

As shown in the above example, for any given closed loop system, the well-posedness can be determined by checking the corresponding conditions derived in the previous sections. Moreover, we can give an explicit characterization of all feedback gains which guarantee the well-posedness of the closed loop systems in question.

For the closed loop system with two modes given by (16) and (17), letting $K \triangleq K_2 - K_1$ and denoting $A + BK_1$ by A again, we have

$$\Sigma_O \begin{cases} \text{mode 1} : \dot{x} = Ax, & \text{if } y = Cx \geq 0 \\ \text{mode 2} : \dot{x} = (A + BK)x, & \text{if } y = Cx \leq 0. \end{cases} \tag{18}$$

For single-input control systems (16), we obtain the following result.

Theorem 3. *Assume that the pair (C, A) is observable and the relative degree for the triple (C, A, B) is $p(\leq n)$ (i.e., $CB = CAB = \ldots = CA^{p-2}B = 0$ and $CA^{p-1}B \neq 0$). Then the following statements are equivalent.*
(i) The system Σ_O is well-posed.
(ii) $K^T \in span\{C^T, (CA)^T, \cdots, (CA^{p-1})^T\}$
$+\{\xi \in \mathcal{R}^n \mid \xi = \gamma(CA^p)^T, \ \gamma CA^{p-1}B > -1\}$.

Remark 6. It follows from Theorem 3 that for $p = n$ the closed loop system is well-posed for any K. Note also that the case $K = \kappa_1 C$ corresponds to the vector field of the closed loop system being Lipschitz continuous. Theorem 3 can be extended to the multi-input case.

6   Conclusion

In this paper, we have given fundamental tools for the bimodal system given by two linear systems to be well-posed in the sense of Carathéodory. The obtained condition are easily checkable, and several extensions to the other systems such as multi-modal systems and bimodal systems with multi-criterions are possible [12]. In addition, the basic approach developed here will be able to be applied to the case of nonlinear vector fields.

Acknowledgment
The first author would like to express his gratitude to the Canon foundation for his performing this work at Faculty of Mathematical Sciences, University of Twente, and also to M. Saeki, Professor of Hiroshima University, for his support. The authors thank the anonymous reviewers for giving some references.

References

1. D.D. Bainov and P.S. Simeonov. Systems with impulse effect. *Ellis Horwood*, 1989.
2. B. Brogliato. Nonsmooth impact mechanics - models, dynamics and control. *Lect. Notes in Contr. Inform. Sci. 220, Springer-Verlag, Berlin*, 1996.
3. C.T. Chen. Introduction to linear system theory. *Holt, Rinehart and Winston*, 1970.
4. A.F. Filippov. Differential equations with discontinuous righthand sides. *Kluwer, Dordrecht*, 1988.
5. B.A. Fleishman. Convex superposition in piecewise-linear systems. *J. Math. Anal. Appl. 6(2)*, 182–189, 1963.
6. W.P.M.H. Heemels, J.M. Schumacher, and S. Weiland. Complementarity problems in linear complementarity systems. *Proc. of American Control Conference*, 706–710, 1998.
7. I.N. Hajj and S. Skelboe. Steady-state analysis of piecewise-linear dynamic systems. *IEEE Trans. on Circuits and Systems, CAS-28(3)*, 1981.
8. Hybrid systems I, II, III, and IV. *Lecture Notes in Computer Science 736, 999, 1066, and 1273, New York, Springer-Verlag*, 1993, 1995, 1996, 1997.
9. M. Johansson and A. Rantzer. Computation of piecewise quadratic Lyapunov functions for hybrid systems. *IEEE Trans. Automatic Control, AC-43*, 555–559, 1998.
10. Y.J. Lootsma, A.J. van der Schaft, and M.K. Çamhbel. Uniqueness of solutions of relay systems. *Memorandum 1406, Dept. of Appl. Math., Twente univ.*, October 1997, to appear in Automatica, special issue on hybrid systems.
11. Special issue on hybrid systems. *IEEE Trans. Automatic Control, AC-43*, 1998
12. J. Imura and A.J. van der Schaft. Characterization of well-posedness of piecewise linear systems. *Memorandum 1475, Fac. of Math. Sci., Twente univ.*, December 1998.
13. M. Tittus and B. Egardt. Control design for integrator hybrid systems. *IEEE Trans. Automatic Control, AC-43*, 491–500, 1998.
14. L. Tavernini. Differential automata and their discrete simulations. *Nonlinear analysis, Theory, Methods, and Applications*, 11(6), 665-683, 1996.
15. A.J. van der Schaft and J.M. Schumacher. The complementary-slackness class of hybrid systems. *Mathematics of Control, Signals, and Systems, 9*, 266–301, 1996.
16. A.J. van der Schaft and J.M. Schumacher. Complementarity modeling of hybrid systems. *IEEE Trans. Automatic Control, AC-43*, 483–490, 1998.
17. D. Zwillinger. Handbook of Differential Equations. *Academic Press Inc.*, 1989.

A New Class of Decidable Hybrid Systems

Gerardo Lafferriere[1], George J. Pappas[2], and Sergio Yovine[3]

[1] Department of Mathematical Sciences
Portland State University, PO Box 751
Portland, OR 97207
gerardo@mth.pdx.edu
[2] Department of Electrical Engineering and Computer Sciences
University of California at Berkeley
Berkeley, CA 94720
gpappas@eecs.berkeley.edu
[3] VERIMAG
Centre Equation, 2 Avenue de Vignate
38610 Gieres, France
Sergio.Yovine@imag.fr

Abstract. One of the most important analysis problems of hybrid systems is the reachability problem. State of the art computational tools perform reachability computation for timed automata, multirate automata, and rectangular automata. In this paper, we extend the decidability frontier for classes of *linear hybrid systems*, which are introduced as hybrid systems with linear vector fields in each discrete location. This result is achieved by showing that any such hybrid system admits a finite bisimulation, and by providing an algorithm that computes it using decision methods from mathematical logic.

1 Introduction

Hybrid systems are roughly discrete event systems with differential equations in each discrete location. A modeling approach to hybrid systems extends finite state machines to incorporate simple dynamics. One of the most important problems for hybrid systems is the *reachability problem* which asks whether some unsafe region is reachable from an initial region. For purely continuous systems, the reachability problem is known to be a very difficult problem with a few exceptions. Optimal control [14] and game theoretic [11] approaches have been used among others to calculate reachable sets for some systems.

The main tool for obtaining classes of hybrid system for which the reachability problem is decidable, is given by the concept of *bisimulation*. Bisimulations are simply reachability preserving quotient systems. If an infinite state hybrid system has a finite state bisimulation, then checking reachability for the hybrid system can be equivalently performed on the finite, discrete, quotient graph. Since the quotient graph is finite, the algorithm will terminate. If in addition, each step of the algorithm can be encoded and implemented by a computer

F.W. Vaandrager and J.H. van Schuppen (Eds.): HSCC'99, LNCS 1569, pp. 137–151, 1999.
© Springer-Verlag Berlin Heidelberg 1999

program, then the problem is decidable. The first successful application of this approach was for the model of timed automata [2], and was then extended to multirate [1] and rectangular [6, 9] automata.

Unfortunately, the above decidable classes of hybrid systems have limited modeling power for most control applications, where systems with complicated continuous dynamics are frequently encountered. *In this paper, we extend the decidability frontier to capture classes of hybrid systems with linear dynamics in each discrete location.* This broadens the applicability potential of the approach given the wide use of linear systems in control theory. In addition, this result is achieved by using new mathematical and computational techniques which may bring other benefits to control theorists and practitioners. The notion of *o-minimality* [12] from *model theory* is used to define a class of hybrid systems, called *o-minimal hybrid systems*. In [7], it is shown that all o-minimal hybrid systems admit finite bisimulations. In order to make the bisimulation algorithm computationally feasible, we use the framework of *mathematical logic* as the main tool to symbolically represent and manipulate sets. The main computational tool for symbolic set manipulation in this context is *quantifier elimination*. Since quantifier elimination is possible for the theory of reals with addition and multiplication [10, 15], we either find or transform subclasses of o-minimal hybrid systems which are definable in this theory. This immediately leads to new decidability results. In addition, the framework presented in this paper, provides a unifying platform for further studies along this direction.

In Section 2 we review bisimulations of transition systems which are applied in Section 3 to transition systems generated by a class of hybrid systems. After a brief introduction to mathematical logic and model theory in Section 4, Section 5 transforms the reachability problem for various classes of linear vector fields into a quantifier elimination problem in the decidable theory of the reals as an ordered field. This leads to a computational bisimulation algorithm whose termination is guaranteed in Section 6, using the notion of o-minimality. The main decidability result is contained in Theorem 3.

2   Bisimulations And Decidability

A transition system $T = (Q, \Sigma, \rightarrow, Q_O, Q_F)$ consists of a (not necessarily finite) set Q of states, an alphabet Σ of events, a transition relation $\rightarrow \subseteq Q \times \Sigma \times Q$, a set $Q_O \subseteq Q$ of initial states, and a set $Q_F \subseteq Q$ of final states. A transition $(q_1, \sigma, q_2) \in \rightarrow$ is denoted as $q_1 \overset{\sigma}{\rightarrow} q_2$. The transition system is finite if the cardinality of Q and $\rightarrow$ is finite, and it is infinite otherwise. A region is a subset $P \subseteq Q$. Given $\sigma \in \Sigma$ we define the predecessor $Pre_\sigma(P)$ of a region P as

$$Pre_\sigma(P) = \{q \in Q \mid \exists p \in P : q \overset{\sigma}{\rightarrow} p\} \tag{1}$$

One of the main problems for transition systems is the reachability problem which can be used to formulate many safety verification problems.

Problem 1 (Reachability Problem). Given a transition system T, is a state $q_f \in Q_F$ reachable from a state $q_0 \in Q_O$ by a sequence of transitions?

The complexity of the reachability problem is reduced using special quotient transition systems. Given an equivalence relation $\sim \subseteq Q \times Q$ on the state space one can define a quotient transition system as follows. Let $Q/\sim$ denote the quotient space. For a region P we denote by $P/\sim$ the collection of all equivalence classes which intersect P. The transition relation $\to_\sim$ on the quotient space is defined as follows: for $Q_1, Q_2 \in Q/\sim$, $Q_1 \overset{\sigma}{\to}_\sim Q_2$ iff there exist $q_1 \in Q_1$ and $q_2 \in Q_2$ such that $q_1 \overset{\sigma}{\to} q_2$. The quotient transition system is then $T/\sim = (Q/\sim, \Sigma, \to_\sim, Q_O/\sim, Q_F/\sim)$.

Given an equivalence relation $\sim$ on Q, we call a set a $\sim$-block if it is a union of equivalence classes. The equivalence relation $\sim$ is a *bisimulation* of T iff Q_O, Q_F are $\sim$-blocks and for all $\sigma \in \Sigma$ and all $\sim$-blocks P, the region $Pre_\sigma(P)$ is a $\sim$-block. In this case the systems T and $T/\sim$ are called *bisimilar*. We will also say that a partition is a bisimulation when its induced equivalence relation is a bisimulation. A bisimulation is called finite if it has a finite number of equivalence classes. Bisimilar transition systems generate the same language [5].

Recently, the above bisimulation methodology has been applied to hybrid systems which combine discrete and continuous dynamics. The main reason for doing this is that if T is a transition system with an infinite state space and $T/\sim$ is a finite bisimulation, then the reachability problem for hybrid systems can be converted to an equivalent reachability problem on a finite graph. If, in addition, this can be performed in a computationally feasible way, then one obtains classes of hybrid systems for which the reachability problem is *decidable*. This approach has successfully resulted in various decidable classes of hybrid systems, including timed automata [2], multirate automata [1], and initialized rectangular automata [6, 9].

These results are based on the following geometric characterization of bisimulations. If $\sim$ is a bisimulation, it can be easily shown that if $p \sim q$ then

B1 $p \in Q_F$ iff $q \in Q_F$, and $p \in Q_O$ iff $q \in Q_O$
B2 if $p \overset{\sigma}{\to} p'$ then there exists q' such that $q \overset{\sigma}{\to} q'$ and $p' \sim q'$

Based on the above characterization, given a transition system T, the following algorithm computes increasingly finer partitions of the state space Q. If the algorithm terminates, then the resulting quotient transition system is a finite bisimulation. The state space $Q/\sim$ is called a bisimilarity quotient.

Algorithm 1 (Bisimulation Algorithm for Transition Systems)
Set $Q/\sim = \{Q_O, Q_F, Q \setminus (Q_O \cup Q_F)\}$
while $\exists\, P, P' \in Q/\sim$ and $\sigma \in \Sigma$ such that $\emptyset \neq P \cap Pre_\sigma(P') \neq P$
    set $P_1 = P \cap Pre_\sigma(P')$, $P_2 = P \setminus Pre_\sigma(P')$
    refine $Q/\sim = (Q/\sim \setminus \{P\}) \cup \{P_1, P_2\}$
end while

In order for a transition system to have a finite bisimulation, the above algorithm must terminate after a finite number of iterations. If, in addition, each step of the algorithm is constructive, then the reachability problem for the transition system is *decidable*. This requires that we have computational methods to represent sets, perform set intersections and complements, check whether a set is empty, and compute $Pre_\sigma(P)$ for any set P and any $\sigma \in \Sigma$.

3   Hybrid Systems

In this paper, we focus on transition systems generated by the following class of hybrid systems.

Definition 1. *A hybrid system is a tuple $H = (X, X_O, X_F, F, E, I, G, R)$ where*

- $X = X_D \times X_C$ *is the state space with* $X_D = \{q_1, \ldots, q_m\}$ *and* X_C *a manifold.*
- $X_O \subseteq X$ *is the set of initial states.*
- $X_F \subseteq X$ *is the set of final states.*
- $F : X \longrightarrow TX_C$ *assigns to each discrete location* $q \in X_D$ *a vector field* $F(q, \cdot)$.
- $E \subseteq X_D \times X_D$ *is the set of discrete transitions.*
- $I : X_D \longrightarrow 2^{X_C}$ *assigns to* $q \in X_D$ *an invariant of the form* $I(q) \subseteq X_C$.
- $G : E \longrightarrow X_D \times 2^{X_C}$ *assigns to* $e = (q_1, q_2) \in E$ *a guard of the form* $\{q_1\} \times U$, $U \subseteq I(q_1)$.
- $R : E \longrightarrow X_D \times 2^{X_C}$ *assigns to* $e = (q_1, q_2) \in E$ *a reset of the form* $\{q_2\} \times V$, $V \subseteq I(q_2)$.

Trajectories of the hybrid system H originate at any $(q, x) \in X_O$ and consist of concatenations of continuous evolutions and discrete jumps. Continuous trajectories keep the discrete part of the state constant, and the continuous part evolves according to the continuous flow $F(q, \cdot)$ as long as the flow remains inside the invariant set $I(q)$. If the flow exits $I(q)$, then a discrete transition is *forced*. If, during the continuous evolution, a state $(q, x) \in G(e)$ is reached for some $e \in E$, then discrete transition e is *enabled*. The hybrid system may then instantaneously jump from (q, x) to any $(q', x') \in R(e)$ and the system then evolves according to the flow $F(q', \cdot)$. Notice that even though the continuous evolution is deterministic, the discrete evolution may be nondeterministic. The discrete transitions allowed in our model are slightly more restrictive than those in initialized rectangular automata.

Every hybrid system H generates a transition system $T = (Q, \Sigma, \rightarrow, Q_O, Q_F)$ by setting $Q = X$, $Q_O = X_O$, $Q_F = X_F$, $\Sigma = E \cup \{\tau\}$, and $\rightarrow = (\cup_{e \in E} \overset{e}{\rightarrow}) \cup \overset{\tau}{\rightarrow}$ where

Discrete Transitions $(q, x) \overset{e}{\rightarrow} (q', x')$ for $e \in E$ iff $(q, x) \in G(e)$ and $(q', x') \in R(e)$

Continuous Transitions $(q_1, x_1) \overset{\tau}{\rightarrow} (q_2, x_2)$ iff $q_1 = q_2$ and there exist $\delta \geq 0$ and a curve $x : [0, \delta] \longrightarrow X_C$ with $x(0) = x_1$, $x(\delta) = x_2$ and for all $t \in [0, \delta]$ it satisfies $\frac{dx}{dt} = F(q_1, x(t))$ and $x(t) \in I(q_1)$.

The continuous τ transitions are time-abstract transitions, in the sense that the time it takes to reach one state from another is ignored. Having defined the continuous and discrete transitions $\overset{\tau}{\rightarrow}$ and $\overset{e}{\rightarrow}$ allows us to formally define $Pre_\tau(P)$ and $Pre_e(P)$ for $e \in E$ and any region $P \subseteq X$ using (1). Furthermore, the structure of the discrete transitions allowed in our hybrid system model results in

$$Pre_e(P) = \begin{cases} \emptyset & \text{if } P \cap R(e) = \emptyset \\ G(e) & \text{if } P \cap R(e) \neq \emptyset \end{cases} \tag{2}$$

for all discrete transitions $e \in E$ and regions P. Therefore, if the sets $R(e)$ and $G(e)$ are blocks of any partition of the state space, then no partition refinement is necessary in the bisimulation algorithm due to any discrete transitions $e \in E$. This fact, in a sense, decouples the continuous and discrete components of the hybrid system. In turn, this implies that the initial partition in the bisimulation algorithm should contain the invariants, guards and reset sets, in addition to the initial and final sets. This allows us to carry out the algorithm independently for each location.

More precisely, define for any region $P \subseteq X$ and $q \in X_D$ the set $P_q = \{x \in X_C \mid (q, x) \in P\}$. For each location $q \in X_D$ consider the finite collection of sets

$$A_q = \{I(q), (X_O)_q, (X_F)_q\} \cup \{G(e)_q, R(e)_q \mid e \in E\} \qquad (3)$$

which describes the initial and final states, guards, invariants and resets associated with location q. Let S_q be the coarsest partition of X_C compatible with the collection A_q (by compatible we mean that each set in A_q is a union of sets in S_q). The (finite) partition S_q can be easily computed by successively finding the intersections between each of the sets in A_q and their complements. We define (q, S_q) to be the set $\{\{q\} \times P \mid P \in S_q\}$. These collections (q, S_q) will be the starting partitions of the bisimulation algorithm. In addition, since by definition $Pre_\tau(P)$ applies to regions $P \subseteq X$, but not to its continuous projection P_q, we define for $Y \subseteq X_C$ the operator $Pre_q(Y) = (Pre_\tau(\{q\} \times Y))_q$. The general bisimulation algorithm for transition systems then takes the following form for the class of hybrid systems that are considered in this paper.

Algorithm 2 (Bisimulation Algorithm for Hybrid Systems)
Set $X/\sim \; = \bigcup_q (q, S_q)$
for $q \in X_D$
    while $\exists \; P, P' \in S_q$ such that $\emptyset \neq P \cap Pre_q(P') \neq P$
        Set $P_1 = P \cap Pre_q(P')$; $P_2 = P \setminus Pre_q(P')$
        refine $S_q = (S_q \setminus \{P\}) \cup \{P_1, P_2\}$
    end while
end for

It is clear from the structure of the bisimulation algorithm that, the iteration is carried out independently for each discrete location. In order for the above algorithm to terminate, the partition refinement process must terminate for each discrete location $q \in X_D$. It therefore suffices to look at one continuous slice of the hybrid system at a time and see whether we can construct a finite bisimulation that is consistent with all relevant sets of each location q as well as with the continuous flows of the vector field $F(q, \cdot)$. Since we focus on each continuous slice at a time, we will drop the q subscript from $Pre_q(Y)$, which will be denoted from now on by $Pre(Y)$.

It is now clear that the decidability of the reachability problem amounts to solving the following two problems.

Problem 1 (Computability) In order for the bisimulation algorithm to be *computational*, we need to effectively

1. Represent sets,
2. Perform set intersection and complement,
3. Check emptiness of sets,
4. Compute $Pre(Y)$ of a set Y.

Problem 2 (Finiteness) Determine whether the bisimulation algorithm terminates in a finite number of steps.

A natural platform for solving the above computational issues is provided by mathematical logic where sets would be represented as formulas of first-order logic over the real numbers. In the next section we introduce the necessary notions of mathematical logic and model theory that will provide the means for representing and manipulating sets defined by first-order formulas (Section 5) as well as for ensuring the termination of the algorithm (Section 6).

4   Mathematical Logic and Model Theory

4.1   Languages and formulas

A *language* is a set of symbols separated in three groups: relations, functions and constants. The sets $\mathcal{P} = \{<, +, -, 0, 1\}$, $\mathcal{R} = \{<, +, -, \cdot, 0, 1\}$, and $\mathcal{R}_{\exp} = \{<, +, -, \cdot, 0, 1, \exp\}$ are examples of languages where $<$ (less than) is the relation, $+$ (plus), $-$ (minus), $\cdot$ (product) and $\exp$ (exponentiation) are the functions, and 0 (zero) and 1 (one) are the constants.

Let $V = \{x, y, z, x_0, x_1, \dots\}$ be a countable set of *variables*. The set of *terms* of a language is inductively defined as follows. A term θ is a variable, a constant, or $F(\theta_1, \dots, \theta_m)$, where F is a m-ary function and θ_i, $i = 1, \dots, m$ are terms. For instance, $x - 2y + 3$ and $x + yz^2 - 1$ are terms of $\mathcal{P}$ and $\mathcal{R}$, respectively. In other words, terms of $\mathcal{P}$ are linear expressions and terms of $\mathcal{R}$ are polynomials with integer coefficients. Notice that integers are the only numbers allowed in expressions (they can be obtained by adding up the constant 1).

The *atomic formulas* of a language are of the form $\theta_1 = \theta_2$, or $R(\theta_1, \dots, \theta_n)$, where θ_i, $i = 1, \dots, n$ are terms and R is an n-ary relation. For example, $xy > 0$ and $x^2 + 1 = 0$ are terms of $\mathcal{R}$.

The set of *(first-order) formulas* is recursively defined as follows. A formula ϕ is an atomic formula, $\phi_1 \wedge \phi_2$, $\neg\phi_1$, $\forall x : \phi_1$ or $\exists x : \phi_1$, where ϕ_1 and ϕ_2 are formulas, x is a variable, $\wedge$ (conjunction) and $\neg$ (negation) are the boolean connectives, and $\forall$ (for all) and $\exists$ (there exists) are the quantifiers.

Examples of $\mathcal{R}$-formulas are $\forall x \, \forall y : xy > 0$, $\exists x : x^2 - 2 = 0$, and $\exists w : xw^2 + yw + z = 0$. The occurrence of a variable in a formula is *free* if it is not inside the scope of a quantifier; otherwise, it is *bound*. For example, x, y, and z are free and w is bound in the last example. We often write $\phi(x_1, \dots, x_n)$ to indicate that $x_1, \dots, x_n$ are the free variables of the formula ϕ. A *sentence* of $\mathcal{R}$ is a formula with no free variables. The first two examples are sentences.

4.2 Models

A *model* of a language consists of a non-empty set S and an interpretation of the relations, functions and constants. For example, $(\mathbb{R}, <, +, -, \cdot, 0, 1)$ and $(\mathbb{Q}, <, +, -, \cdot, 0, 1)$, are *models* of $\mathcal{R}$ with the usual meaning of the symbols.

We say that a set $Y \subseteq S^n$ is *definable* in a language if there exists a formula $\phi(x_1, \ldots, x_n)$ such that $Y = \{(a_1, \ldots, a_n) \in S^n \mid \phi(a_1, \ldots, a_n)\}$. For example, over $\mathbb{R}$, the formula $x^2 - 2 = 0$ defines the set $\{\sqrt{2}, -\sqrt{2}\}$. Two formulas $\phi(x_1, \ldots, x_n)$ and $\psi(x_1, \ldots, x_n)$ are *equivalent* in a model, denoted by $\phi \equiv \psi$, if for every assignment $(a_1, \ldots, a_n)$ of $(x_1, \ldots, x_n)$, $\phi(a_1, \ldots, a_n)$ is true if and only if $\psi(a_1, \ldots, a_n)$ is true. Equivalent formulas define the same set.

4.3 Theories

A *theory* is a subset of sentences. Any model of a language defines a theory: *the set of all sentences which hold in the model.* We denote by $\mathsf{Lin}(\mathbb{R})$ the theory defined as the formulas of $\mathcal{P}$ that are true over $(\mathbb{R}, <, +, -, 0, 1)$, i.e., $\mathsf{Lin}(\mathbb{R})$ is the theory of linear constraints (polyhedra). We denote by $\mathsf{OF}(\mathbb{R})$ the theory obtained by interpreting $\mathcal{R}$ over $(\mathbb{R}, <, +, -, \cdot, 0, 1)$. In other words, $\mathsf{OF}(\mathbb{R})$ is the set of all true assertions about the set of real numbers when viewed as an *ordered field*. The theory $\mathsf{OF}_{\exp}(\mathbb{R})$ is the extension of the ordered field of real numbers with the exponentiation.

4.4 Decidability and quantifier elimination

Given a theory, it is important to determine the sentences of the language that belong to the theory. Tarski [10] showed that $\mathsf{OF}(\mathbb{R})$ is *decidable*, i.e., there is a computational procedure that, given any $\mathcal{R}$-sentence ϕ, decides whether ϕ belongs to $\mathsf{OF}(\mathbb{R})$. The decision procedure is based on the elimination of the quantifiers. Over $\mathbb{R}$, every formula $\phi(x_1, \ldots, x_n)$ of $\mathcal{R}$ is equivalent to a formula $\psi(x_1, \ldots, x_n)$ without quantifiers. Moreover, there is an algorithm that transforms ϕ into ψ by *eliminating the quantifiers*. For example, the formula $\exists w : xw^2 + yw + z = 0$ is equivalent to $4xz - y^2 \leq 0$.

Quantifier elimination implies that every $\mathcal{R}$-definable set $Y \subseteq \mathbb{R}^n$ is definable without quantifiers. Moreover, the decidability of $\mathsf{OF}(\mathbb{R})$ implies that the algorithm for eliminating the quantifiers also provides a computational procedure (that terminates in a finite number of steps) for checking whether Y is empty: $Y = \{(y_1, \ldots, y_n) \in \mathbb{R}^n \mid \phi(y_1, \ldots, y_n)\} = \emptyset$ if and only if the sentence $\exists y_1 \ldots \exists y_n : \phi(y_1, \ldots, y_n)$ is equivalent to the (quantifier-free) formula *false*. There are different methods to perform quantifier elimination, e.g., [3, 15]. All the examples considered in this paper have been solved using the tool REDLOG [4].

Therefore, the theory $\mathsf{OF}(\mathbb{R})$ provides the means for representing sets as well as performing boolean operations and checking for emptiness. All that remains in order to make the bisimulation algorithm computational, is to compute

$Pre(Y)$ for any definable set Y. Computing $Pre(Y)$ for a linear vector field generally includes formulas involving the exponential function which are definable in $\mathsf{OF}_{\exp}(\mathbb{R})$. This theory does not admit elimination of quantifiers. Nevertheless, in the next section, we identify several subsets of $\mathcal{R}_{\exp}$ where quantifiers can be eliminated and the resulting quantifier-free formula is in $\mathcal{R}$, yielding a decision procedure.

5   Linear Hybrid Systems

For subclasses of hybrid systems, like multirate automata and rectangular automata [1], where the subsets of $\mathbb{R}^n$ obtained by the application of the bisimulation algorithm are polyhedral sets, i.e., sets definable in the language of linear constraints $\mathcal{P}$, the computation of $Pre(Y)$ relies on the decidability of the theory $\mathsf{Lin}(\mathbb{R})$ via the elimination of the quantifiers.

In this section we identify several classes of hybrid systems with linear vector fields where the ability of computing $Pre(Y)$ depends on the decidability of $\mathsf{OF}(\mathbb{R})$.

Definition 2 (Linear Hybrid Systems). *A hybrid system* $H = (X, X_O, X_F, F, E, I, G, R)$ *is a linear hybrid system if*

- $X_C = \mathbb{R}^n$.
- *for each* $q \in X_D$ *the vector field* $F(q, x) = A_q x$, *where* $A_q \in \mathbb{Q}^{n \times n}$.
- *for each* $q \in X_D$ *the family of sets* $\mathcal{A}_q = \{I(q), (X_O)_q, (X_F)_q\} \cup \{G(e)_q, R(e)_q \mid e \in E\}$ *is definable in* $\mathsf{OF}(\mathbb{R})$.

As indicated previously, having a computational bisimulation algorithm requires having a procedure for computing $Pre(Y)$ for a definable set Y for each discrete location q. Therefore, we only need to investigate a single location and a single linear vector field $F(x) = Ax$ where the subscript q is dropped for notational convenience. In addition, since the invariant $I(q)$ is a definable set, there exists an $\mathcal{R}$-formula $I(x)$ such that $I(q) = \{x \in \mathbb{R}^n \mid I(x)\}$.

Now let $Y \triangleq \{y \in \mathbb{R}^n \mid P(y)\}$. Then we can write explicitly

$$Pre(Y) = \{x \in \mathbb{R}^n \mid \exists y\, \exists t : P(y) \wedge t \geq 0 \wedge x = e^{-tA}y \wedge$$
$$\forall t' : 0 \leq t' \leq t \implies I(e^{-t'A}y)\}$$

In order to simplify the following presentation, we will assume that $I(x)$ is *true*. In this case, the above definition reduces to

$$Pre(Y) = \{x \in \mathbb{R}^n \mid \exists y\, \exists t : P(y) \wedge t \geq 0 \wedge x = e^{-tA}y\} = \{x \in \mathbb{R}^n \mid \eta(x)\} \quad (4)$$

It will be clear from the following results that more complicated invariant sets can be dealt with by the same techniques. Our goal in this section is to transform formula $\eta(x)$ to an equivalent formula in $\mathsf{OF}(\mathbb{R})$, which is indeed decidable. Based on the eigenstructure of A, we identify several classes of linear vector fields for which this transformation is feasible.

5.1 Nilpotent matrices

We consider first the special case when the vector field is linear with a nilpotent matrix A, that is, $A^n = 0$. Recall that nilpotent matrices can only have zero as an eigenvalue. Another important property of nilpotent matrices is that we can express e^{-tA} explicitly as a finite sum

$$e^{-tA} = \sum_{k=0}^{n-1}(-1)^k \frac{t^k}{k!}A^k \tag{5}$$

Thus, the formula $\eta(x)$ can be rewritten as:

$$\eta(x) \triangleq \exists y \, \exists t : P(y) \wedge t \geq 0 \wedge x = \sum_{k=0}^{n-1}(-1)^k \frac{t^k}{k!}A^k y$$

$$\triangleq \exists y : P(y) \wedge \mu(x, y)$$

Clearly, $\mu(x, y)$ is an $\mathcal{R}$-formula, and so is $\eta(x)$, which implies that the following proposition holds.

Proposition 1. *Let $F(x) = Ax$ be a linear vector field and $A \in \mathbb{Q}^{n \times n}$ a nilpotent matrix, and $Y \subseteq \mathbb{R}^n$ definable in $\mathcal{R}$. Then $Pre(Y)$ is definable in $\mathcal{R}$.*

Therefore, based on the computational procedure for eliminating quantifiers in $OF(\mathbb{R})$, we can compute $Pre(Y)$ for linear vector fields with nilpotent matrices. Note that nilpotent linear vector fields capture integrators which are an extremely important class of linear systems.

5.2 Diagonalizable matrices with rational eigenvalues

In this case we can write $A = TDT^{-1}$ where D is a diagonal matrix with the eigenvalues of A along the diagonal and both T and T^{-1} have rational entries. Then $e^{-tA} = [f_{ij}(t)]$ where $f_{ij}(t) = \sum_{k=1}^{n} a_{ijk}e^{-\lambda_k t}$ with $a_{ijk} \in \mathbb{Q}$ for all i, j, k, and $\{\lambda_k\}$ are the eigenvalues of A. Moreover, $x = e^{-tA}y$ can be written component-wise as $x_i = \sum_{k=1}^{n} \psi_{ik}(y)e^{-\lambda_k t}$, with ψ_{ik} being $\mathcal{R}$-formulas. Therefore, $\eta(x)$ can be rewritten as

$$\eta(x) \triangleq \exists y \, \exists t : P(y) \wedge t \geq 0 \wedge \bigwedge_{i=1}^{n} x_i = \sum_{k=1}^{n} \psi_{ik}(y)e^{-\lambda_k t}$$

$$\triangleq \exists y : P(y) \wedge \varphi(x, y)$$

Since the formula for Y, $P(y)$, is already in $\mathcal{R}$, we will concentrate on studying $\varphi(x, y)$. First we reparameterize the time t to reduce the problem to integers in the exponent. More precisely, for each $k = 1, \dots, n$ let d_k denote the denominator of λ_k and let $d_0 = \prod d_k$. We assume that the λ_k are in reduced form, with positive denominators. Then $d_0 > 0$ and for each $k = 1, \ldots, n$ we write $r_k = \lambda_k d_0$. Then we have that $\varphi(x, y) \equiv \varphi_{\mathbb{Z}}(x, y)$ where

$$\varphi_{\mathbb{Z}}(x, y) \triangleq \exists s : s \geq 0 \wedge \bigwedge_{i=1}^{n} x_i = \sum_{k=1}^{n} \psi_{ik}(y) \, e^{-r_k s} \tag{6}$$

Still, $\varphi_{\mathbb{Z}}$ is an $\mathcal{R}_{\exp}$-formula. We consider a second formula $\zeta(x, y)$ which does not involve the exponential function:

$$\zeta(x, y) \triangleq \exists z : 0 < z \leq 1 \wedge \bigwedge_{i=1}^{n} x_i = \sum_{k=1}^{n} \psi_{ik}(y) \, z^{r_k} \tag{7}$$

The following lemma holds.

Lemma 1. *Formulas $\varphi_{\mathbb{Z}}(x, y)$ and $\zeta(x, y)$ are equivalent.*

Proof. The equivalence follows from the change of variables $z = e^{-s}$.

The third step eliminates negative polynomial powers. It consists of grouping the indices $1, \ldots, n$ according to the sign of the corresponding eigenvalue. Let $I^+ = \{k \mid r_k > 0\}$, $I^- = \{k \mid r_k < 0\}$, and $I^0 = \{k \mid r_k = 0\}$. Consider now the following formula:

$$\nu(x, y) \triangleq \exists w_1 \, \exists w_2 : \tag{8}$$
$$w_1 > 0 \wedge w_2 > 0 \wedge w_1 w_2 = 1$$
$$\wedge \bigwedge_{i=1}^{n} x_i = \sum_{k \in I^+} \psi_{ik}(y) \, w_1^{r_k} + \sum_{k \in I^-} \psi_{ik}(y) \, w_2^{-r_k} + \sum_{k \in I^0} \psi_{ik}(y)$$

Clearly, $\nu(x, y)$ is an $\mathcal{R}$-formula. The following lemma holds.

Lemma 2. *The formulas $\zeta(x, y)$ and $\nu(x, y)$ are equivalent.*

Proof. The equivalence follows from the change of variables $w_1 = z$, $w_2 = 1/z$.

The combination of the above lemmas gives the following proposition which implies that we have a computational procedure for computing reachable sets for diagonalizable linear vector fields with rational eigenvalues.

Proposition 2. *Let $F(x) = Ax$ be a linear vector field and $A \in \mathbb{Q}^{n \times n}$ a diagonalizable matrix with rational eigenvalues, and $Y \subseteq \mathbb{R}^n$ definable in $\mathcal{R}$. Then $Pre(Y)$ is definable in $\mathcal{R}$.*

Example 1. Consider the linear vector field

$$F(x) = \begin{bmatrix} 2 & 0 \\ 0 & -1 \end{bmatrix} \cdot \begin{bmatrix} x_1 \\ x_2 \end{bmatrix} \tag{9}$$

Let $Y = \{(y_1, y_2) \in \mathbb{R}^2 \mid y_1 = 4 \wedge y_2 = 3\}$. Let ψ be such that $Pre(Y) = \{(x_1, x_2) \in \mathbb{R}^2 \mid \psi(x_1, x_2)\}$. Applying the previous lemmas we have that

$$
\begin{aligned}
\psi(x_1, x_2) &\triangleq \exists y_1 \, \exists y_2 \, \exists t : y_1 = 4 \wedge y_2 = 3 \wedge t \geq 0 \wedge x_1 = y_1 e^{-2t} \wedge x_2 = y_2 e^t \\
&\equiv \exists y_1 \, \exists y_2 \, \exists z : y_1 = 4 \wedge y_2 = 3 \wedge 0 < z \leq 1 \wedge x_1 = y_1 z^{-2} \wedge x_2 = y_2 z \\
&\equiv \exists y_1 \, \exists y_2 \, \exists w_1 \, \exists w_2 : y_1 = 4 \wedge y_2 = 3 \wedge w_1 > 0 \wedge w_2 > 0 \wedge w_1 w_2 = 1 \\
&\qquad \wedge x_1 = y_1 w_1^2 \wedge x_2 = y_2 w_2 \\
&\equiv x_1 x_2^2 - 36 = 0 \wedge x_2 > 0
\end{aligned}
$$

5.3 Pure imaginary eigenvalues

In this case, we assume the matrix A is similar to a matrix in a special block-diagonal form, a real Jordan form. First, the number of rows (and columns) of A, is even. Second, there exist D and T such that $A = TDT^{-1}$, T invertible, and D is block diagonal with each block of size 2×2 and of the form

$$
\begin{bmatrix} 0 & b \\ -b & 0 \end{bmatrix}
$$

where b is the imaginary part of an eigenvalue of A. In this case we say that A has a *diagonal real Jordan form*. Moreover, if each eigenvalue is of the form ir with $r \in \mathbb{Q}$, then the entries of D, T, and T^{-1} are all rational.

Using a similar approach as in the case of real eigenvalues, and relying on the fact that $\cos^2 s + \sin^2 s = 1$ and $\cos(ms)$, $\sin(ms)$ can be written as polynomials in $\cos(s), \sin(s)$, results in the following proposition.

Proposition 3. *Let $F(x) = Ax$ be a linear vector field and $A \in \mathbb{Q}^{n \times n}$ a matrix with pure imaginary eigenvalues of the form ir with $r \in \mathbb{Q}$, and with a real Jordan form. If $Y \subseteq \mathbb{R}^n$ is definable in $\mathcal{R}$, then $Pre(Y)$ is definable in $\mathcal{R}$.*

Proposition 3 implies that we have a computational procedure for the reachability problem of a class of linear vector fields with pure imaginary eigenvalues of the form ir with $r \in \mathbb{Q}$.

We have presented above three classes of linear vector fields for which $Pre(Y)$ can be computed for sets Y definable in $OF(\mathbb{R})$. The computational results obtained in the section are now summarized by the following theorem.

Theorem 1. *Let H be a linear hybrid system where for each discrete location $q \in X_D$ the vector field is of the form $F(q, x) = Ax$ where*

- *$A \in \mathbb{Q}^{n \times n}$ is nilpotent or*
- *$A \in \mathbb{Q}^{n \times n}$ is diagonalizable with rational eigenvalues or*
- *$A \in \mathbb{Q}^{n \times n}$ has pure imaginary eigenvalues of the form ir, $r \in \mathbb{Q}$, with diagonal real Jordan form.*

Then the reachability problem for H is semidecidable.

To obtain a decision procedure, we need to guarantee that the bisimulation algorithm terminates. In the next section we use the notion of o-minimality in order to guarantee termination of the bisimulation algorithm.

6   O-Minimal Hybrid Systems And Decidability

In order for the bisimulation algorithm to terminate, the partition of the state space resulting from the bisimulation algorithm should have a finite number of equivalence classes. It is therefore important that during the partition refinement process, the intersection of the predecessor of an equivalence class with any other equivalence class has a finite number of connected components. The search for such desirable finiteness properties of definable sets has lead to the notion of *o-minimality*. While this concept applies to any theory, we consider here only theories over the real numbers. Let $\mathcal{L}$ be a language and $Th(\mathbb{R})$ be a theory of the reals.

Definition 3. $Th(\mathbb{R})$ *is* o-minimal *("order minimal") if every definable subset of $\mathbb{R}$ is a finite union of points and intervals (possibly unbounded).*

The class of o-minimal theories is quite rich. Quantifier elimination implies that $\mathsf{Lin}(\mathbb{R})$ and $\mathsf{OF}(\mathbb{R})$ are o-minimal. In addition, even though $\mathsf{OF}_{\exp}(\mathbb{R})$ does not admit elimination of quantifiers, such theory is indeed o-minimal (see [16]). Another extension of $\mathsf{OF}(\mathbb{R})$ is obtained by adding to $\mathcal{R}$ a symbol $\hat{f}$ for every function defined by

$$\hat{f}(x) = \begin{cases} f(x) & \text{if } x \in [-1,1]^n \\ 0 & \text{otherwise} \end{cases}$$

where f is a real-analytic function in a neighborhood of the cube $[-1,1]^n \subset \mathbb{R}^n$. The resulting theory denoted $\mathsf{OF}_{\mathrm{an}}(\mathbb{R})$ is then an extension of $\mathsf{OF}(\mathbb{R})$ which is also o-minimal. The theory $\mathsf{OF}_{\mathrm{an}}(\mathbb{R})$ includes subanalytic sets as definable sets. In addition, it captures periodic trajectories of linear systems as the sine function restricted to a period is definable. Finally, the theory $\mathsf{OF}_{\mathrm{an},\exp}(\mathbb{R})$ obtained by adding both symbols $\hat{f}$ and exp is also o-minimal (see [13]). The following table summarizes the o-minimal theories that are of interest in this paper along with examples of sets and flows that are definable in these theories.

Table 1 : O-Minimal Theories			
Theory	Model	Definable Sets	Definable Flows
$\mathsf{Lin}(\mathbb{R})$	$(\mathbb{R}, +, -, <, 0, 1)$	Polyhedral sets	Linear flows
$\mathsf{OF}(\mathbb{R})$	$(\mathbb{R}, +, -, \cdot, <, 0, 1)$	Semialgebraic sets	Polynomial flows
$\mathsf{OF}_{\mathrm{an}}(\mathbb{R})$	$(\mathbb{R}, +, -, \cdot, <, 0, 1, \{\hat{f}\})$	Subanalytic sets	Polynomial flows
$\mathsf{OF}_{\exp}(\mathbb{R})$	$(\mathbb{R}, +, -, \cdot, <, 0, 1, \exp)$	Semialgebraic sets	Exponential flows
$\mathbb{R}_{\exp,\mathrm{an}}$	$(\mathbb{R}, +, -, \cdot, <, 0, 1, \exp, \{\hat{f}\})$	Subanalytic sets	Exponential flows

Based on the notion of o-minimality, the concept of o-minimal hybrid systems is introduced as hybrid systems whose relevant sets and flows are definable in an o-minimal theory.

Definition 4. *A hybrid system* $H = (X, X_O, X_F, F, E, I, G, R)$ *is said to be* o-minimal *if*

- $X_C = \mathbb{R}^n$.
- for each $q \in X_D$ the flow of $F(q, \cdot)$ is complete (exists for all time).
- for each $q \in X_D$ the family of sets $A_q = \{I(q), (X_O)_q, (X_F)_q\} \cup \{G(e)_q, R(e)_q \mid e \in E\}$ and the flow of $F(q, \cdot)$ are definable in the same o-minimal theory.

For various classes of o-minimal hybrid systems, the reader is referred to [7], where the following property of o-minimal hybrid systems is proven.

Theorem 2. *Every o-minimal hybrid system admits a finite bisimulation. In particular, the bisimulation algorithm terminates for o-minimal hybrid systems.*

We can now combine the semidecision result of Theorem 1 and the termination result of Theorem 2 in order to obtain the desired decidability result.

Theorem 3. *Let H be a linear hybrid system where for each discrete location $q \in X_D$ the vector field is of the form $F(q, x) = Ax$ where*

- $A \in \mathbb{Q}^{n \times n}$ is nilpotent or
- $A \in \mathbb{Q}^{n \times n}$ is diagonalizable with rational eigenvalues or
- $A \in \mathbb{Q}^{n \times n}$ has purely imaginary eigenvalues of the form ir, $r \in \mathbb{Q}$, with diagonal real Jordan form.

Then the reachability problem for H is decidable.

Proof. All relevant sets of linear hybrid systems are by definition definable in $\mathsf{OF}(\mathbb{R})$ and the flows of linear vector fields are complete. Therefore, given the semidecision result of Theorem 1, all we have to show is that the flow of the linear vector field Ax is definable in an o-minimal theory. Then Theorem 2 would guarantee termination of the bisimulation algorithm. If A is nilpotent then the flow is also definable in $\mathsf{OF}(\mathbb{R})$ which is o-minimal. If A is diagonalizable then the flow is definable in $\mathsf{OF}_{\exp}(\mathbb{R})$ which is also o-minimal. If A has purely imaginary eigenvalues, then the flow contains the functions sin and cos which are not definable in any of the o-minimal theories of Table 1. However, o-minimality of the flow is only used in the proof of Theorem 2 to show o-minimality of the *Pre* operator. Even though the flow of this vector field is not definable, the *Pre* operator corresponding to these periodic flows is still definable, as all we need is the restriction of sin and cos on $[0, 2\pi]$. These restrictions are indeed definable in $\mathsf{OF}_{\mathrm{an}}(\mathbb{R})$ which is also o-minimal. Therefore in all cases the relevant objects are definable in the same o-minimal theory, $\mathsf{OF}_{\mathrm{an,exp}}(\mathbb{R})$

Theorem 3 is the first decidability result in the area of hybrid systems that provides the modeling expressiveness to capture relatively complex continuous dynamics. In addition, Theorem 3 contains in it a purely continuous version of reachability analysis for linear systems under state constraints, a problem which is known to be very difficult. As a result, its potential application to analyze various realistic hybrid systems using computational methods is significant.

7 Conclusions

In this paper, we presented a new class of hybrid system with a decidable reachability problem. This new class captures classes of linear vector fields in each discrete location. In addition, this extension is obtained using techniques from mathematical logic and model theory. The mathematical machinery presented in this paper provides a natural and unified platform for pursuing further research along this direction.

Issues for further research include the incorporation of linear vector fields with inputs in each discrete location. This will allow to model significant modeling disturbances as well as provide us with a framework for doing symbolic controller synthesis. Preliminary results along this direction indicate very delicate nonresonance conditions between the control inputs and the eigenvalues of the systems [8]. In addition, more complicated discrete transitions are also of interest.

Another direction of research includes complexity analysis and reduction of the proposed algorithms as well as their implementation into a computational tool whose kernel will be a quantifier elimination engine.

Acknowledgment: This research is supported by the Army Research Office under grants DAAH 04-95-1-0588 and DAAH 04-96-1-0341. The work of the third author has been partially supported by California PATH, University of California at Berkeley.

References

1. R. Alur, C. Coucoubetis, N. Halbwachs, T.A. Henzinger, P.H. Ho, X. Nicolin, A. Olivero, J. Sifakis, and S. Yovine. The algorithmic analysis of hybrid systems. *Theoretical Computer Science*, 138:3–34, 1995.
2. R. Alur and D.L. Dill. A theory of timed automata. *Theoretical Computer Science*, 126:183–235, 1994.
3. D. S. Arnon, G. E. Collins, and S. McCallum. Cylindrical algebraic decomposition I: The basic algorithm. *SIAM Journal on Computing*, 13(4):865–877, November 1984.
4. A. Dolzman and T. Sturm. REDLOG : Computer algebra meets computer logic. *ACM SIGSAM Bulletin*, 31(2):2–9, June 1997.
5. T.A. Henzinger. Hybrid automata with finite bisimulations. In Z. Fülöp and F. Gécseg, editors, *ICALP 95: Automata, Languages, and Programming*, pages 324–335. Springer-Verlag, 1995.
6. T.A. Henzinger, P.W. Kopke, A. Puri, and P. Varaiya. What's decidable about hybrid automata? In *Proceedings of the 27th Annual Symposium on Theory of Computing*, pages 373–382. ACM Press, 1995.
7. G. Lafferriere, G. J. Pappas, and S. Sastry. O-minimal hybrid systems. Technical Report UCB/ERL M98/29, University of California at Berkeley, Berkeley, CA, April 1998.
8. G. Lafferriere, G.J. Pappas, and S. Yovine. Reachability computation of linear hybrid systems. In *Proc. of 14th IFAC World Congress*. Elsevier Science Ltd., 1999. To appear.

9. A. Puri and P. Varaiya. Decidability of hybrid systems with rectangular differential inclusions. In *Computer Aided Verification*, pages 95–104, 1994.
10. A. Tarski. *A decision method for elementary algebra and geometry.* University of California Press, second edition, 1951.
11. C. Tomlin, G. J. Pappas, and S. Sastry. Conflict resolution for air traffic management : A study in multi-agent hybrid systems. *IEEE Transactions on Automatic Control*, 43(4):509–521, April 1998.
12. L. van den Dries. *Tame Topology and o-minimal structures.* Cambridge University Press, 1998.
13. L. van den Dries and C. Miller. On the real exponential field with restricted analytic functions. *Israel Journal of Mathematics*, 85:19–56, 1994.
14. P. Varaiya. Reach set computation using optimal control. 1997. preprint.
15. V. Weispfenning. A new approach to quantifier elimination for real algebra. Technical Report MIP-9305, Universität Passau, Germany, July 1993.
16. A. J. Wilkie. Model completeness results for expansions of the ordered field of real numbers by restricted pfaffian functions and the exponential function. *Journal of the American Mathematical Society*, 9(4):1051–1094, Oct 1996.

Synthesis of Control Software in a Layered Architecture from Hybrid Automata*

Man Lin

Department of Computer and Information Science
Linköping University
S–581 83 Linköping, Sweden
linma@ida.liu.se
URL: http://www.ida.liu.se/~linma

Abstract. This paper deals with the synthesis of control software for hybrid systems specified as hybrid automata. Instead of generating the software from scratch, the synthesis is based on a generic layered software architecture which supports both periodic and event-triggered computations. The use of the layered software architecture as the framework for implementing hybrid controllers is motivated in the paper.

An automatic code generator HA2LS (from Hybrid Automata to Layered Systems) is introduced. HA2LS reads a specification in terms of hybrid automata and generates intermediate code that can be processed by the tools provided by the layered architecture. The generated software provides a clean interface for a control engineer to plug in the control algorithms. With externally supplied control algorithms and IO procedures, the synthesis of executable hybrid controllers can be completed. The generated code can also be used for simulation purposes if it is generated from the specification of the complete system including the plant. The code generator HA2LS together with the software architecture substantially shorten the time to implement hybrid controllers from hybrid automata.

1 Introduction

By the term "embedded system" one usually refers to a computer program embedded in a physical system. Such an embedded system usually be used to control a mechanical, chemical or other kind of plant. The design and implementation of embedded systems usually involve contributions from both control and computer engineers.

In a complex embedded system, the plant can not be simply described as a set of differential equations. Instead, it can be described by a family of *modes*:

---

* The author would like to thank Dr. J. Malec, Dr. D. Driankov, Dr. S. Nadjm-Tehrani and the anonymous referees for the valuable comments on the paper and thank Volvo Research Foundation, Volvo Educational Foundation, the Center for Industrial Information Technology (CENIIT), and the Swedish Research Council for Engineering Sciences for supporting this research.

F.W. Vaandrager and J.H. van Schuppen (Eds.): HSCC'99, LNCS 1569, pp. 152–164, 1999.
© Springer-Verlag Berlin Heidelberg 1999

each mode given as a set of differential equations. The control is switched among modes depending on the state of the plant. Such a systems is called *hybrid system* since it consists of *continuous* subsystems which are governed by the equations and *discrete* subsystems (the mode switches). The controller for a hybrid system is called hybrid controller. To implement this, the control engineers need to provide the algorithms for each control mode and the computer engineers need to implement the continuous control algorithms, the discrete controllers to control the mode switches, and the interaction between the continuous and the discrete control.

Although there exist many modeling languages [2–4, 10] for modeling the behaviour of hybrid systems, little work has been done to bridge the gap of high level specification and implementation of hybrid controllers. One of the most popular specification languages for hybrid systems is hybrid automata [2]. The problem to be solved in this paper is: given hybrid automata as the specification of a hybrid system, can the software for a hybrid controller be systematically generated? The aim is to reduce or even eliminate the work of the computer engineer in the process of developing a hybrid controller.

It is not easy to implement hybrid controllers using languages like C or ADA since they give very poor support for algorithms expressed in a state machine fashion. Software support and computer aided tools are therefore needed to assist the implementation process. In this paper, we motivate a generic layered software architecture [9] for implementing hybrid controllers. The languages and tools developed within the architecture support periodic computations and discrete computations needed by the hybrid controllers. We provide a systematic method to synthesize hybrid controllers in the generic layered architecture from high level specifications in terms of hybrid automata. The synthesis procedure is semi-automatic. It includes automatic code generation with a tool called HA2LS and the integration of the generated code with the user-supplied procedures. A user-supplied procedure is either a control procedure which implements the control algorithm for the continuous dynamics or an interface procedure which deals with communication from sensors or to actuators. There are two uses for the methodology. One is to generate code for the controller. Another is to generate software for the complete system (including the plant) for simulation purposes.

The paper is organized as follows. First we provide a very brief description of hybrid automata in section 2. Then, we introduce the generic layered architecture and motivate its use for the implementation of hybrid controllers in section 3. A method to synthesize the hybrid controller is provided in section 4. Finally, the conclusions are given. The railway crossing system is used as an example throughout the paper to illustrate the methodology.

2   Hybrid automata

A hybrid automata specification consists of a set of hybrid automata. Basically, a hybrid automaton contains a set of locations and transitions among them. Each location contains a set of linear differential equations to describe one mode of the hybrid system. The transitions describe the mode switching. The components of a hybrid automaton are the following (see [2]):

- a finite set of *variables* $X = \{x_1, x_2, \ldots, x_n\}$;
- a finite set of vertices V called *locations*: each location is associated a *continuous activity* and an *invariant*;
- a finite set of *transitions* E: each transition has a source location, a target location, a guard and a discrete action;
- a finite set of *synchronization events*.

The following is a graphical representation of hybrid automata for a railway crossing system. There are three automata: `Controller`, `Train` and `Gate` in the railway crossing system (see Fig. 1). Each automaton has several locations, e.g. automaton `Gate` has four locations: `raising`, `lowering`, `open` and `closed`. The continuous activities are represented at the locations. For example, the continuous dynamics of `lowering` location of automaton `Gate` is $dg = -9$ (Note that dg denotes the derivative of g, the angle of the gate.). The corresponding invariant is $g \geq 0$ which is easy to understand since the controller should not lower the gate if the gate already reaches its lowest angle. The transitions of the automata are the edges and labels among the locations.

The grammar that we use for describing hybrid automata is taken from the following URL: http://www-cad.eecs.berkeley.edu/~tah/HyTech. The grammar is also used as the input to HyTech [1] which can verify some properties of hybrid systems. The only modification that we make is that we provide the possibility to distinguish whether a synchronization event is used as a guard or is issued as an action. This distinction is necessary since a controller may react to the incoming event by sending out some signals. Without the distinction, a controller can not tell whether this is the event it should react to or the event it should issue.

3   A generic layered software architecture

The layered architecture [9] developed at Linköping University defines a framework for the development of real-time autonomous systems. The architecture provides a clear separation between high level planning, discrete event response and periodic control. The idea is to combine plan-guided actions and situation-driven actions. The architecture contains three layers: *the Analysis Layer, the Rule Layer* and *the Process Layer*. The lowest layer (the Process Layer) hosts periodic computations. The middle layer (the Rule Layer) hosts discrete event response computations. The highest layer (the Analysis Layer) is concerned with planning activities.

Most industrial controllers nowadays do not consider planning activities. Therefore, only the lower two layers are interesting in this context. In the rest of this paper, when we speak about the layered architecture, it refers only to the Rule Layer and the Process Layer.

One of the most important concepts in software engineering is reusability. Reusable code and software structure will help to improve the efficiency and correctness of the software development. The reusability idea is strongly exploited

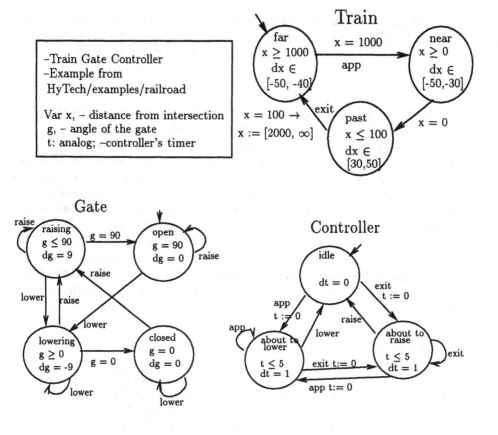

Fig. 1. The railway crossing system

in the construction of software tools and languages in the layered architecture. The idea is to extract the common features for every application as part of the "executive" for the layers while providing a language for expressing application dependent features. The executives for the Process Layer and the Rule Layer are called *PLX (Process Layer Executive)* and *RLX (Rule Layer Executive)* respectively. The corresponding languages are called *PLCL (Process Layer Configuration Language)* and *RL (Rule Language)* respectively.

As will be seen later, these executives and languages can be used to implement hybrid controllers since the two layers match perfectly the computations needed by the hybrid controllers. Next, we provide short descriptions of the two layers.

3.1 The Process Layer

The Process Layer facilitates periodic computations. These computations transform data. A feedback control algorithm can be thought of as such a transformation function. A function takes the data observed via some sensors from the plant and computes the output to send to the actuators as the control input to the plant. In what follows, we will describe in detail PLCL and PLX.

PLCL

The PLCL provides the language constructs for defining data as *vectors* and the constructs for defining *processes* to perform the computations on vectors.

A vector is realized as a dual state vector where two storage locations are used for each data item: one represents the value for kT_i and one for $(k+1)T_i$ where T_i is the flip period for the vector. Each data item in a vector is called a *frame*. Due to the hardware configuration and the difference of physical devices, the procedures to input a frame and output a frame might be different. PLCL associates with a frame three frame procedures: *INIT*, *GET* and *PUT*, each of which encapsulates some device-dependent operations. Either *GET* or *PUT* may be omitted corresponding to either a pure sensor or a pure actuator. The gateV vector below shows the state information for a gate. The SELECT frame gate_mode represents the mode selector for the process associated with the gateV vector.

```
VECTOR gateV is
  INTERVAL 1;
  SELECT gate_mode : tSwitch;     INIT registerSelector[];
                                  GET  receiveFrame [];
  FRAME  g      : int;            INIT registerInt [];
                                  PUT  sendFrameg [];
  FRAME  dg     : int;            INIT registerInt [];
                                  GET  receiveFrame [];
END -- vector: gateV
```

A process structure contains the descriptions of different *modes* that the vector will operate in. The ACTIVE construct defines the initial active mode for a process. Each mode is realized by a sequence of transformations called *modules*. A module is the basic element of computation in the process layer. Using the same mechanism as the frame procedures one module is an encapsulation of a single algorithm in a specific context. The code of the algorithm is obtained from a library (supplied externally). When writing a PLCL program, one only supplies the specification of the function names and the input-output arguments for the modules. The following is an example of a mode definition:

```
MODE closed is
MODULE    module1 ( IN *gateV#dg : int, OUT gateV#g : int )
          USE gate_closed_func [];
```

The mode named closed contains only one module called module1. The actual transformation function is called gate_closed_func. The module has one input argument: dg belonging to the vector gateV and one output argument g belonging also to the vector gateV.

PLX

The PLX manages the Process Layer tick counter, *flip* and *dump* to each dual state vector and execute the processes at appropriate time points. The PLX algorithm is as follows:

```
INIT vectors;
LOOP
    WAIT FOR tick;
    FLIP;
    EXECUTE processes;
    DUMP;
END
```

The *flip* operation copies the current state to the previous state and acquires a new sample for a frame if it has a *GET* procedure. The *dump* operation outputs a frame if it has a *PUT* method. The execution of processes is done by activating the module functions associated with active modes.

3.2 The Rule Layer

The Rule Layer reacts to discrete events. Whenever some important event occurs, the system starts to respond by evaluating rules. In the context of hybrid controllers, the effect of the Rule Layer is to select different control algorithms (mode switching) and provide parameters for the control algorithms. The selection of a different control algorithm is done by changing the **SELECT** frame in the Process Layer. The parameters are changed by changing frames of the Process Layer accessible to the library functions. In what follows, we describe in detail RL and RLX.

RL

The Rule Layer deals with finite domain variables called *slots*. An RL program contains the description of slots and rules defined over them. RL also provide the possibility to specify the communication channels.

A rule contains WHEN, IF and THEN parts. The WHEN part contains the activation conditions for the rules and the IF part is the additional condition for the rules to be able to fire. The THEN part contains actions of the rules: either changing the value of a slot or sending signals to specific channels.

The slots of the Rule Layer are realized as dual vectors containing the current values and the previous values. The primitive condition can query both the current value and the previous value of a slot. This is done by the comparison operators defined in RL.

– $x|{=}v$ is used to compare the previous value of x with v.
– $x{=}|v$ is used to compare the current value of x with v.

- $x*=v$ is used to check whether x has changed to v.
- All the other operators $>, <, \geq, \leq$ will take the current value of x for comparison. The comparison will follow the obvious mathematical interpretation.

The rules must be interpreted declaratively. The order of writing the rules should not influence the semantics of the rules. A response of the rule-based system is the firing of a sequence of rules. To impose a consistent view of the state and the response, we have defined a semantics for the desired response of a rule-based system and correctness criteria for RL programs. Static checker has also been developed. For details, see [6].

RLX

The executive of the Rule Layer serves as an inference engine for the rules. Basically, the inference engine starts to evaluate rules whenever a stimulus comes. If there is a rule enabled, then the engine fires the rule. The fired rule might generate new changes which, in turn, activate other rules. The inference continues until there are no rules to be fired. The details about how the engine's implementation works can also be found in [6].

3.3   The integration of the Process Layer and the Rule Layer

One way to integrate the two layers is to have a communication link between them. We have developed a socket communication model for the two layers in the UNIX environment.

One important problem is the study of the timing issue for the integrated system. We assume that all the inputs to the Rule Layer must come from the Process Layer. Thus there is at most one discrete response for each smallest period of the Process Layer since the Process Layer *dumps* the vectors only once for each smallest period. We also have to assume that the period of a process is smaller than the sampling rate of the inputs that are to be used by the process. Therefore, the environment won't change significantly during the processing by the system.

To make sure that the mode switch can happen immediately when a condition is satisfied, we assume (and require) that the Rule Layer reacts timelessly, meaning that the response of the Rule Layer is short enough such that the Process Layer can always receive the mode switching command at the beginning of the next tick. This assumption can be checked with the help of timing analysis of the code.

3.4   Motivation for developing hybrid control systems using this framework

We summarize the benefits of developing hybrid control systems in the framework of the layered architecture:

- The layered architecture separates the computations in a clean way. Periodic computations and event-triggered computations are put into different layers but integrated in an interleaving manner. The integration maintains the sampling rate while mode switching can be performed immediately when the conditions are satisfied.
- It reduces the work for the software engineer since most of the common functions are encoded in the executives which are reusable for every application.
- The architecture and the languages developed provide a good interface for the control engineer. To develop a hybrid controller in the layered architecture, the control engineer only needs to provide the following:
  - a PLCL program with the definitions for the vectors and processes;
  - an RL program with the rules for mode switching;
  - a set of control algorithms whose implementation can be written in the C language;
  - specification of communication channels when necessary.

  The automatic code generator HA2LS (from Hybrid Automata to Layered Systems) to be described in the next section can generate most of the code needed for a hybrid controller directly from hybrid automata.
- Some analysis [6–8] performed on PLCL programs and RL programs can help a user to identify whether a program is logically correct and if the timing constraints can be fulfilled.

Ravn et al. have used the idea of the software architecture to implement a mode-switching controller for a hydraulic cylinder [5].

4   Automatic code generator: HA2LS

In this section, we introduce the tool HA2LS which automatically generates layered software from hybrid automata and give the guidelines for generating the final executable hybrid controllers.

Fig. 2 illustrates the translation and assembly process for hybrid controllers. Currently, the software for the executives of layered architecture is available on two execution environments: a real-time version running under $PSOS^+$ on a MC68020 microprocessor and a simulated version running under UNIX.

HA2LS can generate a PLCL program, an RL program and partial codes for the control algorithms in C. Next, we present how each individual program is generated with an example: the railway crossing control system (see Fig. 1). We will focus on the code generation for Gate automaton in the discussion.

4.1   Generation of PLCL program

For each automaton of the hybrid automata specification, we generate one vector and one process description.

Each vector contains an interval indicating the period of *flip* and a set of frames. The generated period is set to 1. It can be changed by the user once

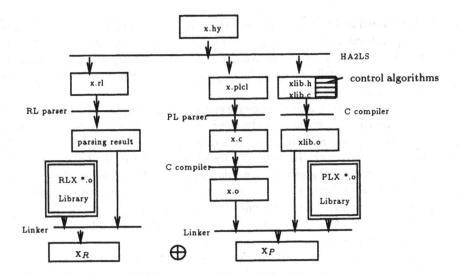

Fig. 2. The translation and assembly process

the sampling rate for the frame is decided, which depends on the environment of the controller and the speed requirement for the control. Three types of frames: *mode-selector-frame*, *value-frame* and *dot-frame* are generated for each vector. Frame **gate_mode**, **y**, **dy** (see the following PLCL code) are examples of *mode-selector-frame*, *value-frame* and *dot-frame*, respectively. The **VALUE** construct provides the initial value for a frame and the **ALIAS** construct provides the possibility to refer to the same frame with different names.

```
      ---- PLCL: vector description for automaton Gate
VECTOR gateV is
   INTERVAL 1;
   SELECT gate_mode : tSwitch;        INIT registerSelector[];
                                      GET   receiveFrame [];
   FRAME  g      : int;  VALUE  0;    INIT registerInt [];
                                      PUT   sendFrameg [];
                                      ALIAS   gr;
   FRAME  dg     : int;  VALUE  0;    INIT registerInt [];
                                      GET   receiveFrame [];
END -- vector: gateV
```

Each process contains the mode descriptions and the active mode. Each location in the automaton will be translated to a mode description. The mode description for **raising, lowering, open, closed** for **Gate** automaton can be found in the following generated PLCL code. The active mode (**open**) is the initial location of the corresponding automaton.

```
PROCESS gatePØgateV IS
  MODE closed is
    MODULE    module1  (IN *gateV#dg : int,  OUT gateV#g : int)
              USE gate_closed_func [];
    END           OR
  MODE lowering is
    MODULE    module1  (IN *gateV#dg : int,  OUT gateV#g : int)
              USE gate_lowering_func [];
    END           OR
  MODE open is
    MODULE    module1  (IN *gateV#dg : int,  OUT gateV#g : int)
              USE gate_open_func [];
    END           OR
  MODE raising is
    MODULE    module1  (IN *gateV#dg : int,  OUT gateV#g : int)
              USE gate_raising_func [];
    END
  ACTIVE open;
END -- process gateP
```

4.2 Generation of RL program

The generation of RL program includes the generation of *slots* and *rules*.

Four types of slots: *value-slot*, *dot-slot*, *mode-slot* and *synchronization-slot* will be generated. Each *value-slot* corresponds to a continuous variable of the hybrid automaton. For each *value-slot*, there is a *dot-slot* indicating the change rate of the slot. The generation of domains for *value-slot* and *dot-slot* is tricky since we need to partition infinite domains into finite ones. What we have done is to abstract intervals to points. A *mode-slot* is used to indicate the mode of an automaton. Each mode corresponds to one location of the automaton. A *mode-slot* in the Rule Layer has a counter part in the Process Layer called *mode-selector-frame*. Changing a *mode-slot* results in switching mode of the Process Layer. Each synchronization event in the hybrid automaton is translated into an EVENT-typed slot whose value will be reset after each response. Different types of generated slots are shown below.

```
    --   VAR SLOT
SLOT                g    :{-1, 0, 89, 90,91 }   := 90;

    --   DOT SLOT
SLOT           dg   :{-10, -9, -1, 0, 8, 9,10 }   := 0;

    --   SYNCHRONIZATION SLOT
SLOT                app   :{0, 1}   := 0  TYPE EVENT;
SLOT                exit  :{0, 1}   := 0  TYPE EVENT;
SLOT                raise :{0, 1}   := 0  TYPE EVENT;
```

```
SLOT                  lower   :{0, 1}   := 0  TYPE EVENT;
  --    MODE SLOT
SLOT   gate_mode :{open, closed, raising, lowering}  := open;
```

Each rule in the rule program corresponds to a discrete transition in a hybrid automaton. A rule in the rule program contains three parts: trigger, condition and assignment. The trigger and condition parts are generated from the guard of the corresponding transition. The assignment part is derived from the action part of the transition.

The criterion to split the guard into trigger and condition parts is the following: if the guard contains a synchronization event e, then we will generate $e*=1$ as the trigger of the rule. Otherwise, we will generate a TRUE trigger for the rule. Other parts of the guard of a transition except the synchronization event will be translated into the condition part of a rule. The translation is straight-forward. Note that the comparative operator $=$ will be translated into $=|$ which is to test the current value of the slot appearing before the operator. One more condition to be generated is the source-location condition. For each transition of an automaton A, A |= loc is generated where loc is the the name of the location from which the transition originates.

In a hybrid automaton, the effect of a transition is to switch to the next location and to reset some values of the variables, e.g., to reset the clocks. The first effect can be easily translated as an assignment: A := loc, where loc stands for the next location. As the reset of values takes effects only at the beginning of the next continuous activity, the new values will not affect the evaluation of the transitions leaving the current location. However, the semantics of RL language adopts the synchrony hypothesis. Any assignment in one response is seen as occurring simultaneously with all others. To make an assignment take effect after the response, we use a trick. If $x:=v$ appears in the action part of a transition, we will generate $xr:=v$ in the assignment part of the rule where xr is an alias for x meaning the slot assigned by the rule layer. The alias construct in the process layer will put the values of xr into the frame x of the process layer. Since xr only appears in the assignment part, such assignment will not lead to further rule firings while, at the same time, it can change the value of frame x in the process layer.

The following shows the RL rules for the Controller automaton.

```
     ---- RULEs for automaton gate
WHEN lower *= 1 IF   gate |= closed THEN  dg := 0, gate := closed;
WHEN raise *= 1 IF   gate |= closed THEN  dg := 9, gate := raising;
WHEN raise *= 1 IF   gate |= lowering THEN dg := 9, gate := raising;
WHEN lower *= 1 IF   gate |= lowering THEN  dg := -9, gate := lowering;
WHEN TRUE IF y   =| 0 AND gate |= lowering  THEN dg := 0, gate := closed;
WHEN lower *= 1 IF   gate |= open THEN  dg := -9, gate := lowering;
WHEN raise *= 1 IF   gate |= open THEN  dg := 0, gate := open;
WHEN lower *= 1 IF   gate |= raising THEN  dg := -9, gate := lowering;
WHEN raise *= 1 IF   gate |= raising THEN  dg := 9, gate := raising;
WHEN TRUE IF y   =| 90 AND gate |= raising THEN dg := 0, gate := open;
```

4.3   Generation of library functions

Partial code for the library of module functions for control algorithms are also generated. These include the declarations of the functions and the interface of the functions. The comments about the range of the DOT-VALUEs (expected control objectives) are also generated. Only the function bodies to achieve the control objectives need to be changed later. What follows is the module function to lower the gate which only contains the effect of expected control algorithm: the angle of the gate g is lowered with the rate $dg = -9$. The real control algorithm has to be plugged in later.

```
void gate_lowering_func (int dg, int *g)
{ *g = *g + dg;           /**    -9 <= dg <= -9        **/
  return; }
```

4.4   What is left to make the control software complete?

There are two uses for the methodology. One is to generate code for the controller. Another is to generate software for the complete system (including the plant) for simulation purposes.

Usually, we start with the simulation version. Given the model of a control system and its environment, we generate a simulated closed loop system where the control system and its environment are coupled using shared variables. The plant behaviours are simulated by dedicated processes. The simulation version enables simulation of various proposed controller algorithms. This is done by changing the module functions. For the railway crossing system, if code is also generated from the **train** automaton, then we have a simulated close loop system.

To derive the control software from the simulation version, the following needs to be done:

- Replace the simulated plant with the real plant coupled by sensors and actuators. This includes:
  - specify the input and output procedures for the input and output frames;
  - delete the processes simulating the behaviour of input frames;
  - plug in the algorithms for the controllers into the body of the generated module functions for the processes simulating the output frames;
- Specify the channels for the communication with sensors and actuators if there are any;
- Change the sampling rate of vectors if needed;

For example, to couple the railroad control system with its environment, one needs to do the following: Since the control system needs to read the distance of the incoming train from the intersection periodically, the input procedure for the frame x should be provided. The process together with its module functions simulating the behaviour of the train will be deleted. The output procedure for the angle of the gate g is needed. The control algorithms for raising the gate and lowering the gate with the given rate should be plugged into the module functions gate_raising_func and gate_lowering_func.

5 Conclusions

In this paper, we have discussed how to bridge the gap between high-level specification and an implementation of hybrid systems. An automatic code generator HA2LS has been introduced to generate layered systems in a generic layered software architecture from hybrid automata. The generated software provides a good interface for the control engineer to plug in real control algorithms. With the help of control engineer supplying the control algorithms and the language and tools developed for the layered architecture, a complete hybrid controller can be conveniently constructed. The hybrid controller can be run in UNIX environment and on $PSOS^+$-based real-time system. With a reasonable amount of work, the software can be ported to other platforms. This work greatly reduces the work of a computer engineer in implementing hybrid controllers.

References

1. R. Alur, C. Courcoubetis, N. Halbwachs, T.A. Henzinger, P.-H. Ho, A. Olivero, J. Sifakis, and S. Yovine. The algorithmic analysis of hybrid systems. *Theorectical Computer Science*, pages 3–34, 1995.
2. R. Alur, C. Courcoubetis, T.A. Henzinger, and P.-H. Ho. Hybrid automata: an algorithmic approach to the specification and verification of hybrid systems. In R.L. Grossman, A. Nerode, A.P. Ravn, and H. Rischel, editors, *Hybrid Systems, Lecture Notes in Computer Science 736*, pages 209–229. Springer-Verlag, 1993.
3. M.S. Branicky, V.S. Borkar, and S. K. Mitter. A unified framework for hybrid control. In *Proceedings of 33rd Conference of Decision and Control*, pages 4228–4234. IEEE, 1994.
4. Z. Chaochen, A. P. Ravn, and M. R. Hansen. An Extended Duration Calculus for Hybrid Real-Time Systems. In R.L. Grossman, A. Nerode, A.P. Ravn, and H. Rischel, editors, *Proc. Workshop on Theory of Hybrid Systems, October 1992, LNCS 736*, pages 36–59. Springer Verlag, 1993.
5. T. J. Eriksen, S. T. Heilmann, M. Holdgaard, and A. P. Ravn. Hybrid systems: a real-time interface to control engineering. In *Proceedings of 8th Euromico Workshop on Real-Time Systems*, pages 114–120. IEEE, 1996.
6. M. Lin. *Formal Analysis of Reactive Rule-based Programs*. Licentiate thesis, Linköping University, Linköping University, Sweden, 1997.
7. M. Lin and J. Malec. Timing analysis of reactive rule-based programs. *Control Engineering Practice*, 6:403–408, 1998.
8. M. Morin. Predictable cyclic computations in autonomous systems: A computational model and implementation. Licenciate thesis 352, Department of Computer and Information Sciences, Linköping University, 1993.
9. M. Morin, S. Nadjm-Tehrani, P. Real-time hierarchical control. *IEEE Software*, 9(5):51–57, September 1992.
10. S. Nadjm-Tehrani. *Reactive Systems in Physical Environments: Compositional Modelling and Framework for Verification*. PhD thesis, Dept. of Computer and Information Science, Linköping University, March 1994. Dissertation No. 338.

An Overview of Hybrid Simulation Phenomena and Their Support by Simulation Packages

Pieter J. Mosterman*

Institute of Robotics and System Dynamics, DLR Oberpfaffenhofen
P.O. Box 1116, D-82230 Wessling, Germany
Pieter.J.Mosterman@dlr.de
http://www.op.dlr.de/~pjm

Abstract. Hybrid systems combine continuous behavior evolution specified by differential equations with discontinuous changes specified by discrete event switching logic. Numerical simulation of continuous behavior and of discrete behavior is well understood. However, to facilitate simulation of mixed continuous/discrete systems a number of specific hybrid simulation issues must be addressed. This paper presents an overview of phenomena that emerge in simulation of hybrid systems, reported in previously published literature. They can be classified as (i) event handling, (ii) run-time equation processing, (iii) discontinuous state changes, (iv) event iteration, (v) comparing Dirac pulses, and (vi) chattering. Based on these phenomena, numerical simulation requires the implementation of specific hybrid simulation features. An evaluation of existing simulation packages with respect to these features is presented.

1 Introduction

Continuous system dynamics can be described by, possibly large, systems of differential equations. These can be either ordinary differential equations (ODEs) or contain algebraic constraints as well to form differential and algebraic equations (DAEs). Complex systems, such as aircraft, often operate in different *modes* of continuous operation and when mode changes occur, the continuous dynamics change abruptly. Even small physical components may operate in different modes, e.g., a diode can operate as a short or open circuit, requiring abrupt discrete changes in the system of equations.

Simulation of pure continuous and pure discrete systems is well-understood. ODEs and DAEs are a common representation for continuous systems from which numerical simulation algorithms generate behaviors. Variable step size approaches may be applied to ensure the numerical grid is sufficiently dense with respect to some error measure given the dynamic behavior of the system. Discrete simulation is often based on the particular description formalism, such as Petri nets [21] and finite state machines [12]. Typical is the use of random

* Pieter J. Mosterman is supported by a grant from the DFG Schwerpunktprogramm KONDISK.

F.W. Vaandrager and J.H. van Schuppen (Eds.): HSCC'99, LNCS 1569, pp. 165–177, 1999.

distribution functions to model, e.g., queue processing and to facilitate discrete phenomena such as nondeterminism and parallelism.

Recently, there is a growing interest in *hybrid systems*, i.e., systems with mixed continuous/discrete behavior. This interest is driven by (i) the increasing need for comprehensive analysis of systems where discrete controllers operate on a continuous process, and (ii) efficient handling of otherwise stiff continuous equations. In hybrid simulation, typically, the continuous model part generates discrete events when continuous signal variables cross threshold values. These discrete events may affect continuous behavior evolution by changing active model components and discontinuously changing the continuous state variables.

Though individually the continuous and discrete formalisms can be treated well, their interaction causes a number of unique problems in simulation. Based on these phenomena, numerical simulation requires the implementation of specific hybrid simulation features. This paper discusses and illustrates these issues and refers to possible solutions. It identifies which of the features specific to hybrid simulation are incorporated by a number of simulation packages. It does not evaluate robustness or quality of the implemented solutions (for additional information refer to [13]). The evaluation shows that some of the hybrid simulation features are incorporated in a number of simulation packages, whereas a number of other features are only implemented by a few or none. The aim of this paper is to categorize issues in hybrid simulation to evaluate the quality of simulation packages, and to refer to possible methods of implementation.

2   Generating the Simulation Model

This section describes how ODE and DAE forms are achieved and defines a number of terms required to describe the numerical simulation problem.

1. Complex dynamic system models are composed of declarative submodels specified by noncausal equations [6]. After the complete system of equations is *compiled* a *sorting* procedure assigns computational causality. In matrix form, this corresponds to a lower triangular form. Circular dependencies between variables cause blocks of dependent equations (which can be solved symbolically or numerically) to be kept together, resulting in a block lower triangular (BLT) matrix structure of the sorted equations.

2. After sorting, the complexity of the simulation problem, indicated by the *index*, needs to be sufficiently low [5]. Higher-index problems contain algebraic constraints on time derivative variables, and the system of equations needs to be *solved* to arrive at a lower index, e.g., by Pantelides' algorithm [22] and the *dummy derivative* approach [15].

3. Next, consistent initial values of the state variables need to be calculated from user specified values. For example, to start simulation from steady state, all time derivative values can be set to 0. The corresponding variable values are then computed based on these conditions.

4. Once the consistent initial values of the system of equations are computed, a numerical solver evolves the system behavior over time. Numerical solvers

vary from an explicit fixed step Euler to complex implicit integration schemes with variable step size and error control such as DASSL.

3 Hybrid Simulation Phenomena

When discrete event changes occur, equations may become (de)activated, which requires a number of specific problems to be solved. This section presents small benchmark examples that embody these phenomena.

3.1 Newton's Cradle

Consider the three colliding bodies in Fig. 1 where m_1 has an initial velocity, $v_1 = v$. Initially, m_1 moves towards m_2 with constant velocity while m_2 and m_3 are at rest:

$$\begin{bmatrix} m_1 & 0 & 0 \\ 0 & m_2 & 0 \\ 0 & 0 & m_3 \end{bmatrix} \begin{bmatrix} \dot{v}_1 \\ \dot{v}_2 \\ \dot{v}_3 \end{bmatrix} = \begin{bmatrix} 1 & 0 & 0 \\ 0 & 1 & 0 \\ 0 & 0 & 1 \end{bmatrix} \begin{bmatrix} F_1 \\ F_2 \\ F_3 \end{bmatrix} \tag{1}$$

with algebraic equations

$$\begin{bmatrix} 1 & 0 & 0 \\ 0 & 1 & 0 \\ 0 & 0 & 1 \end{bmatrix} \begin{bmatrix} F_1 \\ F_2 \\ F_3 \end{bmatrix} = \begin{bmatrix} 0 \\ 0 \\ 0 \end{bmatrix} \tag{2}$$

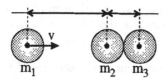

Fig. 1. A sequence of collisions.

Upon collision, $(x_1 \geq x_2) \wedge (v_1 > v_2)$, momentum is instantaneously transferred to m_2. The first condition specifies that in case of point masses there is contact between the two bodies and the second condition that there is a collision. For m_2 and m_3 this condition is not satisfied because $v_2 = v_3$, and, therefore, no collision occurs. A collision event is generated when $x_1 \geq x_2$ that needs to be (i) detected and (ii) located within a small error tolerance to improve precision. In general, continuous variables generate discrete events when they cross threshold values. If time is the continuous variable, the time of occurrence can be located exactly. Otherwise, a root finding mechanism is required, which often relies on less efficient iteration [7].

Characteristic 1 (state events) *Events that are generated when continuous variables cross threshold levels need to be detected and the time of occurrence has to be located.*

Characteristic 2 (time events) *The time of occurrence of events that are generated because of time reaching a threshold value is known in advance, and, therefore, these events can be handled efficiently.*

For the colliding bodies, the change of the velocities immediately before collision, v_i, to their values immediately after collision, v_i^+, is governed by Newton's collision rule

$$v_2^+ - v_1^+ = -\epsilon(v_1 - v_2), \tag{3}$$

Characteristic 3 (explicit reinitialization) *Constraints on initial values for continuous state vector variables after a configuration change may be functions of the final values in the previous configuration.*

The values of v_1 and v_2 are known as their final value when $x_1 \geq x_2$ and continuous behavior was halted. However, both v_1^+ and v_2^+ are unknown and cannot be solved with one equation. To solve for the new velocities, a least mean square fit can be applied but in certain cases this may result in incorrect transfer of physical quantities across discontinuities. In case of the colliding bodies, upon collision the forces F_1 and F_2 are equal, which changes the system of equations by replacing Eq. (2) with

$$\begin{bmatrix} 1 & -1 & 0 \\ 0 & 0 & 0 \\ 0 & 0 & 1 \end{bmatrix} \begin{bmatrix} F_1 \\ F_2 \\ F_3 \end{bmatrix} = \begin{bmatrix} 0 \\ 0 \\ 0 \end{bmatrix} \tag{4}$$

Characteristic 4 (changing simulation model equations) *Discrete events may change the set of equations that describes system behavior by adding and removing equations.*

Using Eq. (1) to solve the equation in the top row of Eq. (4) yields $m_1 \dot{v}_1 = m_2 \dot{v}_2$ and this can be integrated to

$$m_1(v_1^+ - v_1) = m_2(v_2^+ - v_2), \tag{5}$$

which embodies the physical conservation of momentum constraint. Combined with Eq. (3) it can be uniquely solved for v_1^+ and v_2^+.

The use of a conservation law based on physics is a general characteristic that needs to be accounted for in hybrid simulation of physical system models. This can be automated for semi-explicit systems with linear constraints by integrating the system of differential equations [17].

Characteristic 5 (conservation constraints) *Computing initial values for the continuous state vector variables in a new configuration is governed by physical conservation constraints.*

When the new velocities of the colliding bodies are updated ($v_i \leftarrow v_i^+$), $v_2 > v_3$ and an immediate further configuration change occurs that models the collision between m_2 and m_3.

Characteristic 6 (state updating event iteration) *After updating the state vector, new initial continuous state variable values may generate discrete events that immediately cause further configuration changes.*

Again Newton's collision rule and conservation of momentum applies and the new velocities of m_2 and m_3 are computed. In case $m_1 = m_2 = m_3$, after momentum is transferred to m_3, no further configuration changes occur and m_3 starts to move continuously in time.

3.2   The Falling Rod

Consider the rigid rod sliding on a rough surface in Fig. 2. In case of Coulomb friction, the friction force F_f depends on the normal force F_N by a constant coefficient μ, i.e., $F_f = \mu|F_N|$ [14], active in the direction opposite to $v_{A,x}$, the velocity of the contact point at the surface. F_y is the kinetic force exerted by the center of mass in the vertical direction. Combined with the gravitational force, F_g, this yields the normal force $F_N = F_y - F_g$. Velocity v_x is the horizontal velocity of the center of mass, M, v_y the vertical velocity, and ω the angular velocity. The system has three inertial, energy storing, components, *viz.*, the linear inertias m_x and m_y and the rotational inertia J.

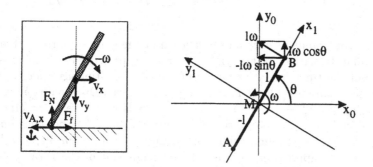

Fig. 2. Coulomb friction.

If initially the rod is falling freely towards the floor,

$$\begin{bmatrix} m_x & 0 & 0 \\ 0 & m_y & 0 \\ 0 & 0 & J \end{bmatrix} \begin{bmatrix} \dot{v}_x \\ \dot{v}_y \\ \dot{\omega} \end{bmatrix} = \begin{bmatrix} 1 & 0 & 0 \\ 0 & 1 & 0 \\ 0 & 0 & 1 \end{bmatrix} \begin{bmatrix} F_x \\ F_y \\ F_\omega \end{bmatrix} \quad (6)$$

with algebraic constraints

$$\begin{bmatrix} 1 & 0 & 0 \\ 0 & 1 & 0 \\ 0 & 0 & 1 \end{bmatrix} \begin{bmatrix} F_x \\ F_y \\ F_\omega \end{bmatrix} = \begin{bmatrix} 0 \\ F_g \\ 0 \end{bmatrix}. \quad (7)$$

Upon collision, Fig. 2 shows that the linear velocities v_x and v_y are constrained to move with respect to ω according to

$$\begin{cases} v_x = -l\omega \sin\theta \\ v_y = l\omega \cos\theta \end{cases} \tag{8}$$

Therefore, the initial state variables v_x, v_y, and ω become dependent by algebraic constraints. This causes higher index problems that require differentiation of equations to arrive at an equivalent lower index system that can be simulated [22].

Characteristic 7 (simulation model index reduction) *Adding and removing equations may result in higher index problems, requiring run-time index reduction.*

Furthermore, the forces in the x and y direction relate to the torque F_ω as

$$-l\sin\theta F_x + l\cos\theta F_y + F_\omega - l\cos\theta F_g = 0. \tag{9}$$

Therefore, upon collision a state event (Characteristic 1) causes Eq. (7) to be replaced by

$$\begin{bmatrix} 1 & 0 & l\sin\theta & 0 & 0 & 0 \\ 0 & 1 & -l\cos\theta & 0 & 0 & 0 \\ 0 & 0 & 0 & -l\sin\theta & l\cos\theta & 1 \end{bmatrix} \begin{bmatrix} v_x \\ v_y \\ \omega \\ F_x \\ F_y \\ F_\omega \end{bmatrix} = \begin{bmatrix} 0 \\ 0 \\ l\cos\theta F_g \end{bmatrix} \tag{10}$$

These constraints require reinitialization of the state vector, i.e., the linear and angular velocities, by applying conservation principles (Characteristic 5). Because of the time-derivative nature of forces, discontinuous velocity changes may cause an impulsive force $F_y = m_y \dot{v}_y$. Here, $\dot{v}_y$ is a Dirac pulse, δ, with area $v_y^+ - v_y$, $\dot{v}_y = \delta[v_y^+ - v_y]$. F_x is also of an impulsive nature, $\dot{v}_x = \delta[v_x^+ - v_x]$. Since $F_N = F_y - F_g$, the condition for sliding, $|F_x| > \mu F_N$, becomes $|m_x \dot{v}_x| > \mu(m_y \dot{v}_y - F_g)$ which requires evaluating $|m_x \delta[v_x^+ - v_x]| > \mu(m_y \delta[v_y^+ - v_y] - F_g)$. By numerically approximating the infinite magnitude of the Dirac pulses, F_g may inadvertently affect the comparison.

Characteristic 8 (Dirac pulse comparison) *Discontinuous changes in continuous variables may cause Dirac pulses on derivative variables. These pulses have to be treated separate from non-impulsive variables.*

A possible solution uses the areas of the Dirac pulses instead of magnitudes of forces. So, if upon collision $m_x|v_y^+ - v_y| > \mu m_y(v_y^+ - v_y)$, the rod starts to slide and another set of algebraic constraints becomes active

$$\begin{bmatrix} 0 & 0 & 0 & 1 & \mu & 0 \\ 0 & 1 & -l\cos\theta & 0 & 0 & 0 \\ 0 & 0 & 0 & -l\sin\theta & l\cos\theta & 1 \end{bmatrix} \begin{bmatrix} v_x \\ v_y \\ \omega \\ F_x \\ F_y \\ F_\omega \end{bmatrix} = \begin{bmatrix} 0 \\ 0 \\ l\cos\theta F_g \end{bmatrix} \tag{11}$$

Again, consistent initial values have to be computed from the current state variable values before continuous integration can resume. Note that in this case v_x is not constrained, and, therefore, is not required to change discontinuously. To generate physically meaningful behavior, the new state variable values should not be calculated from the values that were obtained by solving the initialization problem in the stuck mode that is active upon collision, because this mode was instantaneously departed, i.e., before updating the state vector. Instead, the values should be derived from the final values in the last mode of continuous behavior, i.e., where the rod was falling freely [16].

Characteristic 9 (state invariant event iteration) *New initial continuous state variable values may generate events that immediately cause further configuration changes before the original state vector values are updated.*

To illustrate, consider the situation where $\mu = 0$. Upon collision with the floor, in the stuck mode, the linear and angular velocities are subject to the algebraic constraint in Eq. (8) and this causes $v_x^+ \neq 0$. Because $\mu = 0$, it is immediately inferred that the rod starts sliding and the algebraic constraints are replaced by those in Eq. (10). These do not enforce a value $v_x^+ \neq 0$. In fact, during the entire process there is no force in the horizontal direction because the friction coefficient, μ, is 0. Therefore, $v_x = 0$ should hold. But, if the state vector would have been updated in the intermediate stuck mode ($v_x \leftarrow v_x^+$) this would be violated. For details refer to [16].

3.3   An Evaporator Vessel

In a fast breeder reactor, an evaporator vessel stores hot sodium and warms water that flows through a helical coil inside the vessel. As a safety mechanism, an overflow that connects to the sodium sump may become active when a specific fluid level is reached, see Fig. 3.

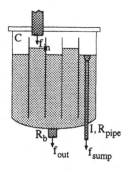

Fig. 3. An evaporator vessel.

A model of this system may consist of the vessel with capacity, C, and outflow resistance, R_b. The overflow may be modeled by its flow resistance, R_{pipe}, and fluid inertia, I. When the fluid level is below the overflow level,

$$\begin{bmatrix} I & 0 \\ 0 & C \end{bmatrix} \begin{bmatrix} \dot{f} \\ \dot{p} \end{bmatrix} = \begin{bmatrix} -R_{pipe} & 0 \\ 0 & -\frac{1}{R_b} \end{bmatrix} \begin{bmatrix} f \\ p \end{bmatrix} + \begin{bmatrix} 0 \\ f_{in} \end{bmatrix}, \tag{12}$$

a steady state is achieved such that the inflow, f_{in}, equals the outflow f_{out}. If R_b requires a liquid level higher than the overflow level, a state event (Characteristic 1) is generated that activates this mechanism,

$$\begin{bmatrix} I & 0 \\ 0 & C \end{bmatrix} \begin{bmatrix} \dot{f} \\ \dot{p} \end{bmatrix} = \begin{bmatrix} -R_{pipe} & 1 \\ -1 & -\frac{1}{R_b} \end{bmatrix} \begin{bmatrix} f \\ p \end{bmatrix} + \begin{bmatrix} 0 \\ f_{in} \end{bmatrix}, \tag{13}$$

and another steady state level is achieved such that $f_{in} = f_{out} + f_{sump}$. For specific parameters (e.g., $R_b = 1, R_{pipe} = 0.5, I = 0.5, C = 15, f_{in} = 0.25, \Delta T = 0.025$), this may cause the liquid level to fall below the overflow level and the mechanism becomes inactive but now the liquid level rises and the overflow becomes active again after an infinitesimal short period of time, causing the system to *chatter* between modes.

In general, simulation across discontinuities may cause large errors for a fixed integration step, see Fig. 4(a). Continuous time integration may be executed until within a small tolerance of the discontinuity, see Fig. 4(b). However, in case of chattering, the integration step size is repeatedly reduced to its minimal value, Fig. 4(c), and simulation times become excessively long. Efficient hybrid simulation needs to detect this behavior and possibly apply an equivalence relation to eliminate the fast chattering motion, but preserve the dynamics of the slow motion along the chattering surface. For details refer to [20].

Characteristic 10 (chattering) *When simulation switches back and forth between configurations, the fast chattering behavior needs to be removed while preserving the slower dynamics along the switching surface.*

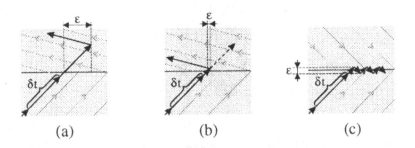

(a)   (b)   (c)

Fig. 4. Chattering may slow down simulation.

3.4   Summary

The following hybrid simulation phenomena can be identified.

- **state events:** If events occur because of continuous system variables crossing threshold values [2, 7, 23],
  - the event needs to be **detected**, and
  - its time of occurrence needs to be **located**.
- **time events:** If time crosses a threshold value, the time of occurrence is known *a priori*, which can be treated efficiently [7].
- **simulation model:** The system of equations that describes model behavior may change.
  - Blocks of sorted and solved equation may simply appear or disappear (e.g., a vehicle entering and leaving a highway), and, therefore, can be dynamically **added/removed**.
  - In some cases equations can be replaced by others, changing computational causality, and the system of equations may have to be **sorted** again.
  - In other cases, algebraic constraints between state variables may become active and the system of equations needs to be **solved** again to reduce the index of the system [22] (e.g., the rod making contact with the floor).
- **reinitialization:** There may be a discontinuous change in state variable values.
  - This change may be **explicitly** specified by the user by a new initial state equation (e.g., Eq. (3) of the colliding bodies).
  - The system of equations may have to be **integrated** to derive physically consistent initial values for a new mode. This ensures that physical conservation principles hold [17].[1]
- **event iteration:** When an event occurs, new system variable values may immediately trigger a further event. Two types of event iteration exist [18],
  - the state vector is **invariant** across the entire iteration (e.g., the falling rod that immediately starts to slide), and
  - the state vector is **updated** after each iteration step (e.g., the sequence of colliding bodies).
- **Dirac pulses:** Discontinuous changes in continuous variables may cause Dirac pulses to occur. If their magnitudes are numerically approximated, comparison may be affected by non-Dirac type variables (e.g., the sliding condition for the falling rod). To ensure numerically precise treatment, Dirac pulse values should be distinguished from non-Dirac pulse values and evaluation of Dirac pulses can be based on their areas [16].
- **chattering:** If the system chatters between modes, root finding to locate the exact time of occurrence of the event causes continuous integration to become excessively slow. An equivalence relation eliminates the fast chattering motion, but preserve the dynamics of the slow motion along the chattering surface [20].

---

[1] There may be instantaneous dissipation of energy.

4   Packages Evaluation

The following software packages were investigated with respect to their support of the described hybrid features, see Table 1.

- χ is a simulation environment initially developed for modeling and simulation of manufacturing plants at the University of Eindhoven [10].
- *ABACUSS* is a derivative work of the gPROMS software [1]. It is developed at the Massachusetts Institute of Technology.
- *BaSiP* is developed at the University of Dortmund for simulation of recipe-driven production in complex multi-purpose batch plants [26].
- *DOORS* is a prototype distributed real-time simulator for mechatronic design developed at the University of Magdeburg [11].
- *Dymola* provides a powerful object oriented modeling and simulation environment for education and the professional engineer [9].
- *gPROMS* was initially developed at Imperial College, London for process modeling, simulation and optimization [3]. It is now commercially available.
- HYBRSIM is an experimental hybrid bond graph modeling and simulation tool based on physical principles developed at the DLR Oberpfaffenhofen [19].
- *Omola* is developed at the Lund Institute of Technology for modeling and simulation of continuous time and discrete event dynamic systems [2].
- SHIFT is a programming language for describing dynamic networks of hybrid automata, developed at the University of California, Berkeley [8].
- SIMULINK is a block diagram based modeling and simulation environment of the MathWorks [24]. This package has similar characteristics as MATRIX$_X$-SYSTEMBUILD.
- *Smile* is a simulator for energy systems of GMD FIRST, Berlin [25].
- *20-SIM* ("Twente Sim") is a modeling and simulation program developed at the University of Twente [4].

Table 1 shows that both state and time events are typically handled by these packages, though the implementation may vary [23]. Also, they facilitate adding and removing equations that do not change causality. Furthermore, the use of noncausal numerical simulation algorithms supports conditional equations that do not affect the DAE structure, i.e., those that would only require a new sorting stage in case explicit numerical integration routines are used.

Solving the system of equations during run-time is more complicated. This may require index reduction and new initial values of the state variables may have to be calculated. This has been implemented in HYBRSIM for index 2 DAEs by built-in physical conservation constraints. However, HYBRSIM is an interpreted simulator, and, therefore, not as efficient. ABACUSS provides some support for run-time solving, but this is still under development.

Event iteration is a crucial part of hybrid simulation. Most packages only implement event iteration by halting continuous time and updating the state vector values at every discrete event step. The SIMULINK semantics differ in this respect since it may abort event iteration before the discrete event model has

Table 1. Tools evaluation.

		χ	ABACUSS	BaSiP	DOORS	Dymola	gPROMS	HYBRSIM	Omola	SHIFT	SIMULINK	Smile	20sim
state events	detection	√	√	√	√	√	√	√	√	√	√	√	√
	location	√	√	√	√	√	√	√	√			√	√
time events		√	√		√	√	√		√		√	√	√
simulation model	add/remove	√	√	√	√	√	√	√	√	√	√	√	
	re-sort	√	√		√	√	√	√	√			√	
	re-solve							√					
reinitialization	explicit	√	√		√	√	√	√	√	√	√	√	√
	integration							√					
event iteration	invariant							√					
	update	√	√	√	√	√	√	√	√	√		√	
Dirac pulses								√					
chattering													

converged (i.e., further discrete state changes may still be possible) and progress time over one *integration* step before continuing the event iteration process. Chattering is at present not addressed in any general purpose simulator.

5   Conclusions

In simulation of hybrid systems a number of phenomena that require specific facilities can be classified: (i) event handling, (ii) run-time equation processing, (iii) discontinuous state changes, (iv) event iteration, (v) comparing Dirac pulses, and (vi) chattering. This papers presents a set of small examples that exhibit these phenomena and identifies their support by simulation packages.

If a simulation package does not support these phenomena, the model needs to be processed into a different form that can be handled. Often this requires global knowledge that is explicitly added to the model. For example, in case of the colliding bodies, the conservation of momentum constraint can be explicitly added to the model, thus circumventing the need for integrating the $m_1\dot{v}_1 = m_2\dot{v}_2$ equation but requiring additional modeling effort.

In other cases, simulation can only be performed at the cost of precision. For example, chattering behavior can be handled when the integration step is fixed. However, this results in a larger simulation error. Moreover, the fixed simulation step needs to be sufficiently small to numerically solve fast continuous gradients. The entire trajectory needs to be simulated with the smallest required integration time step.

Furthermore, the discrete model structure can be extended by incorporating *ad hoc* knowledge about the simulation scenario. For example, in case of the falling rod, the condition for sliding, $F_x > \mu F_N$, can be explicitly modeled to be evaluated as $F_x > \mu F_y$, in case of discontinuous changes in the velocities, thus eliminating the effect of the F_g component. Again, this requires additional modeling effort and hampers model re-use.

6 Acknowledgment

Thanks to the following people for their discussion and providing their expertise:
Jan F. Broenink, Georgina Fábián, Martin Fritz, Martin Otter, Clemens Klein-
Robbenhaar, Olaf Stursberg, Hubertus Tummescheit, and Andreas Wolf.

References

1. ABACUSS. http://yoric.mit.edu/abacuss/abacuss.html, 1995. Massachussets Institute of Technology.
2. M. Andersson. *Object-Oriented Modeling and Simulation of Hybrid Systems*. PhD dissertation, Department of Automatic Control, Lund Institute of Technology, Lund, Sweden, 1994.
3. P. I. Barton. *The Modelling and Simulation of Combined Discrete/Continuous Processes*. PhD dissertation, University of London, 1992.
4. Jan F. Broenink. Modelling, simulation and analysis with 20-sim. *Journal A*, 38(3):22–25, January 1998. Special CACSD issue.
5. Pawel Bujakiewicz. *Maximum weighted matching for high index differential algebraic equations*. PhD dissertation, TU Delft, Delft, Netherlands, 1994. ISBN 90-9007240-3.
6. F. E. Cellier, H. Elmqvist, and M. Otter. Modelling from physical principles. In W.S. Levine, editor, *The Control Handbook*, pages 99–107. CRC Press, Boca Raton, FL, 1996.
7. François E. Cellier. *Combined Continuous/Discrete System Simulation by Use of Digital Computers: Techniques and Tools*. PhD dissertation, ETH, Zurich, Switzerland, 1979.
8. Akash Deshpande, Aleks Göllü, and Luigi Semenzato. *SHIFT Programming Language and Run-Time System for Dynamic Networks of Hybrid Automata*. California PATH, UC Berkeley.
9. H. Elmqvist, D. Brück, and M. Otter. *Dymola — User's Manual*. Dynasim AB, Research Park Ideon, Lund, 1996.
10. G. Fábián, D. A. van Beek, and J. E. Rooda. Integration of the discrete and the continuous behaviour in the hybrid Chi simulator. In *Proceedings 1998 European Simulation Multiconference*, pages 252–257, Manchester, 1998.
11. R. Kasper and W. Koch. Object-oriented behavioural modeling of mechatronic systems. In 3rd *Conference on Mechatronics and Robotics '95*, Paderborn, Germany, October 1995.
12. Zvi Kohavi. *Switching and Finite Automata Theory*. McGraw-Hill, Inc., New York, 1978.
13. Stefan Kowalewski, Martin Fritz, Holger Graf, Jörg Preussig, Silke Simon, Olaf Stursberg, and Heinz Treseler. A Case Study in Tool-Aided Analysis of Discretely Controlled Continuous Systems: the Two Tanks Problem. In *Fifth International Conference on Hybrid Systems*, Notre Dame, Indiana, September 1997.
14. P. Lötstedt. Coulomb friction in two-dimensional rigid body systems. *Z. angew. Math. u. Mech.*, 61:605–615, 1981.
15. Sven Erik Mattsson and Gustaf Söderlind. A new technique for solving high-index differential-algebraic equations. In *Proceedings of the 1992 Symposium on Computer-Aided Control System Design*, pages 218–224, Napa, California, March 1992.

16. Pieter J. Mosterman. *Hybrid Dynamic Systems: A hybrid bond graph modeling paradigm and its application in diagnosis.* PhD dissertation, Vanderbilt University, 1997.

17. Pieter J. Mosterman. State Space Projection onto Linear DAE Manifolds Using Conservation Principles. Technical Report #R262-98, Institute of Robotics and System Dynamics, DLR Oberpfaffenhofen, P.O. Box 1116, D-82230 Wessling, Germany, 1998.

18. Pieter J. Mosterman and Gautam Biswas. Principles for Modeling, Verification, and Simulation of Hybrid Dynamic Systems. In *Fifth International Conference on Hybrid Systems*, pages 21–27, Notre Dame, Indiana, September 1997.

19. Pieter J. Mosterman, Gautam Biswas, and Martin Otter. Simulation of Discontinuities in Physical System Models Based on Conservation Principles. In *SCS Summer Simulation Conference*, pages 320–325, Reno, Nevada, July 1998.

20. Pieter J. Mosterman, Feng Zhao, and Gautam Biswas. Sliding mode model semantics and simulation for hybrid systems. In *Hybrid Systems V*. Springer-Verlag, 1998. Lecture Notes in Computer Science.

21. Tadao Murata. Petri nets: Properties, analysis and applications. *Proceedings of the IEEE*, 77(4):541–580, April 1989.

22. Constantinos C. Pantelides. The consistent initialization of differential-algebraic systems. *SIAM Journal of Scientific and Statistical Computing*, 9(2):213–231, March 1988.

23. Taeshin Park and Paul I. Barton. State event location in differential-algebraic models. *ACM Transactions on Modeling and Computer Simulation*, 6(2):137–165, April 1996.

24. SIMULINK. *Dynamic System Simulation for Matlab.* The MathWorks, January 1997.

25. Smile. http://gargleblaster.cs.tu-berlin.de/~smile. TU Berlin.

26. K. Wöllhaf, M. Fritz, C. Schulz, and S. Engell. BaSiP - Batch process simulation with dynamically reconfigured process dynamics. *Proceedings of ESCAPE-6, Supplement to Comp. & Chem. Engineering*, 20(972):1281–1286, 1996.

Building Hybrid Observers for Complex Dynamic Systems Using Model Abstractions

Pieter J. Mosterman[1*] and Gautam Biswas[2**]

[1] Institute of Robotics and System Dynamics DLR Oberpfaffenhofen
P.O. Box 1116, D-82230 Wessling, Germany
Pieter.J.Mosterman@dlr.de
[2] Knowledge Systems Laboratory, Department of Computer Science
Stanford University, Stanford, CA 94305, U.S.A.
biswas@ksl.stanford.edu

Abstract. Controllers for embedded dynamic systems require models with continuous behavior evolution and discrete configuration changes. These changes may cause fast continuous transients in state variables. *Time scale* and *parameter* abstractions simplify the analysis of these transients, causing discontinuities in the state variables. The two abstraction types have a very different impact on the analysis of system behavior. We have developed a systematic modeling approach that introduces formal semantics for behavior generation. This paper discusses the implementation of this scheme in a hybrid observer designed to track embedded system behavior. The resultant observer is based on piecewise simpler continuous models with mode transitions defined between them. Actual mode transitions in the system are provided by a digital controller and directly obtained from measuring physical variables.

1 Introduction

The drive to achieve more optimal and reliable performance on complex systems such as aircraft and nuclear plants while meeting rigorous safety constraints is necessitating detailed modeling and analysis of the embedded controllers for these systems. In embedded systems, the continuous physical process interaction with digital control signals requires modeling schemes that facilitate the analysis of mixed continuous and discrete, i.e., *hybrid* behavior. Discrete phenomena may also occur when modeling abstractions are applied to simplify fast nonlinear continuous process behavior.

Consider the primary aerodynamic control surfaces of an airplane in Fig.1 [19]. Modern avionics systems employ electronic signals generated by a digital computer, which are transformed into the power domain by electro-hydraulic actuators. The primary flight control system exemplifies the need for hybrid modeling

---

* Pieter J. Mosterman is supported by a grant from the DFG Schwerpunktprogramm KONDISK.
** Gautam Biswas is on leave from the Department of Computer Science, Vanderbilt University, Nashville, TN.

F.W. Vaandrager and J.H. van Schuppen (Eds.): HSCC'99, LNCS 1569, pp. 178–192, 1999.
© Springer-Verlag Berlin Heidelberg 1999

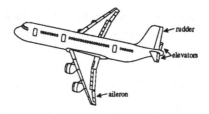

Fig. 1. Aerodynamic control surfaces.

in embedded control systems. At the lowest level in the control hierarchy, positioning of the rudder, elevators, and ailerons is achieved by continuous PID control. Desired set point values are generated directly by the pilot or by a supervising control algorithm implemented on a digital processor. Digital control may mandate *mode* changes at different stages of a flight plan (e.g., *take-off*, *cruise*, and *go-around*). Detection of component failures may lead to discrete changes in system configuration. Model simplification by discretizing fast nonlinear transients also results in discontinuous variable changes.

We have developed a hybrid modeling paradigm that encompasses analysis of embedded systems and modeling abstractions in physical systems [13, 16, 17]. The methodology for abstracting complex transients has been developed into compositional hybrid automata models with formal semantics for computing the discontinuous changes in the system state vector [15].

Observer schemes form a key component in the design and implementation of controller and diagnosis schemes for dynamic systems [10, 18]. In this paper, our focus is on applying our hybrid modeling methodology to develop effective observers for complex dynamic systems. We illustrate the approach by building a hybrid observer, i.e., an observer that includes mode change effects, for the elevator positioning subsystem of the primary flight control system of aircraft. The resultant observer simplifies complex nonlinear models to simpler piecewise linearized models with discrete mode transitions. There exist well-defined robust schemes for constructing observer models for linear systems [4, 10]. Formal model semantics enable us to compute the discontinuous changes in the state vector across mode transitions.

2 Hybrid Modeling of Physical Systems

Hybrid modeling paradigms [1, 7, 17] supplement continuous system description by mechanisms that model discrete state changes resulting in discontinuities in the field description and the continuous state variables. In previous work, we [12, 16] have formulated a systematic approach to hybrid modeling of dynamic physical systems based on a local switching mechanism implemented as finite state automata. The dynamically generated topology in a mode is used to translate the switching specifications to conditions based on state variables. Switching conditions may be expressed in terms of the values immediately before switching occurred, (*a priori* values), or in terms of the values computed by solving the initial value problem for the newly activated mode (*a posteriori* values).

2.1 Definitions

Differential equations form a common representation of continuous system behavior. The system is described by a state vector, x. Behavior over time is specified by field f. Discrete systems, modeled by a state machine representation, consist of a set of discrete modes, α. Mode changes caused by events, σ, are specified by the *state transition function* ϕ, i.e., $\alpha_{i+1} = \phi(\alpha_i)$. A transition may produce additional discrete events, causing further transitions. A mode change from α_i to α_{i+1}, may result in a field definition change from f_{α_i} to $f_{\alpha_{i+1}}$. Discontinuous changes in the state vector are governed by an algebraic function g, $x^+ = g_{\alpha_i}^{\alpha_{i+1}}(x)$. Discrete mode changes are caused by an *event generation function*, γ, associated with the current active mode, α_i, $\gamma_{\alpha_i}(x) \leq 0 \rightarrow \sigma_j$.

2.2 Abstractions in Physical System Models

On a macroscopic level, physical systems are continuous but phenomena may occur at multiple temporal and spatial scales. To simplify system models, small, parasitic, dissipation and storage parameters are abstracted away to simplify the system model causing discontinuous changes in system behavior. Time scale abstractions collapse the end effect of phenomena associated with very fast time constants to a point in time. Parameter abstractions remove small and large parameter values (parasitic dissipative and storage elements) from the model [14, 16].

Time Scale Abstraction. Consider filling a cylinder with oil by pulling a piston (Fig. 2). Fig. 2(a) shows a loose object in the oil moving towards the connection between the cylinder and the pipe that provides the oil supply. If the object is rigid, it obstructs oil flow when it reaches the orifice (Fig. 2(b)). Therefore, the oil flow rate becomes 0, but at this point if the external force is insufficient to keep the object in place, it drops down, and the orifice opens partially. As a result flow resumes, and the force exerted by the oil on the object causes it to topple over and move through the orifice. If all these phenomena occur at time constants much smaller than the normal oil flow rate, the oil flow rate seems to change at a very fast rate from a nonzero value to a zero value, before oil flow resumes and builds up a new velocity, v_{new}. Therefore, the new configuration, α_3, is reached immediately after the state vector is updated.

The hybrid model imposes an algebraic constraint on the oil flow velocity in the *obstruct* mode, α_2, and the switching specifications are such that this mode is departed immediately after the state vector is updated, $x = x^+$. The intermediate mode α_2 is called a *pinnacle*.

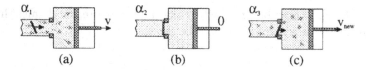

Fig. 2. An object may block an orifice.

Parameter Abstraction. If the object is flexible, it bends by the force of the pulling piston when it blocks the orifice. Almost immediately the object pops through the connection (Fig. 3(c)). A hybrid modeling approach may be adopted to avoid specifying this detailed bending behavior. When reasoning with this model, one has to analyze whether the force on the stuck object is sufficient to pull it through the orifice. This is done by computing the force generated by the instantaneous change of oil velocity from a finite value to 0. If the computed force is large enough, the model switches to a new mode where the orifice is not blocked, and the oil continues to flow with an initial value that equals the value of oil flow velocity just before the object got stuck in the orifice.

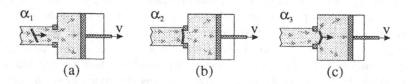

Fig. 3. An object may be pulled through an orifice.

Formally, the consecutive mode switch from α_2 to α_3 has to occur before the state vector is updated to its *a posteriori* value, $x = x^+$. otherwise the velocity would be 0. The intermediate mode, α_2, between α_1 and α_3 is a so-called *mythical* mode [16].

3   The Elevator System

Attitude control in an aircraft is achieved by the elevator control subsystem [6, 19]. This system may consist of two mechanical elevators (Fig. 1) which are positioned by electro-hydraulic actuators. When a failure occurs, redundancy management may switch actuator systems to ensure maximum control. Continuous feedback control drives the elevator to its desired set point, while higher level redundancy management selects the active actuator.

Fig. 4 shows the operation of an actuator. The continuous PID control mechanism for elevator positioning is implemented by a servo valve. The output of the servo valve controls the direction and speed of travel of the piston in the cylinder by means of a spool valve mechanism, illustrated in Fig. 5. The piston and connected elevator flap constitute the load. In the servo mechanism, the feedback signal may be provided by the fluid pressure, mechanical linkage, electrical signals, or a combination of the three.

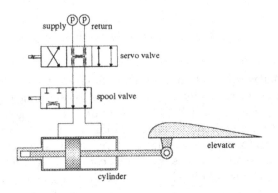

Fig. 4. Hydraulics of one actuator.

A typical spool valve (Fig. 5) consists of a piston that moves in a cylinder. A number of cylinder ports connect the supply and return part of the hydraulic system with the load. Cylindrical blocks called lands, connected to the piston, can be placed at different positions to render the servo mechanism and actuator *active* or *passive*. Fig. 5(a) and (c) show two possible oil flow configurations of the actuator. In Fig. 5(a) the control signal passes through the spool valve to the load, i.e., the actuator is *active*. In Fig. 5(c) the spool valve causes damping behavior, i.e., the actuator is *passive*. When the actuator is active, the spool valve is in its *supply* mode and the control signal generated by the servo is transferred to the cylinder that positions the elevator. The direction of the transmitted signal depends on which port is connected to the supply. When the actuator is *passive*, the spool valve is in its *loading* mode, and control signals cannot be transferred to the cylinder. However, oil flow between the chambers is possible through a loading passageway, as shown in Fig. 5(c), otherwise the cylinder would block movement of the elevator, canceling control signals from the redundant *active* actuator. When moving between *supply* and *loading*, the spool valve passes through the *closed* configuration where oil flow is blocked, as shown in Fig. 5(b).

Consider a scenario where a sudden pressure drop is detected in the left elevator actuator. Redundancy control moves the spool valve of this actuator from *supply* to *loading* and the spool valve of the other actuator from *loading* to *supply*. This causes transients that are studied in greater detail below.

4   Modeling the Elevator System

We employ *parameter* and *time scale* abstractions to design a simpler but adequate model of the elevator subsystem for control purposes. We show how these different abstraction types relate back to physical parameters in the real system.

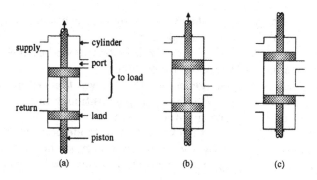

Fig. 5. A typical spool valve.

4.1   Mode Changes in the Spool Valve

To facilitate analysis of transient behavior, four modes of operation are modeled for the spool valve:

α_0 *loading:* The valve operates as a load. Pressure changes generated by the servo valve are blocked. Oil flow between chambers of the elevator positioning cylinder occurs through the loading passageway (Fig. 5(c)).

α_1 *closed:* The spool valve is closed. Pressure changes generated by the servo valve are blocked. Oil flow between the chambers of the elevator positioning cylinder is not possible as shown in Fig. 5(b).

α_2 *opening:* The valve is opening. While its lands move past the ports, fluid inertia effects may become active. Depending on the physical construction of the valve, these may have significant effects on transient behavior.

α_3 *supply:* The spool valve is opened and supplies control power. Pressure changes generated by the servo valve are transferred to the cylinder that positions the elevator. Flow of oil into and out of this cylinder is possible. This corresponds to the configurations in Fig. 5(a).

The modes α_1 and α_2 are transitional modes between α_0 and α_3. Mode changes of the spool valve are controlled by the redundancy management module which monitors a number of critical system variables. In the fault scenario, a sensor reading in actuator1 generates the failure event, σ_f. In response, the redundancy management reconfigures control by generating a sequence of discrete control signals that cause a switch of actuators. The combined state α_{ij} indicates the state of actuator2, α_i, and actuator1, α_j.

1. A control event is generated that causes the piston in the spool valve of actuator1 to move from its *supply* to *loading* position at a constant rate of change. Along the trajectory, a number of physical events occur when the displacement of the spool valve cylinder with respect to its center point, Δx, reaches valve specific values:

    (a) $\Delta x > -\lambda \to \sigma_{close} \Rightarrow \alpha_1$, the overall system mode becomes α_{01} (actuator2 is loading, and actuator1 is closed).

(b) $\Delta x > \lambda \to \sigma_{open} \Rightarrow \alpha_2$, the overall system mode becomes α_{02} (actuator2 is loading, and actuator1 is opening).

(c) $\Delta x > x_{th} \to \sigma_{load} \Rightarrow \alpha_0$, the overall system mode becomes α_{00} (actuator2 is loading, and actuator1 is loading).

2. A second control event is generated that moves the piston in the spool valve of actuator2 from its *loading* to *supply* position with a constant rate of change, causing the following physical events:

(a) $\Delta x < \lambda \to \sigma_{close} \Rightarrow \alpha_1$, the overall system mode becomes α_{10}

(b) $\Delta x < -\lambda \to \sigma_{open} \Rightarrow \alpha_2$, the overall system mode becomes α_{20}

(c) $\Delta x < -x_{th} \to \sigma_{supply} \Rightarrow \alpha_3$, the overall system mode becomes α_{30}.

Actuator1 is completely deactivated by switching to its *loading* mode α_0 before actuator2 is activated. The values of λ and x_{th} are based on physical parameters of the valve, e.g., the shape of ports and lands [6,11]. An *overlapped* or *closed center* valve has a small nonzero λ value, whereas a a *zero lap* or *critical center* type valve has $\lambda = 0$ [6].

4.2   Model Assumptions

When an actuator moves to its *closed* mode, oil flow into and out of the positioning cylinder is blocked. This implies that the cylinder piston that controls elevator position cannot move, and the elevator stops moving as well. In more detail, internal dissipation and small elasticity parameters of the oil cause the elevator velocity to change continuously during the transition. The continuous transient behavior between *supply* and *closed* is shown in Fig. 6. How quickly the system reaches 0 velocity in the *closed* mode depends on the elasticity and internal dissipation parameters of the oil.

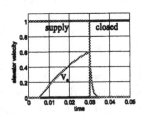

Fig. 6. Continuous transients: *closed* mode.

After a short time in the *closed* mode, the actuator moves to the *opening* mode, and inertial effects become active. Fig. 7 illustrates the continuous transients in this transition. The fluid inertia parameter of the clearance determines the final elevator velocity, v_e. In the *opening* mode, the inertial effect decreases as the clearance between port and land increases eventually becoming negligible, and the actuator operates as a simple load (*loading* mode). This is shown in Fig. 8 for two different values of the fluid inertia parameter. This also shows that the redundant actuator takes over elevator positioning control (mode α_{30}).

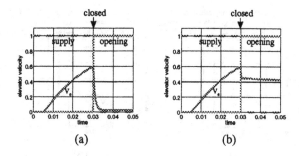

Fig. 7. Continuous transients: *opening* mode. Fluid inertia (a) $I = 1$ and (b) $I = 100$.

The continuous transients described above are not of much interest to the modeler for analysis and control (see Fig. 8 where the transients in the opening mode are still clearly visible but the continuous transients in the closed mode are not). Model simplification results in removal of small elasticity and inertia effects, but Fig. 8 illustrates that depending on their magnitude, they may have a distinct impact on the overall system behavior.

4.3   Applying Abstractions to Achieve Model Simplification

We apply model simplification by abstraction to analyze the elevator system.

Time Scale Abstraction. In the *opening* mode, fluid inertia and dissipative effects in the clearance between land and port cause a second order build-up of fluid flow. Though the fluid flow velocity and its time derivative are 0 initially, the velocity of the elevator and the driving piston are not. This results in a pressure build-up in the cylinder governed by the elasticity coefficient of the oil which causes a rapid increase of fluid flow through the land/port clearance. The pressure also causes the elevator velocity to decrease rapidly, resulting in the transient in Fig. 7(a). The initial transient from moving into the *closed* mode is replaced by the transient moving into the *opening* mode. The difference is best seen by comparing Fig. 6 with Fig. 7(a). The final value of the velocity after this transient depends on the dissipative effects and starting point and duration of the *opening* mode.

If the elastic and inertial effects are abstracted away, the *closed* and *opening* modes are traversed instantaneously in sequence into the *loading* mode. However, the inertia element has a distinct effect on system behavior, and the influence occurs over a small time interval. This is an example of time scale abstraction, where mode change phenomena are expressed at *a point* in time. An important implication is that the state vector has to be modified through the sequence of mode changes. An algebraic relation is derived to compute the elevator velocity to correspond to the fast transient behavior in the mode transitions (Fig. 6 and Fig. 7(a)).

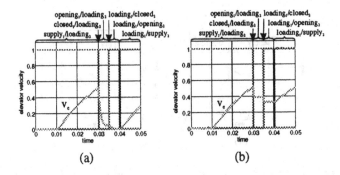

Fig. 8. Continuous transients: *loading* mode. Fluid inertia (a) $I = 1$ and (b) $I = 100$.

Parameter Abstraction. When dissipation in the land/port clearance domi-
nates the inertial effect, a much faster response in fluid flow velocity occurs because dissipation does not introduce a time derivative effect. The flow of oil into
and out of the cylinder is fast, and the pressure build-up in the cylinder is small.
As a result, elevator velocity remains almost unchanged as the model switches
from *closed* to *opening* (Fig. 7(b)). Small parameter values are abstracted away,
and the transitions through the *closed* and *opening* modes are instantaneous
(no time derivative effects are present). For small parameter values (Fig. 6 and
Fig. 7(b)), the transients to *opening* (Fig. 7) may result in very different behavior from transients into *closed* (Fig. 6). When a discontinuous jump occurs, the
eventual elevator velocity is not computed by first executing the jump to *closed*
and then to *opening*, but immediately to *opening*. Otherwise, *closed* would have
set the velocity to 0, which would also be the value in the *opening* mode. For
parameter abstractions these intermediate steps are completely abstracted away.

5 A Hybrid Observer

Modern controllers and fault isolation systems are based on accurate estimation
of the state vector of the system under consideration. Typically, a limited number
of physical measurements are made on the system, and functional redundancy
methods are utilized to derive values of other system variables [8]. These approaches rely on accurate numerical models that can be used to reconstruct the
internal system state from the observed variables by means of an observer [10].
Modeling uncertainties often require that temporal sequences of a number of
variable values be available to make accurate estimates of the system state.

For linear continuous processes this form of system identification is well understood, and some techniques exist for nonlinear continuous systems [5]. However, complex embedded systems, such as the elevator control subsystem, operate
in multiple modes, and the continuity constraint is often violated in the *models*
that define the transitions between the modes of operation. As a result, mode
transitions are often accompanied by discontinuities in the system state vector.
Conventional observer schemes cannot be applied in these situations. Because

a mode change may cause a large deviation in the estimated state vector (e.g., Fig. 7), convergence of the observer may require much more time. Incorporating the semantics of time scale and parameter abstractions into the observer scheme may mitigate a number of problems introduced by the large state changes when mode changes occur.

5.1   The Observer for Continuous Behavior

We investigate this conjecture by building a hybrid observer for the elevator velocity of the elevator control system using a standard observer scheme depicted in Fig. 9. The input to the elevator actuators includes the PID control pressure for actuator1 and actuator2:

$$u = [p_{in,1} \; p_{in,2}].$$ (1)

The measured variables associated with the actuator hydraulics are:

$$y = [q_{oil,1} \; q_{oil,2} \; p_{oil,1} \; p_{oil,2}],$$ (2)

where q_{oil} represents the oil flow rate into the servo valve, and p_{oil} represents the internal pressure in the positioning cylinder. The state vector of the system is

$$x = [p_{oil,1} \; p_{oil,2} \; v_e],$$ (3)

where v_e is the elevator velocity. We focus on reconstructing v_e from the measured variables in the observer system.

Fig. 9. A general observer setup.

A traditional Luenberger observer scheme for computing the estimated velocity, $\hat{v}_e$, during normal operation uses an observer model that is identical to the process model. We present a simplified observer structure, without accounting for scaling factors, derived for actuator1 in its *active* mode and actuator2 in its *passive* mode. This corresponds to the overall system mode α_{03}.

The estimate of v_e is computed from the estimated oil flow into the active hydraulics system, which is $\hat{q}_{in,1}$ when actuator1 is active. The value of $\hat{q}_{in,1}$ is a function of the pressure difference across the servo and spool valve, $p_{in,1} - \hat{p}_{oil,1}$, and the fluid resistance in this path, R_{hy} (equal in value for both actuator1 and actuator2):

$$\hat{q}_{in,1} = \frac{p_{in,1} - \hat{p}_{oil,1}}{R_{hy}}.$$ (4)

The pressure $p_{in,1}$ is a known control signal. Using $\hat{q}_{in,1}$ the oil flow into the positioning cylinder capacity, C_{hy}, can be calculated by subtracting the amount of oil required to move the piston in the cylinder at the estimated velocity v_e. Consequently, the estimated pressure change in the cylinder is given by

$$\dot{\hat{p}}_{oil,1} = \frac{1}{C_{hy}}(\hat{q}_{in,1} - \hat{v}_e + K_p(p_{oil,1} - \hat{p}_{oil,1})), \qquad (5)$$

where K_p is the gain factor used to determine observer convergence.

The change of pressure in the inactive cylinder is computed similarly, the difference being that there is no external oil flow into the cylinder. Instead, there is an oil flow through its loading passageway. This yields

$$\dot{\hat{p}}_{oil,2} = \frac{1}{C_{hy}}(-\frac{\hat{p}_{oil,2}}{R_{leak}} - \hat{v}_e + K_p(p_{oil,2} - \hat{p}_{oil,2})), \qquad (6)$$

where K_p is the gain factor for observer convergence (same as in Eq. (5)).

Finally, the change of elevator velocity is a function of the oil pressures generated by both cylinders, $\hat{p}_{oil,1}$ and $\hat{p}_{oil,2}$, and a convergence term based on the difference in actual oil flow into the system and the estimated oil flow, i.e.,

$$\dot{\hat{v}}_e = \frac{1}{m_e}(\hat{p}_{oil,1} + \hat{p}_{oil,2} + K_v(q_{in,1} - \hat{q}_{in,1})). \qquad (7)$$

Convergence is a function of the gain factor, K_v. The dynamics for mode α_{30} can be computed similarly. Since the system is well behaved and linear, and given that the input values and output measurements listed in Eq. (1) and (2) are known for the above observer, $\hat{v}_e$ easily converges to v_e.

5.2   The Hybrid Observer Scheme

As a next step, mode changes as a result of system failure are included. The resulting elevator control system model now includes a number of modes α_{ij} described earlier, where fast continuous transients occur. These behaviors contain complex nonlinearities and are hard to model in detail. This makes it difficult to build a robust and efficient observer.

Instead of modeling complex nonlinearities directly, we apply our model abstraction schemes to transform the behavior into a set of piecewise simpler (possibly linear) behaviors with a set of discontinuous mode transitions incorporated into the observer model. Fig. 10 shows the discrete event switching structure of the hybrid observer. Because the complex continuous transitions between modes are abstracted into discontinuous changes, the complex ODEs can be replaced by piecewise simpler ODE models (sODE) [15]. The states are indexed as sODE$_{ijk}$, where i determines whether the observer feedback is active ($i = 1$) or not ($i = 0$), j corresponds to α_j of actuator2, and k corresponds to α_k of actuator1. When a failure occurs and the redundancy management controller starts to switch states of actuator1, it generates event $\sigma_{close,1}$. This event changes the observer mode to the one representing both actuator1 and actuator2 in the *loading* mode. In this setup, there can be significant discrepancies between the actuator system model and the observer model. Since the fast continuous transients are not modeled in the observer, the error feedback is deactivated when these transitions occur.

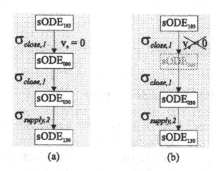

Fig. 10. Hybrid automata specifying the observer discrete event structure.

Since the detailed continuous transients are abstracted into discontinuous changes in the observer model, an immediate further change occurs, with the spool valve of actuator2 going into its *supply* mode. To achieve this consecutive change, the same event $\sigma_{close,1}$ is modeled to trigger the transition. Because the fast continuous transients are still active, the observer feedback is still disabled. Given the valve specifications, the fast continuous transients settle when $\Delta x < -x_{th}$, and this generates the physical event $\sigma_{supply,2}$ that causes the observer to restart the estimation process.[1] Note that the transition through sODE$_{000}$ is crucial for time scale abstraction because it brings about the corresponding discontinuous jump in v_e (see Fig. 10(a)). However, removing states based on this global knowledge destroys the compositional characteristics of the hybrid automata [15].

The resulting observer model contains the two continuous modes α_{03} and α_{30} described above, where it may or may not have its error feedback activated. The complete observer configuration is illustrated in Fig. 11.

After mode changes are completed, the continuous observer is reactivated, and this includes setting the estimated value of p_{oil} to its measured value. However, this information is not available for $\hat{v}_e$ and its initial value has to be computed using the applicable semantics derived from the mode transition definitions. To ensure a short convergence time after the continuous observer is reactivated, it is important that $\hat{v}_e$ estimated from discontinuous transition semantics be close to the actual value of v_e after the fast continuous transients.

---

[1] In general, estimating mode transitions, and the actual mode of the system in itself can be a very complex task.

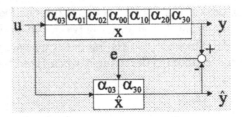

Fig. 11. A hybrid observer configuration.

5.3   Applying Abstraction Semantics

Depending on the type of spool valve used to execute the mode change, the elevator velocity may quickly be forced to zero, or it may change little from its current value (see Fig. 7). Fig. 12 shows by simulation the estimated v_e for two types of spool valves. The grayed bars indicate the time interval during which the observer feedback mechanism is disabled because discrete mode changes occur. In Fig. 12(a), time scale abstraction explains the fast continuous opening/closing behavior of the valve (note the discontinuous change in $\hat{v}_e$ at 0.01). Because of the time derivative behavior, this response takes more time than a parameter abstraction which has semantics that explain the behavior shown in Fig. 12(b). Both results show quick convergence of $\hat{v}_e$ after mode changes have occurred. A small initial difference between v_e and $\hat{v}_e$ demonstrates the convergence process of the observer during continuous behavior.

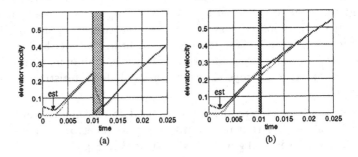

Fig. 12. Elevator velocity estimate for (a) time scale and (b) parameter abstraction.

To make the need for applying the correct modeling abstractions more explicit, we implemented an observer where time scale and parameter abstraction semantics were intentionally switched. Fig. 13 shows the results in terms of estimate convergence. In both cases the initial estimate of $\hat{v}_e$ after mode changes deviates much more from the actual value, and, therefore, convergence takes much longer in the continuous mode. This may result in inefficient or even erroneous behavior by the system.

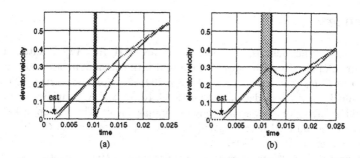

Fig. 13. Elevator velocity estimate for incorrect (a) time scale and (b) parameter abstraction.

6   Conclusions

The hybrid observer for the elevator system demonstrates how observer design can be made computationally simpler without sacrificing accuracy and convergence characteristics. In other work [15], we have implemented combined discrete transitions with continuous behavior evolution as hybrid automata, where discrete transitions cause changes in the system behavior model, but discontinuous changes in the state vector values may also occur. Time scale abstraction and parameter abstraction along with associated semantics that govern discontinuous changes in behavior specification are incorporated into the hybrid automata framework. An important feature of our work is that these abstractions relate back to physical parameters of the physical system that cause fast continuous transients. The hybrid automata scheme can form the basis for designing observer schemes in the manner illustrated in the last section.

Further extensions to this approach will involve extending our modeling schemes to represent complex nonlinear behaviors as piecewise simpler behaviors with discrete transitions between modes, e.g., the transition from *closing* to *supply* may involve regions of turbulent and laminar flow. The system model can then be decomposed into piecewise components each corresponding to a different regime of operation, and controllers synthesized for the resulting hybrid models. There has been work in this area by [2,3], but it would be very interesting to combine such approaches with modeling abstractions discussed in this paper, and apply them to controller design tasks. In other work we have started looking at the use of singular perturbation techniques [9] to automate the generation of simpler models and the mode switching conditions with the correct abstraction semantics.

References

1. R. Alur, C. Courcoubetis, T.A. Henzinger, and P.H. Ho. Hybrid automata: An algorithmic approach to the specification and verification of hybrid systems. *Lecture Notes in Computer Science*, vol. 736, pp. 209–229. Springer-Verlag, 1993.

2. A. Balluchi, M.Di Benedetto, C. Pinello, C. Rossi, and A. Sangiovanni-Vincentelli. Hybrid control for automotive engine management: The cut-off case. *Lecture Notes in Computer Science, Hybrid Systems: Computation and Control*, pp. 13–32, Springer-Verlag, Berlin, 1998.

3. A. Beydoun, L.Y. Wang, J. Sun, and S. Sivasankar. Hybrid control for automotive powertrain systems: A case study. *Lecture Notes in Computer Science, Hybrid Systems: Computation and Control*, pp. 33–48, Springer-Verlag, Berlin, 1998.

4. K. Brammer and G. Siffling. *Kalman-Bucy Filters*. Artech House, Norwood, MA, 1989.

5. R.S. Bucy and J.M.F. Moura. *Nonlinear Stochastic Problems*. Reidel, Dordrecht, 1983.

6. W.L. Green. *Aircraft Hydraulic Systems*. John Wiley & Sons, Chichester, UK, 1985.

7. J. Guckenheimer and S. Johnson. Planar hybrid systems. In P. Antsaklis, W. Kohn, A. Nerode, and S. Sastry, eds., *Hybrid Systems II*, vol. 999, pp. 202–225. Lecture Notes in Computer Science, Springer-Verlag, 1995.

8. R. Isermann. A review on detection and diagnosis illustrate that process faults can be detected when based on the estimation of unmeasurable process parameters and state variables. *Automatica: IFAC Journal*, 20(4):387–404, 1989.

9. P. Kokotovic, H.K. Khalil, and J. O'Reilly. *Singular Perturbation Methods in Control*. Academic Press, London, UK, 1986.

10. D.G. Luenberger. *Introduction to Dynamic Systems: Theory, Models, & Applications*. John Wiley, New York, 1979.

11. H.E. Merritt. *Hydraulic Control Systems*. John Wiley, New York, 1967.

12. P.J. Mosterman. *Hybrid Dynamic Systems: A hybrid bond graph modeling paradigm and its application in diagnosis*. PhD dissertation, Vanderbilt University, 1997.

13. P.J. Mosterman and G. Biswas. Formal Specifications for Hybrid Dynamical Systems. *IJCAI-97*, pp. 568–573, Nagoya, Japan, Aug. 1997.

14. P.J. Mosterman and G. Biswas. Principles for Modeling, Verification, and Simulation of Hybrid Dynamic Systems. *5th Intl. Conf. on Hybrid Systems*, pp. 21–27, Notre Dame, IN, Sep. 1997.

15. P.J. Mosterman and G. Biswas. Hybrid Automata for Modeling Discrete Transitions in Complex Dynamic Systems. *IFAC Intl. Symp. on AI in Real-Time Control*, Grand Canyon National Park, AZ, Oct. 1998.

16. P.J. Mosterman and G. Biswas. A theory of discontinuities in dynamic physical systems. *J. of the Franklin Institute*, 335B(3):401–439, Jan. 1998.

17. P.J. Mosterman, G. Biswas, and J. Sztipanovits. A hybrid modeling and verification paradigm for embedded control systems. *Control Engg. Practice*, 6:511–521, 1998.

18. H. Schneider and P.M. Frank. Observer-based supervision and fault detection in robots using nonlinear and fuzzy logic residual evaluation. *IEEE Trans. on Control Systems Technology*, 4(3):274–282, May 1996.

19. J. Seebeck. *Modellierung der Redundanzverwaltung von Flugzeugen am Beispiel des ATD durch Petrinetze und Umsetzung der Schaltlogik in C-Code zur Simulationssteuerung*. Diplomarbeit, TU Hamburg-Harburg, 1998.

Integration of Analog and Discrete Synchronous Design

Simin Nadjm–Tehrani

Dept. of Computer & Information Science,
Linköping University
S-581 83 Linköping, Sweden
simin@ida.liu.se

Abstract. The synchronous family of languages (Lustre, Esterel, Signal, Statecharts) provide a great deal of support for verifying a control program at the design and compilation stage. However, a common aspect of embedded systems is that significant properties of the system can not be verified by formally analysing the controller (software) on its own. To analyse the system one requires to state and document assumptions on the environment. Furthermore, proving timeliness properties necessitates justifying a sampling interval and relating the synchronous step to metric time. Support for these activities is generally missing from current formal methods tools.

In this paper we exploit simulation models – based on physical modelling of the environment – together with theorem proving, to prove properties of a closed loop system. We report on the work in progress on a case study provided by Saab Aerospace where deductive tools such as NP-Tools and simulation environments such as MATRIXx-SystemBuild are jointly used for verifying designs in Statecharts or programs in Lustre. The case study treats temperature and flow control in a climatic chamber.

1 Introduction

Many applications of formal methods in system development are in the requirements specification phase – often formalising a subset of requirements corresponding to functional behaviour of the system [11,7]. In embedded systems, these requirements commonly refer to the component which is under design – typically the controller for some physical devices (realised either as software or electronics).

One approach to verification of embedded systems concentrates on the *design*. This approach while providing a deep understanding of the system subject to design, can not be taken further when it comes to implemented code (except in few reported cases e.g. [13]). Nevertheless, several misconceptions can be identified while analysing early requirements (design) documents [4].

A different approach corresponds to verifying the very *control program* implemented in a language natural for this purpose. Here, the languages in the synchronous family for programming (Lustre, Esterel, Signal and Statecharts)

F.W. Vaandrager and J.H. van Schuppen (Eds.): HSCC'99, LNCS 1569, pp. 193–208, 1999.

play an important role [8, 10]. This family of languages have the benefit of a formal semantics (as synchronous I/O machines or Mealy automata) and an intuitive appeal within the engineering community.

One reason for choosing such languages is the support provided in the development environments: the controller can be analysed to eliminate causal inconsistencies, and to detect nondeterminism in the reactive software. The clock calculi in Lustre and Signal, as well as constructive semantics in Esterel can be seen as verification support provided directly by the compiler (comparable to several properties verified by model checking in [4]). Most of the works reported within this community, however, apply verification techniques to check the controller on its own.

To attack the class of properties arising as a result of interaction between the controller and the controlled environment, we need to explicitly model those aspects of the environment which are relevant to the property in question. This approach is common in control engineering. However, the analysis tools within this field primarily provide support for continuous system simulation, and are less adequate for proving properties of programs with discrete mode changes and (or) complex non-linear dynamics in the plant.

In this paper we present an approach whereby modelling tools used for analysis of analog systems can be used to substantiate the properties of the environment when formally verifying a synchronous controller. We use a case study provided by Saab Aerospace within the Esprit project SYRF on SYnchronous Reactive Formalisms, and present some verifications performed so far.

In the framework of the project we are able to use algorithmic (symbolic model checking techniques) as well as deductive (theorem proving) methods for verification. The former is possible with hybrid mathematical models where the variation of continuous variables is linear in time [9]. The latter approach uses the propositional theorem prover NP-Tools which is based on Stålmarck's method [3]. This tool has now been extended so that the models may include both boolean and integer variables, and the enhanced proving power with integer arithmetic is subject to test in the case studies. However, irrespective of the method used, we need constraints on the behaviour of the physical environment. This requires a modelling activity for which none of the above tools are suited.

With regard to the controller model, the idea is to combine this family of languages in the same environment so that appropriate parts of each application can be modelled according to the preferrable style (state-based, dataflow, textual or graphical). Thus, the verification techniques should be insensitive to the application language and operate on a common intermediate format (see [5]).

Our approach can therefore be summarised as follows. For safety properties of the closed loop system we decompose the property into conjuncts which can either be discharged based on the assumptions on the operating environment, verified on the control program by theorem proving, or verified in the physical environment model using analysis of a continuous mathematical model and (or) simulation. For timeliness properties our work is still explorative. We propose using the simulation models of the environment for estimating an adequate sampling interval, and combining this information with the bounded (step) response in the discrete setting. This can be achieved if discrete time models of the environment are translated to a logical representation via a two step translation process (via mode-automata and Lustre to NP-Tools format). Work on implementing these compilers is in progress in other parts of the project. For the time being we are working on NP-Tools models as if they were the result of automatic translation.

The rest of the paper is structured as follows. In section 2 we set out our framework for modelling and analysis. Section 3 is devoted to description of the case study. Section 4 presents the informal requirements document as presented to us by the industrial partners. Section 5 describes our understanding of the control goals of the system and a selected list of properties to prove. In section 6 we present a mathematical model of the environment based on specific assumptions. Section 7 uses this model to present some simulation examples and indicate a reasonable sampling interval. In section 8 we refine some of the selected properties and explain the verification steps performed so far, combining simulation and theorem proving. Section 9 covers our concluding remarks and related works.

2   Analog-digital synchronous design

Given a controller developed as a synchronous program, the following further activities are essential to verification of the embedded system.

1. Identifying and formalising the properties to prove – i.e. those critical to correct functioning of the controlled system based on informal requirements specifications.
2. Specifying physical properties of the environment.
3. Applying abstraction techniques to obtain "expressive enough" mathematical models.
4. Composing the controller and environment models to obtain the model of the closed loop system.
5. Applying algorithmic or deductive methods for verifying the properties specified under 1.

Step 1 is certainly the most cognitive step. Due to lack of a common methodology for applying formal methods in industry, the requirements documents typically consist of prescriptive texts at the implementation level. Declarative goals for the global behaviour of the system are often implicit. We therefore began by eliciting from the requirements document the properties to check by a formal proof. This is the first difference between working with a real case study and an academic example aiming to test a particular theory or method. The case study also illustrates that the "step" is performed iteratively: first the properties to prove are identified and stated informally, then the detailed design and modelling in step 2 is used to formalise the requirements.

Even step 2 turned out to require more information than that presented to us. We have stated certain assumptions and formalised a model for the environment based on these assumptions.

Adopting the deductive method of proof in NP-Tools, step 4 is straightforward. First, a model of the program in Lustre (or Statecharts) can be readily translated to the format required by NP-Tools (translators for this purpose are being developed in other parts of the project). Secondly, the composition of the environment model and the Lustre (or Statechart) program simply amounts to a conjunction of the environment constraints (as logical formulas) and the translated controller formulas in NP-Tools. One goal of the named case study is to explore whether necessary properties of the environment can be abstracted and represented in this verification environment.

In this article we treat the first two steps and the insights gained for the third step through the case study. We also report on current work under step 5, identifying certain restrictions for step 3.

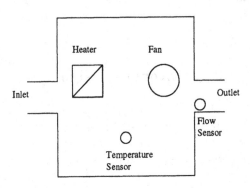

Fig. 1. The hardware components of the air control system.

3   The climatic chamber

The case study consists of a climatic chamber. A control system regulates and monitors the flow and the temperature of air which is circulating in the chamber. Originally, it was developed as a demo system which can be published freely, and demonstrates the kind of problems appearing in developing realistic subsystems such as the ventilation system in the JAS 39 Gripen aircraft. It was presented to the project partners in terms of a 4 page textual specification and an implemented code for a controller in the ASA development environment. ASA allows a top down breakdown of a functional description as a block diagram, where the lowest levels of the description are finite state machines[1].

---

[1] Currently, Saab is evaluating the tool Statemate for design specification as a replacement for ASA. The approach is equally applicable to any synchronous formalism with an interface to NP-Tools.

Figure 1 presents the component model of the chamber, while Figure 2 shows the interface between the system and the operator.

The chamber is to be ventilated through the inlet and outlet and has a given volume. It has two sensors for measuring the internal air temperature and the air flow. The external interface primarily consists of an on-off button, two analog knobs for setting the values for required temperature and flow, as well as warning signals in terms of a light and a sound. It also includes lights for showing some of its internal modes of operation (wait, work, or block). This may appear to be outside the realms of a component and interface specification (mixing the design related mode names) but has been kept for compatibility with the provided specification.

Unfortunately, the original architectural specification says nothing about the actuators. In the functional part of the document one can see that the heater and fan can be turned on and off. It is also implicit that their output effect can be adjusted in an analog manner, but there is no other information about the nature of control signals or the range of values. Section 6 attempts to fill this gap by making certain assumptions.

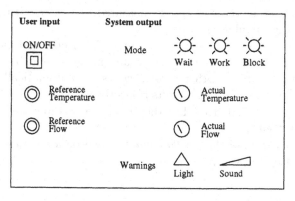

Fig. 2. The external interface to the system.

4   Functional specification

Here we present the original specification as presented to the project (only some variable names were changed to comply with the mathematical notation used in the physical model of section 6).

This informal (and incomplete) specification associates a number of operations with four discrete modes in the controller. The controller code or state machine design document is too large to be presented here. The description below should give an overall picture of the system and illustrate the large step needed before formal methods can be applied to a control system documented by informal text.

temperature control

- At start up, a requested temperature T_{Ref} is selected and the heater is turned on. It is possible to choose T_{Ref} in the interval $[T_{Refmin}, T_{Refmax}]$, where T_{Refmin} is greater than the incoming air temperature T_{in}. Any other chosen values outside the above interval are automatically adjusted up (down) to the lower (upper) limit respectively.
- Let $T_{min} = T_{Ref} - 2\Delta_{temp}$ and $T_{max} = T_{Ref} + 2\Delta_{temp}$. Then the system is considered to be in the **wait** mode while the actual chamber temperature (T_{chamb}) is not within those limits.
- The system will be in the **sol** mode (read solution mode) from the time the temperature hits the region $[T_{min}, T_{max}]$. The system makes a transition to the **work** mode when $|T_{chamb} - T_{Ref}| \le \Delta_{temp}$ is fulfilled. The time taken for the system to enter and leave the **sol** mode is expected to be a given fixed time called *solution_time*.

ventilation control

- The ventilation or the flow rate is measured by the time it takes to replace the air in the chamber once.
- At start up, a requested flow time $tvent_{Ref}$ is selected and the fan is turned on. This chosen time for the rate of air change should be in the interval $[tventmin, tventmax]$. Other chosen values are automatically adjusted upwards (downwards) to these limits respectively.
- Ventilation is regulated so that the air in the chamber is changed at least once every $tvent_{Ref}$.
- The observed rate of change *flowtime* is based on a measure of the air flow q delivered by the sensor.

monitoring

- The continuously measured values of T_{chamb} and q are to be displayed on the control panel.
- Three lamps indicate being in the system modes **wait**, **work** and **block** respectively[2].
- A warning by light and sound shall be activated whenever $|T_{Ref} - T_{chamb}| > \Delta_{temp}$ after being in the **sol** mode for a duration of *solution_time*.
- The warning light is activated if *flowtime* $> X.tvent_{Ref}$ for some fixed ratio X.
- When in **work** mode, if $|T_{Ref} - T_{chamb}| > 2\Delta_{temp}$, or the derivative of the temperature, based on T_{chamb} exceeds a maximum value $Tgrad$, then the mode shall change to the **block** mode.

---

[2] Apparently no lights should be on when the system is in the solution mode.

- When in **block** mode, the heater is immediately turned off, the warning light and the light indicating the **block** mode are activated. The fan is active for another *tblock* seconds before it is turned off. It shall not be possible to influence the system by changing T_{Ref} or $tvent_{Ref}$. The system must be turned off before restart.
- If T_{Ref} is changed such that $T_{Ref} > T_{chamb} + \Delta_{temp}$ the system makes a transition to the **wait** mode, and the heater is turned on.
- If T_{Ref} is changed such that $T_{Ref} < T_{chamb} - \Delta_{temp}$, the system transits to **wait** mode, the heater is turned off and the fan will be on.

5   Selected Requirements

From the prescriptive requirements document we have identified the following overall goals for the controller.

- Keeping the reference values constant,
  - the work light will be lit within a time bound from the start of the system, and
  - the system will be stable in the work mode.
- Chamber temperature never exceeds a given limit.
- Whenever the reference values are (re)set, the system will (re)stabilise within a time bound or warnings are issued.

Note that these are not properties of the controller on its own. Note also that these global requirements are not dependent on the particular characteristics of the physical environment, the sensors or the actuators. Such detail will be added later when we formalise the conditions under which they will be proved.

Our approach is to apply different proof techniques where they suit best. Although "being stable in the work mode" can be seen as a safety property (the conditions for leaving the mode will not be true), it is most expedient to use control theory methods for proving this property. This is due to the fact that not *all* inputs to the system are kept constant (see the result of the physical modelling step). Hence, it is formulated as a stability property.

Another aspect to point out is on the second (safety) property. Here we look beyond the functional demand on the system to monitor and warn when the temperature falls outside given intervals. We rather attempt to see what is the goal of devising such intervals and mode changes and envisage as a (mode-independent) goal of the system that the air is never heated to a hazardous level (even in the block mode and *after* warnings are issued).

The work on treatment of the third requirement builds on the first requirement and has not been carried out yet.

6   Physical environment

In this section we state some assumptions necessary in our initial model of the physical environment.

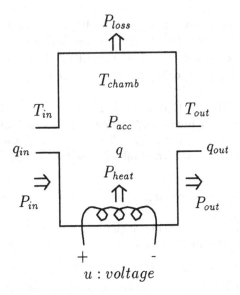

Fig. 3. The variables appearing in the heat model of the chamber.

6.1   Heat-flow inter-relationships

Heat and flow mutually affect each other in non-trivial ways [1, 2]. In a first coarse model of the climatic chamber we include the effects of flow on the heating process but ignore the effects in the opposite direction. The motivation being that the primary concern of the ventilation system is keeping the temperature within given bounds (this may be induced from the nature and extent of warnings).

Next, we make a simplification concerning how flow affects heat. We decide not to consider the air flow effects with respect to the heat transfer factor of the heater (i.e. how velocity of the moving air will affect heating of the air). We only consider the cooling effect as a result of the air exchange caused by the fan with respect to heat generation.

6.2   Assumptions on flow

The flow of mass and volume are related by air density, i.e. $q\,\rho = \dot{m}$, where $q\ [m^3/s]$ denotes volume flow, $\rho\ [kg/m^3]$ is a density factor (normally a function f of temperature T), and $\dot{m}\ [kg/s]$ is the mass flow. Since the design document mostly discusses "the time it takes for exchange of air", it is not clear whether one can measure volume flow or mass flow. No more information about the sensor is available.

The first assumption that we make is that the air pressure in the chamber remains constant relative to the external air pressure. This implies mass balance: that input and output mass flows are equal, $\dot{m}_{in} = \dot{m}_{out}$, which further leads to a similar assumption on volume flow. The latter can, however, be assumed provided that density as a function of temperature can be approximated to a constant factor. That is, assuming $f(T) = k_\rho$, we can further assume that $q_{in} = q_{out}$. Note that the fan can change the value of the chamber flow q, but the assumptions above constrain the in and out flows to stay in a constant relation: more specifically $q = q_{in} = q_{out}$. No further assumptions are made on the input to the flow actuator We simply assume that a given (reference) flow can be obtained placing the emphasis on the physical model of the heat process.

6.3   Model of the heating process

The heater produces heat which is accumulated in the chamber to an extent which depends on the flow. Figure 3 presents the variables appearing in the heat process model.

Based on the earlier assumptions the heat model will have three inputs: the air flow (q), the control signal to the heater (u), and the outside air temperature (T_{in}). The main state variable is the chamber temperature (T_{chamb}) which can be measured by the temperature sensor and delivered to the heat controller. However, to see how the final model is arrived at we start by stating the energy conservation principle:

$$P_{in} + P_{heat} - P_{out} - P_{loss} = P_{acc} \tag{1}$$

where the sum of heat power is stated to be constant, i.e. the incoming power (P_{in}), that resulting from the heater activation (P_{heat}), the outgoing power (P_{out}), and the power lost through the walls of the chamber (P_{loss}) are assumed to give the accumulated heat power in the chamber (P_{acc}).

Making a further assumption that $T_{chamb} = T_{out}$, using the earlier assumption $q = q_{in} = q_{out}$, and using equation 1, we can now simplify the above formula. The P-terms are replaced with their respective definitions, which constrain each element in terms of input/output variables, and some model coefficients (all the k-terms):

$$P_{in} = q_{in}\, T_{in}\, k_{in}$$
$$P_{heat} = u^2\, k_{heat}$$
$$P_{out} = q_{out}\, T_{out}\, k_{out}$$
$$P_{loss} = T_{chamb}\, k_{loss}$$
$$P_{acc} = \dot{T}_{chamb}\, k_{chamb}$$

Thus, we obtain the following equation defining the chamber temperature:

$$\dot{T}_{chamb} = 1/k_{chamb} \tag{2}$$
$$(\, q\, T_{in}\, k_{in} +$$
$$u^2\, k_{heat} -$$
$$T_{chamb}(q\, k_{out} - k_{loss}))$$

Note that despite several simplifying assumptions the model is highly non-linear. So, formally proving the properties of the hybrid model consisting of the synchronous controller and this chamber model, is a challenge for hybrid system verification. For example, significant restrictions remain (to do step 4 in section 2) before a linear hybrid automata for this model is derived [6].

7   Environment constraints

So far we have developed an abstract mathematical model of the environment giving the relationship between the state variables and the inputs. The next step is to study its temporal behaviour. This necessitates a computational model of the physical environment based on equation 2 above.

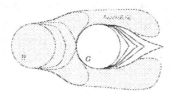

Fig. 4. The dynamics of the heat process with $u = 125V$, $q = 0.05\ m^3/s$, and $T_{in} = 300K$.

Here we have used the modelling and analysis environment provided by the tool MATRIXx with the accompanying toolbox SystemBuild [12]. The translation of the mathematical model into a block diagram is straight forward but there are a number of coefficients and constants which need to be fixed (estimated) Thus, the SystemBuild block diagram is annotated with actual values for the different k-terms in the model above.

Based on these estimates, upper and lower bounds provided for T_{in} and for *flowtime*, we can now simulate the chamber for different T_{in} signals and different input voltages. Figure 4 shows for example the dynamics of the system for a constant air flow, constant temperature control, and a constant environment temperature. A sinusoidal signal for T_{in} with a different value for the control signal gives the simulated trajectory presented in Figure 5.

7.1   Estimating the sampling interval

The first step for checking timing requirements must relate a synchronous step to metric time. Since one of the required properties involves timeliness, we can no longer "ignore" real time as is usual in synchronous programming.

Note that although rules of thumb exist for determining the sampling interval based on model coefficients, these are only rough guesses and reliable enough in the case of linear systems. For non-linear models and where the system is of higher dimension[3] performing selected simulations is the appropriate way to find an adequate step size.

Changing the input signal characteristics in the simulation model provides a feel for how fast the temperature inside the chamber varies in response to factors not governed by the heat controller. For example, the curves mentioned above show that heating up the chamber temperature to required levels takes up to 3000 seconds with the assumed heater.

The series of simulations performed have indicated that the control program provided by Saab has far too high sampling frequency. The program sampling rate is of course not explicit. This is due to the abstraction of time in the synchronous paradigm. However, since the program is supposed to warn for irregularities if the temperature does not stabilise within given time limits, counters are used and incremented at each program cycle.

Deducing time from these counters, one can observe that the program is sampling at 10 Hz where 0.01 Hz is a more reasonable sampling rate. This information was in itself interesting for other work in the project: the test pattern generation team was discovering that to arrive at interesting results test data for 30,000 steps had to be generated.

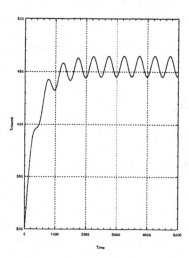

Fig. 5. The dynamics of the heat process with $u = 100$, $q = 0.05$ m^3/s, and T_{in}: 50K amplitude at 0.002 Hz added to 300K bias.

---

[3] Recall that we have assumed the flow to be an input to begin with; adding the flow as a state variable makes the model more complex, and no longer monotonic in presence of a non-zero k_{loss}.

8   Compositional verification

We now revisit the formulated requirements for the case study, and consider how the models presented so far can be used to decompose the requirements into "simpler" requirements. The idea is to break down the overall goals of verification so that suitable methods can be selected for proving different subgoals. In particular, different formal verification techniques or analysis tools from control theory can be used for different subgoals.

8.1   Refinement of requirements

The approach we adopt for performing step 5 (verification of closed loop model) combines formal and informal reasoning as well as continuous analysis. First, we attempt to find sufficient conditions which facilitate proving a property using our knowledge of the system. These auxiliary properties may be of the following kinds:

- an assumption which we discharge informally
- a property of the controller or the environment which we formally prove locally
- another property arising as an interaction of the two, which we further refine by finding further sufficient conditions

Then the system satisfies the top requirement under the informally discharged assumptions.

8.2   Refinement of R_2

Consider the second property which is a safety property. The only actuator in the system causing hazards is the heater which must be shown to heat the air to desired levels but *not* to hazardous levels. Let R_2 express this property.

R_2: The chamber temperature T_{chamb} never exceeds a limit T_H

The aim is to find (strong enough) properties R_{2i} such that $\bigwedge R_{2i}$ is sufficient for proving R_2. We start with the following conditions:

R_{20}: The chamber temperature is equal to T_{in} at start time
R_{21}: The reference temperature T_{Ref} can never exceed $T_{Ref_{max}}$, and
  $T_{Ref_{max}} + 2\Delta < T_H$
R_{22}: Whenever the system is in wait-, solution-, or work-mode, we have
  $T_{chamb} < T_H$
R_{23}: The system is never in block-mode while $T_{chamb} > T_H$

These properties can be discharged informally or proved within the NP-Tools model except for R_{23} which we continue to refine:

R_{231}: $T_{chamb} = T_{Ref} + 2\Delta < T_H$ when entering the block-mode

R_{232}: u = 0 throughout the stay in block-mode

R_{233}: The system leaves the block-mode after *tblock* seconds, and enters the off-mode

R_{234}: The temperature T_{chamb} does not increase while the system is in the block mode

This is sufficient for proving the safety property provided that
$R_{231} \wedge R_{232} \wedge R_{233} \wedge R_{234} \rightarrow R_{23}$.
Properties R_{231} to R_{233} are easily proved using the NP-Tools model of the controller. For the proof of R_{234} we use continuous reasoning based on the simulation models.

8.3   Refinement of R_1

Consider now the first requirement. The stability component of this requirement can only be verified using control theory and exact knowledge of the control algorithm in the work mode. Here, we concentrate on the first component, denoting it by R_1.

R_1: Keeping the reference values constant, the work light will be lit within t_1 from the start of the system

First, we provide sufficient conditions for R_1 to hold in the design model:

R_{11}: The system starts in the wait mode with the chamber temperature equal to T_{in}

R_{12}: While T_{Ref} is constant, the only successor to the wait mode is the solution mode

Given input restrictions R_{10},

R_{13}: The system leaves the wait mode within *wait_time* from the start of the system

R_{14}: the system leaves the solution mode within *solution_time* from entering the mode

R_{15}: While T_{Ref} is constant, the only successor to the solution mode is the work mode, and the work light is turned on whenever work mode is entered

R_{16}: *wait_time* + *solution_time* $\leq t_1$

We initially claim that

$$R_{11} \wedge R_{12} \wedge R_{13} \wedge R_{14} \wedge R_{15} \wedge R_{16} \rightarrow R_1$$

At a later stage we may drop R_{11} and replace it with the assumption that the initial chamber temperature is different from T_{in}. But to begin with, we make the restrictions in R_{10} more explicit, and show that

$$R_{10} \rightarrow R_{12} \wedge R_{13}$$

Here, we have several paths to take, but the choice is guided by the verification techniques we intend to utilise. For example, the following restrictions justify the adoption of a discrete-time model of the environment in a mode-automaton [15] with n discrete modes. Each mode is then governed by a difference equation derived from the continuous model of section 6 in the standard manner (for details see [6]).

$\mathbf{R_{101}}$: q stays constant at Q $[m^3/s]$

$\mathbf{R_{102}}$: u may vary every t_{sample} seconds

$\mathbf{R_{103}}$: T_{in} is piecewise constant taking the values $\{v_1, \ldots, v_n\}$

Note that mode-automata [15] can be seen as a restriction of time-deterministic hybrid transition systems [17] in which changes in state variables in each mode are defined in terms of a Lustre program. Each state variable is thus defined by an equation relating the state variables at the previous (clock) step and the current input. Adopting these restrictions, the verification method would be as follows: using a scheme for compilation from mode-automata to Lustre we obtain a model of the environment in Lustre which can be composed with a controller in Lustre, and further compiled to NP-Tools. In NP-Tools it is possible (but tedious) to show that the number of steps leading to the work light coming on is $\leq N$ for some N (This proves $\mathbf{R_1}$ for a given t_{sample} provided that $t_1 \geq Nt_{sample}$). This is a track we are currently exploring in the project. Note that this is one reason for not choosing a "too short" sampling interval. As well as other reasons which may be associated with oversampling, we would like to prove the bounded response property for as small N as feasible.

9   Concluding remarks

The work we have reported is at a too early stage for making definitive remarks about feasibility of combining "push-botton" theorem provers and simulation environments. More work is also needed to compare the method with "heavy duty" theorem proving in the spirit of [7]. However, some preliminary points for discussion have already emerged. Some of the shortcomings are reminiscent of those reported in [4]: the limitation to integer arithmetic, for example, means that the counter proofs presented by the system are more informative than the safety proofs holding over a limited range. This is, however, compensated in our approach by departing from fully formal proofs and combining with a simulation analysis when (local) reasoning over reals is crucial to the property in question.

There were several obstacles in achieving the goals set out, many of a practical nature, such as childhood problems in translators from e.g. Statecharts to NP-Tools. Since the translators to NP-Tools traditionally encapsulate the model as a circuit with only in- and out-pins visible, it is necessary to recompile the model every time a small error is detected in the original design description. The fact that only a subset of the Statemate languages was covered by the prototype translator also made the Statemate description of the continuous control parts less natural than it could be (c.f. an environment where state-based and data-flow descriptions can be combined). The work should however serve as an illustration of the many steps "in between" going from an informal document and a synchronous program to formal verification of "hybrid" properties.

Our model of the heat process intentionally made several simplifications to fit an early experimental set up. The interested reader may for example refer to a more complex model of heat exchangers in [16] where some of our restrictions are relaxed. The purpose of that paper is the illustration of a rich simulation language and only the plant part of the heat exchanger is subjected to validation by simulation.

To get an impression of the size of the system it should be indicated that the translated NP-Tools version of the full controller with a basic interface to the plant has 96 input variables and 88 output variables (seen as a circuit). The time taken for the proofs has been within fractions of a second for the requirements checked so far. These have included a number of sanity checks (such as determinism in the controller) and the subgoals for the safety property.

It is also interesting to note that the size of the real ventilation subsystem in the same format is 700 input variables and 500 output variables. Despite the seemingly large state space, the size of the reachable states set – as far as required for the types of properties mentioned – is small enough for practical purposes, even in the real system [14].

Acknowledgements

This work was supported by the Esprit LTR project SYRF, the Swedish board of Technical Research (TFR) and the Swedish board of technical development (Nutek). Also, cooperation of the Saab people, specially Ove Åkerlund, is gratefully acknowledged.

References

1. H. Alvarez. *Energiteknik, Del I.* Studentlitteratur, Lund, Sweden, 1990.
2. C.O. Bennett. *Momentum, Heat and Mass Transfer, 3rd Edition.* McGrawHill, 1982.
3. B. Carlson, M. Carlsson, and G. Stålmarck. NP(FD): A Proof System for Finite Domain Formulas. Technical report, Logikkonsult NP AB, Sweden, April 1997. Available from http://www-verimag.imag.fr//SYNCHRONE/SYRF/HTML97/a321.html.
4. W. Chan, R.J. Anderson, P. Beame, S. Burns, F. Modugno, D. Notkin, and J.D. Reese. Model Checking Large Software Specifications. *IEEE Transactions on Software Engineering*, 24:498–519, July 1998.
5. The SYRF Project Deliverables. Work package 2: Combination of Formalisms. Available from http://www-verimag.imag.fr//SYNCHRONE/SYRF/HTML97/a21.html, December 1997.
6. The SYRF Project Deliverables. Work package 7: Integration of Analog and Discrete Synchronous Design. Available from http://www-verimag.imag.fr//SYNCHRONE/SYRF/HTML98/a72.html, December 1998.

7. B. Dutertre and V. Stavridou. Formal Requirements Analysis of an Avionics Control System. *IEEE Transactions on Software Engineering*, 25(5):267–278, May 1997.
8. N. Halbwachs. *Synchronous Programming of Reactive Systems*. Kluwer Academic Publishers, 1993.
9. N. Halbwachs, P. Raymond, and Y.-E. Proy. Verification of Linear Hybrid Systems by means of Convex Approximations. In *In proceedings of the International Symposium on Static Analysis SAS'94, LNCS 864*. Springer Verlag, September 1993.
10. D. Harel. STATECHARTS: A Visual Formalism for Complex Systems. *Science of Computer Programming*, 8:231–274, 1987.
11. M. Heimdahl and N. Leveson. Completeness and Consistency in Heirarchical State-based Requirements. *IEEE transactions on Software Engineering*, 22(6):363–377, June 1996.
12. Integrated Systems Inc. *SystemBuild v 5.0 User's Guide*. Santa Clara, CA, USA, 1997.
13. H. Jifeng, C.A.R. Hoare, M. Fränzle, M. Müller-Olm, E-R. Olderog, M. Schenke, M.R. Hansen, A.P. Ravn, and H. Rischel. Provably Correct Systems. In H. Langmaack, W.-P. de Roever, and J. Vytopil, editors, *Proc. of the 3rd. International Conference on Formal Techniques in Real-time and Fault-tolerant Systems, LNCS 863*, pages 288–335. Springer Verlag, 1994.
14. O. Åkerlund. Application of Formal Methods for Analysis of the Demo System and parts of the Ventilation System of JAS 39 (in swedish). Technical report, Saab Aerospace AB, Linköping, Sweden, January 1997.
15. F. Maraninchi and Y. Rémond. Mode-automata: About modes and states for reactive systems. In *European Symposium On Programming*, Lisbon (Portugal), March 1998. Springer verlag.
16. S.E. Mattsson. On modelling of heat exchangers in modelica. In *Proc. 9th European Simulation Symposium*, Passau, Germany, October 1997. Available through http://www.modelica.org/papers/papers.shtml.
17. S. Nadjm-Tehrani. Time-Deterministic Hybrid Transition Systems. In *Hybrid Systems V, Proceedings of the fifth international workshop on hybrid systems, 1997, LNCS, To appear*. Springer Verlag.

Reachability Analysis of a Class of Switched Continuous Systems by Integrating Rectangular Approximation and Rectangular Analysis

J. Preußig, O. Stursberg, and S. Kowalewski

Process Control Laboratory, Chemical Engineering Department,
University of Dortmund, D-44221 Dortmund (Germany)
Tel. +49.231.755-5128, Fax-5129
{joerg|olaf|stefan}@ast.chemietechnik.uni-dortmund.de

Abstract. The paper presents a concept for the reachability analysis of switched continuous systems in which switching only occurs when the continuous state trajectory crosses thresholds defined by a rectangular partitioning of the state space. It combines an existing approach for approximating such systems by rectangular automata with an existing reachability algorithm for this class of hybrid automata. Instead of creating a complete abstraction of the original system by a rectangular automaton first and then analyzing it, in the presented procedure the flow conditions of the visited locations are determinded on-the-fly during the course of the analysis. The algorithm is illustrated with the help of a simple physical example.

1 Introduction

This paper is concerned with the reachability analysis of systems with continuous dynamics which can switch when the continuous state trajectory crosses rectangular switching manifolds. This class of hybrid systems arises for example in industrial processing plants where logic controllers are used to supervise and enforce operational and safety requirements. Usually, thresholds are defined for single process variables (e.g., alarms for the temperature in a reactor) and the crossing of these thresholds results in a discrete controller action which abruptly changes the continuous dynamics (e.g., switching off the heating of the reactor). An important control objective in these applications is to prevent the process variables from reaching certain undesired or even dangerous ranges, reachability analysis could be a method for checking the correct design of the logic control programs including the choice of the threshold values.

However, reachability analysis is only feasible for very restricted classes of hybrid systems and the appropriate models of logic controlled processing systems (in most cases switched ordinary differential equations) rarely belong to one of them. Therefore, usually a two-step procedure is proposed [6], [10]: First, the considered switched continuous system is approximated (conservatively) by a simpler system for which reachability analysis is possible. Then, in the second

F.W. Vaandrager and J.H. van Schuppen (Eds.): HSCC'99, LNCS 1569, pp. 209–222, 1999.

step, the approximating system is analyzed. In this approach, the effort of approximating the complete original system only pays off if several analysis runs (e.g., for different target or initial regions) have to be performed on the same approximating model. If this not the case and only one or very few scenarios are of interest, it is rarely necessary to find an approximation for everything. Instead it would be sufficient to approximate the continuous dynamics along the paths determined by the reachability algorithm.

Motivated by this idea, we present a reachability algorithm for switched continuous systems which reduces the number of (n-dimensional) rectangles for which the dynamics have to be approximated. It is based on an analysis algorithm for simple rectangular automata [9] and earlier attempts to approximate continuous systems by this class of hybrid automata [10]. The main idea is that the analysis procedure calls the approximation procedure each time when it has determined a new reachable outgoing face on the currently analyzed rectangle. The approximation procedure will then determine the neighboring rectangle and return it together with the corresponding flow conditions (i.e. differential inclusions). The analysis procedure will use this information to compute the possible outgoing faces of the new rectangle, and so on. In other words, the transitions and the flow conditions of the control modes of the approximated rectangular automaton are not given a priori but are determined on demand during the analysis.

The paper is organized as follows. In the next section we define the considered class of switched continuous systems. Section 3 recalls rectangular automata and the analysis algorithm for simple rectangular automata from which the presented procedure was derived. In Sec. 4 the main concepts for approximating switched continuous systems by rectangular automata are presented. In Sec. 5 and 6 we describe the combined analysis/approximation procedure and illustrate it by a simple physical example. Section 7 gives references to related work and in the conclusions we give an outlook on a possible extension of the algorithm.

2   Switched Continuous Systems

Switched continuous systems are a subclass of hybrid systems which is characterized by the property that depending on a discrete-valued input vector and on the actual state vector the dynamics is switched between different sets of ordinary differential equations. We define a switched continuous system by a 5-tupel:

$$SCS = (\boldsymbol{X}, U, L, \Phi, O, out) \tag{1}$$

with the following components:

Continuous state space: For n variables x_j defined on an interval $[x_{j,min}, x_{j,max}]$, $j = \{1, \ldots, n\}$ the continuous state space is given by $\boldsymbol{X} = [x_{1,min}, x_{1,max}] \times \ldots \times [x_{n,min}, x_{n,max}] \subset \mathbb{R}^n$.

Set of input vectors: $U = \{\boldsymbol{u}_1, \ldots, \boldsymbol{u}_l\}$ is the finite set of inputs of SCS where each $\boldsymbol{u}_k$ is defined as an m-dimensional vector $\boldsymbol{u}_k = (u_{k,1}, \ldots, u_{k,m})$ with $u_{k,j} \in \mathbb{R}$, $k = \{1, \ldots, l\}, j \in \{1, \ldots, m\}$.

Sets of landmarks: Each element of the n-tupel $L = \{L_1, \ldots, L_n\}$ denotes an ordered set of *landmarks* which is introduced for the variable x_j: $L_j = \{l_{j,0}, \ldots, l_{j,p_j}\}, j \in \{1, \ldots n\}$. The landmarks correspond to those values of x_j at which either the input is set to a new u or at which a different dynamics becomes valid (see below). The landmarks $l_{j,0}$ and l_{j,p_j} are set to the bounds of the continuous state space, i. e. $l_{j,0} = x_{j,min}$, $l_{j,p_j} = x_{j,max}$. The introduction of L partitions the state space X into a number of $\pi = p_1 \cdot \ldots \cdot p_n$ regions of rectangular geometry. Each of these regions is defined as $X_i = [l_{1,k_1}, l_{1,k_1+1}] \times \ldots \times [l_{n,k_n}, l_{n,k_n+1}], k_j \in \{0, \ldots, p_j - 1\}$, such that $X = \bigcup_{1 \le i \le \pi} X_i$.

Dynamics: The continuous state evolution is given by a set $\Phi = \{f_1, \ldots, f_q\}$. A vector of functions $\dot{x} = f_r(x, u_k), r = \{1, \ldots q\}$ is defined for $x \in X_i$ (or a set of regions) and $u_k \in U$. Each component $f_{r,j}$ is assumed to be a time-invariant, possibly non-linear ODE with a unique and continuous solution over time.

System output and output function: The set of output symbols of SCS is denoted by $O = \{o_1, o_2, \ldots\} \cup \varnothing$. The output function $out : \{X, L\} \to O$ generates a symbol o_i at a point of time at which a variable crosses a landmark: $out : \{X, l_{j,k}\}$ if $x_j = l_{j,k} : out(X, L) = o_i$, else: $out(X, L) = \varnothing\}$. If more than one variable crosses a landmark at the same time instant (i. e. a border or a corner of a rectangular region is reached), out generates the corresponding set of output symbols.

We consider the standard closed-loop-setting of discretely controlled processes, i. e. the switched continuous system is coupled to a controller in the following sense: When the output function of SCS generates a symbol o_i this information is passed on to the controller. The latter computes and sends back an appropriate u_k-signal in order to steer the process into a desired state space region X_i. Since we omit timing functions and external inputs of the controller, and assume that the new u_k is returned without delay, the input of SCS changes only at time instances at which a threshold crossing takes place. For the sake of simplicity, we also assume that chattering does not occur, i. e. the input trajectory $u(t)$ is piece-wise constant over time: $t \in [t_k, t_{k+1}[: u(t) = u_k$ with finitely many switching instances on a bounded interval.

To apply verification techniques to settings which contain switched continuous systems, the dynamics of SCS has to be transformed into a simpler type first. The next chapter describes a class of models with verifiable dynamics as an appropriate target of transformation.

3   Rectangular Automata

3.1   Basic Concepts

We first briefly review basic definitions and concepts related to rectangular automata (RA) [7]. Let $Y = \{y_1, \ldots, y_n\}$ be a set of variables. A *rectangular inequality* over the set Y is an inequality of the form $y_j \sim c$, for some $y_j \in Y$, some relation $\sim \in \{\leq, =, \geq\}^1$ and some rational $c \in \mathbb{Q}$. A *rectangular predicate* over Y is a conjunction of rectangular inequalities over Y. The set of rectangular predicates over Y is denoted $\mathcal{R}(Y)$.

A *rectangular automaton* A is a system $(X, V, inv, flow, init, E, guard, reset\_vars, reset)$ consisting of the following components:

Variables: A finite set $X = \{x_1, \ldots, x_n\}$ of variables.

Control modes: A finite set V of control modes.

Invariant conditions: A function *inv* that maps every control mode v_k to an invariant condition I_k in $\mathcal{R}(X)$. Control of the automaton may remain in a control mode only when its invariant is satisfied.

Flow conditions: A function *flow* that maps every control mode v_k to a flow condition φ_k in $\mathcal{R}(\dot{X})$, where $\dot{X} = \{\dot{x}_1, \ldots, \dot{x}_n\}$ with $\dot{x}_j$ representing the first derivative of x_j with respect to time. While control remains in a given mode, the variables evolve according to the differential inclusion specified by the mode's flow condition.

Initial conditions: A function *init* that maps every control mode to an initial condition in $\mathcal{R}(X)$.

Control switches: A finite multiset E of control switches in $V \times V$. For a control switch (v_k, v_l), we say that v_k denotes the *source mode* and v_l the *target mode*.

Guard conditions: A function *guard* that maps every control switch to a guard condition in $\mathcal{R}(X)$. Intuitively, the guard must be satisfied before the mode switch can be taken.

Resets: A function *reset_vars* that maps every control switch to an *update set* in 2^X, and a function *reset* that maps every control switch e to a reset condition in $\mathcal{R}(X)$. We require that for every control switch e and for every $x \in X$, if $x \in reset\_vars(e)$, then $reset(e)$ implies $x = c$ for some constant c. Intuitively, after the mode switch, the variables must satisfy the reset condition. Variables that appear in the update set must be reset to the fixed value indicated by the reset condition. Furthermore, all other variables must be unchanged.

Our analysis algorithm works on a subclass of rectangular automata, namely simple rectangular automata [9]. A *simple* rectangular automaton has the following properties. Its invariant, initial, flow, guard, and reset conditions represent bounded sets. Its guard conditions include tests for equality for one of the variables' bounding values in the source mode's invariant. Furthermore, if the variable is reset, then it is reset to a bounding value for the target mode's invariant. Finally, if it is not reset, then its value in the guard must be a bounding value in the target mode's invariant.

Simple rectangular automata often arise naturally when approximating more complex hybrid systems. In order to conservatively overapproximate the flow field of nontrivial continuous dynamics, one may partition the state space into rectangular blocks, and for each variable provide constant lower and upper bounds on the flow within each block [6]. A control mode is split into several

---

[1] For simplicity, we consider only nonstrict inequalities.

control modes, one for each block of the partition. Crossing from one block in the state space to another is modeled by mode switches among the blocks, with the guards being tests for equality across common boundaries. For example, a mode v with the invariant $1 \leq x \leq 3$ may be split into two modes — v_1 with the invariant $1 \leq x \leq 2$ and v_2 with the invariant $2 \leq x \leq 3$ — with mode switches between them having the guard $x = 2$.

3.2   Analysis of Rectangular Automata

In our analysis procedure we use parts of an algorithm that has been introduced in [9]. It uses the concept of faces. A *face* is a rectangular predicate with one dimension fixed to a certain value. Our rationale for introducing faces is to use rectangular faces to represent non-rectangular sets. A *face-region* $\mathcal{F}$ is a set $\{F_1, \ldots, F_q\}$ where each F_i is a face. The semantics of $\mathcal{F}$ is the convex hull over its q faces, i.e. $[\![\mathcal{F}]\!] = convexhull\{[\![F_1]\!], \ldots, [\![F_q]\!]\}$. This is shown for an example in Fig.1 where a face-region $\mathcal{F}_1$ is represented by the two faces F_1 and F_2. In practice, the faces of a face-region over n variables are derived from $2n$ constraints of the form $x_j = l_1$ or $x_j = l_2$. In the example, the face F_1 corresponds to $x_1 = 1$ and the face F_2 to $x_2 = 7$, with the empty faces for $x_1 = 7$ and $x_2 = 1$ being omitted.

Our algorithm makes use of the fact that the invariants in a control mode of a rectangular automaton form a rectangular region. So, a reachable face-region within the invariants can be represented by faces that lie on the invariant's bounds. Let $\mathcal{F}_1$ be a reachable face-region in a control mode v_1. Now we want to compute the new face-region $\mathcal{F}_2$ in another control mode v_2 that is adjacent to v_1 in terms of the invariant conditions. Then we can first check if any face of $\mathcal{F}_1$ is within the invariant condition of v_2. In our example this holds for F_2. So, we can use this face to determine a reachable region $\mathcal{F}_2$ in control mode v_2. This is done by determining for each bound l of an invariant of v_2 a face as the part of invariant l that can be reached starting from F_2 according to the possible flow in v_2. Here, only for the bound $x_1 = 7$ a face can be found, namely F_3.

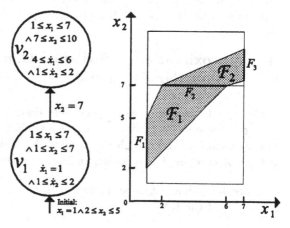

Fig. 1. Analysis of RA

The basic idea how a face can be computed from another faces will be shown with the help of our example in Fig.1 and the computation of F_2 from F_1. First we determine an interval of times in which any point within F_1 can/must be moved to F_2 according to the flow in dimension x_2. The distance between F_1 and F_2 in dimension x_2 ranges between 2(=7-5) and 5(=7-2). With a flow $1 \le \dot{x}_2 \le 2$ in v_1 this distance can/must be cleared within a time interval $T = [1; 5]$. Since the flow in each dimension is independent from the other dimensions, we can now use this time interval to compute how any point in F_1 can/must be shifted in the other dimensions while moving towards F_2. In our example, the only other dimension is x_1 for which we have a fixed flow $\dot{x}_1 = 1$. So in the time interval $T = [1; 5]$ a point starting from $x_1 = 1$ can flow to values ranging from 2 to 6. This yields F_2 with $2 \le x_1 \le 6 \wedge x_2 = 7$.

Intuitively, we can consider F_1 a face that is *ingoing* to $\mathcal{F}_1$ and F_2 as *outgoing*. A complete reachability analysis is performed by considering all outgoing faces of an initial control mode as ingoing faces to adjacent control modes to which control switches exist. For these incoming faces then the outgoing faces within the invariants of the adjacent control modes are computed. In the next step these newly computed faces are considered as ingoing to all adjacent control modes again and so an iteration evolves. This iteration terminates when all reachable faces of a given automaton are found. The termination is guaranteed, since our RAs are always defined over a finite state space and our analysis is approximate. Due to rounding in the approximative analysis there is only a finite number of points considered in the (finite) continuous state space. So, there is also only a finite number of faces that the algorithm can find within this state space. In contrast to this for the the exact reachability analysis of RA termination is not guaranteed, i.e. the reachability problem is not decidable.

4   Approximation of Switched Continuous Systems by Rectangular Automata

To be able to use the analysis procedure described above for the verification of controlled systems according to Sec. 2, switched continuous systems have to be transformed into Rectangular Automata. For this purpose, each element of the 8-tupel A is referred to a corresponding property of SCS (compare to [10]).

- *Variables*: Each state variable x_j of SCS is assigned to one element of the set X of the RA.

- *Control modes*: The set V is formed by assigning a control mode to a sub-region of the state space. The regions X_i obtained in Sec. 2 from hyper-rectangular partitioning have in general not an appropriate size to allow an approximation of sufficient accuracy. Hence, an additional finer partitioning is established by introducing a grid between adjacent landmarks such that a largely regular X-partition results. The number g_j of gridpoints which are introduced for a variable x_j parametrizes the modelling accuracy and the computational effort of analysis. The again rectangular region, which is bounded by pairs of adjacent gridpoints in all coordinates, is called a *cell* below, denoted by $c_k \in C$ where $C = \{c_1, ..., c_\pi\}$ stands for the set of all cells. The rectangular region of c_k is referred to as X_{c_k} in the sequel. A control mode $v_k \in V$ of the RA is assigned to each cell c_k.

- *Invariant conditions*: The invariant condition I_k of a control mode v_k equals the bounds of the region X_{c_k}, i.e. $inv : v_k \to \mathcal{R}(X), I_k = \bigwedge_{1 \leq j \leq n} \min_{c_k} \{x_j\} \leq x_j \leq \max_{c_k} \{x_j\}$.

- *Flow conditions*: The crucial transformation step is the simplification of the dynamics given by Eq. 1 into a flow condition φ_k in $\mathcal{R}(\dot{X})$. The mapping of nonlinear functions f_r into a rate interval is necessarily an approximation which must comply with the requirement of conservativity. Hence, we define the flow condition as intervals including all derivative values of a state variable occuring within the cell under consideration: $flow : v_k \to \mathcal{R}(\dot{X}), \varphi_k = \bigwedge_{1 \leq j \leq n} \min_{c_k} \{\dot{x}_j\} \leq \dot{x}_j \leq \max_{c_k} \{\dot{x}_j\}$. In our implementation of the modelling and analysis procedure, numerical optimization is used to determine the rate interval. For this purpose, we chose *Sequential Quadratic Programming*, a standard solution method for constrained nonlinear optimization [3] where the constraints are given by the invariant conditions I_k. To obtain a conservative approximation, the global minimum / maximum on the cell region X_{c_k} has to be found. Obviously, this is not guaranteed in all cases, namely for arbitrary non-convex functions f_r. Instead of using only the cell center as starting point of the optimization (as implemented so far), a set of starting points which are appropriately distributed on X_{c_k} could improve the probability of computing a conservative flow conditions, but the computational effort would be increased correspondingly.

- *Initial conditions*: The *init* function specifies a set of regions X_{c_k} by means of rectangular predicates $\mathcal{R}(X)$ as the initialisation of all variables.

- *Control switches*: A control switch is introduced into the RA for all pairs (v_k, v_l) of adjacent control modes, which have corresponding cell regions with a shared $(n - i)$-dimensional face $(i \geq 1)$.

- *Guard conditions*: The guard which is assigned to a control switch (v_k, v_l) equals the rectangular predicate $\mathcal{R}(X)$ that describes the $(n-i)$-dimensional face $(i \geq 1)$ which is shared by the corresponding cell regions.

- Resets: The functions $reset_{vars}$ and $reset$ are omitted since the set of considered variables is the same in all control modes, and jumps of the state trajectory given by Eq. 1 are excluded.

Following this scheme, a processing system can be modelled as a RA using the switched continuous system as an intermediate format to capture the relevant physical behavior. To investigate the behavior of the controlled system by verification, the controller has to be modelled as a RA, too. The overall RA model of the controlled processing system is obtained by composition of the process RA and the controller RA.

5 Combining the Approximation and the Analysis Procedure

In this section we describe a combination of the approximaton procedure from Sec. 4 and analysis procedure from Sec. 3.2. This combination is motivated by the fact that in the analysis of a RA often only a minor part of the RA's control modes are found to be reachable. In examples with fine discretizations and several mode switches the reachable part may be a small fraction of the RA's overall control modes. This means that the approximation procedure spends a considerable time on computing subregions that the analysis procedure will never reach and, a more severe problem, that the analysis procedure has to keep an unnecessarily huge automaton structure in memory. Note, that standard on-the-fly techniques for reachability analysis are only a partial solution to the latter problem. These techniques only reduce the number of control modes that are generated by composition of subsystems, whereas our approach also avoids generating unnecessary control modes of the subsystems.

In our approach the analysis procedure does not keep the transition structure and the flow conditions of the control modes in memory, but calls the approximation procedure each time when it has determined a new reachable outgoing face on the currently analyzed rectangle. The approximation procedure will then determine the corresponding transition, i.e. the neighboring rectangle, and return it together with the approximated flow conditions. Based on this information, the analysis procedure computes the possible outgoing faces of the new rectangle and the iteration continues.

To realize this interaction, two elements from the analysis procedure of Sec. 3.2 are needed. First, we must be able to compute the outgoing faces for a given rectangular invariant I, its flow conditions φ, an ingoing face F of this invariant, and the current discrete mode u. In the following pseudo-code description of the reachability algorithm, this is represented as a function **Outfaces**(I, φ, F, u) which returns a list *OutFaceList* consisting of triples $(F, \pm, u)$. F is an outgoing face and $\pm$ provides the information in which direction along the fixed dimension it is actually outgoing, which is needed by the approximation routine. Since the mode u never changes within an invariant this vector is simply copied by the function **Outfaces**.

From the approximation algorithm, we extract the procedure **Approximation**$(F, \pm, u)$ which returns a tripel (I, φ, u). I represents the rectangle adjacent to F in the given direction $\pm$, φ is an approximation of the flow condition in this rectangle, and u is the new discrete input which may have changed by switching when F was crossed. Note that **Approximation** is specific for a given switched continuous system and a controller. The state of the controller is stored in memory between calls of **Approximation**.

Based on these routines, the main body of the reachability algorithm is realized by a recursive procedure **Reach** of the following form, where I is an invariant, φ a flow, F a face, $\pm$ a direction of a face, and u a discrete input.

PROCEDURE **Reach**(I, φ, F, u){
OutFaceList := **Outfaces**(I, φ, F, u)
FOR EACH (F, $\pm$, u) $\in$ *OutFaceList* DO
    IF $\{(F, \pm, u)\} \cap$ *ReachedList* $= \emptyset$
        ReachedList := *ReachedList* $\cup \{(F, \pm, u)\}$
        CALL **Approximation**(F, $\pm$, u)
        READ (I, φ, u)
        Reach(I, φ, F, u)
    END IF
END FOR EACH
}

In words, **Reach** determines from a given face a list of new faces and for each member of this list it checks whether it has been found before (with the same direction pointer and discrete input). If not, it stores the face and calls the approximation procedure to receive the new invariant, flow conditions, and discrete mode. Then the procedure calls itself again to compute the successors of the current face . By this, all the reached faces are processed in a recursive, depth-first-search manner until no more new faces are found. The whole iteration is started by calling **Reach** with a given initial invariant, a flow condition of the kind $\dot{x}_j \in [-1, 1]$ for all continuous variables, an arbitrary face on the boundary of the invariant, and a given initial u. **Reach** will then determine all bounds of the initial invariant as the initial outgoing faces.

6 Example

For illustration of our approximation and analysis procedure we apply it to a simple technical process, a two-tank system which has been considered in [10] and [8] in modified versions before. The two tanks are arranged such that the first vessel is filled by an input flow F_{in} and is emptied into *Tank 2* through a connecting pipe (see Fig. 2). The outflow of *Tank* 2, which is located on a lower level (height difference: H) than *Tank* 1, is denoted by F_{out}. The flow through the system depends on the liquid levels h_1 and h_2 in both tanks, the setting of the valve controlling the flow F_{12}, and naturally the fixed flow F_{in}. For our purposes the following switched continuous system is sufficient do describe the dynamical behavior of the system:

$$\dot{h}_1 = (F_{in} - F_{12})/A_1,$$
$$\dot{h}_2 = (F_{12} - F_{out})/A_2,$$

(2)

$$h_2 < H \ : \ F_{12} = K_1 \cdot \sqrt{h_1},$$
$$h_2 \geq H \ : \ F_{12} = K_1 \cdot \sqrt{h_1 - h_2 + H} \quad \text{if} \quad h_1 \geq h_2 - H,$$
$$F_{12} = 0 \quad \text{else},$$
$$F_{out} = K_2 \cdot \sqrt{h_2},$$
$$valve = \begin{cases} \text{'half-open'} : K_1 = K_1^1 \\ \text{'open'} \quad\quad : K_1 = K_1^2 \end{cases}$$

While the state vector is given by h_1 and h_2, the variable *valve* denotes the input of the system. Changes of the gradient field defined by Eq. 2 occur when either *valve* is switched to another discrete value, or when h_2 exceeds H. The parameters are (units omitted): $A_1 = 1.14 \cdot 10^{-2}$, $A_2 = 1.98 \cdot 10^{-3}$, $H = 0.4$, $F_{in} = 1.11 \cdot 10^{-4}$, $K_1^1 = 1.2 \cdot 10^{-4}$, $K_2^1 = 3.4 \cdot 10^{-4}$, and $K_2 = 1.5 \cdot 10^{-4}$.

Remark: The different cases in the definition of the flow F_{12} for $h_2 \geq H$ constitute a non-orthogonal partitioning of the state space, i. e. Eq. 2 does not correspond to the definition of SCS given in Sec. 2. Our method can nevertheless be applied since the distinction of the two cases is considered when the procedure **Approximation** calls the optimization routine. Note furthermore that Eq. 2 contains the strict inequality $h_2 < H$. For the transformation into RA according to Sec. 3.1, it is replaced by the nonstrict equality $h_2 \leq H$ using the limit $\lim_{h_2 \to H} F_{12}$ for the calculation of the gradients.

We investigate the following scenario in the sequel: It is assumed that the initial liquid heights are given by $h_1 = [0.2, 0.3]$ and $h_2 = [0.2, 0.3]$ and that *valve* = 'half-open' applies. Since F_{12} is smaller than F_{in} at this setting, h_1 will rise. To prevent an overflow of *Tank* 1 the controller switches the value of *valve* to 'open' when it receives the information that h_1 has reached the value $h_{1,S} = 0.8$. As a consequence, h_1 drops immediately and h_2 increases, where the latter effect will be considerably larger since the cross-sectional area A_2 of *Tank* 2 is much smaller than A_1. We will use our analysis procedure to check whether opening the valve can lead to the situation that the range $h_2 > 0.9$ can be reached.

The model according to Eq. 2 as well as a controller model containing the switching logic for the *valve*-variable are implemented in an initialization file for the procedure **Approximation**. Additionally, the accuracy of the optimization algorithm and the partitioning parameters g_j have to be supplied - for the latter we choose to divide the range of h_1 and h_2 into 10 intervals each of equal length. Following the iterative procedure decribed in 5, the reachable region is generated step by step. The result of this procedure is shown in Fig. 3 where the grey-shaded area marks the region which is determined as reachable from the dark-shaded initial region. The plot reveals that the *critical* region with $h_2 > 0.9$ is found to be reachable, i.e. the switching value $h_{1,S}$ was not chosen appropriately to avoid an overflow of *Tank* 2. To provide a better understanding of the analysis result, the continuous trajectories starting at the corners of the initial region are drawn

Fig. 2. Scheme of the two-tank system.

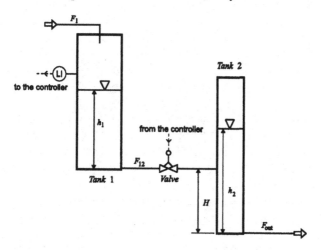

additionally [2]. It is obvious that the shaded area completely contains the actually reachable region (which looks like a single trajectory most of the time) and considerably overestimates it. A more accurate estimation can be obtained by choosing a finer partitioning at the expense of an increased computational effort. *Remark*: The reader might wonder why only complete rectangles are reachable in the left upper part of the reachable regions. This is due to the fact that in this area the gradient field points towards the line on which the trajectories move into the equilibrium point at $(h_1, h_2) = (0.25, 0.55)$. Hence, rate intervals including zero are obtained for these cells such that the whole cell area is reachable. (For the area above the dashed line applies: $F_{12} = 0$.).

The advantage of integrating the rate computation into the analysis routine becomes obvious from the ratio between shaded and non-shaded cells: If the whole RA for the two-tank system is computed in advance and then processed by the analysis tool, the optimization to evaluate the flow conditions has to be carried out 200 times - for each cell twice (since we have two discrete input values). Using the combined procedure, the approximation algorithm is only called 53 times. This advantage becomes even more important if we deal with systems of higher dimension, with a finer partitioning and with a large number of discrete input values.

---

[2] Obviously, in this example the reachable region can easily be determined by simulation. However, for more complex systems exhaustive simulation may become impossible.

Fig. 3. Analysis results.

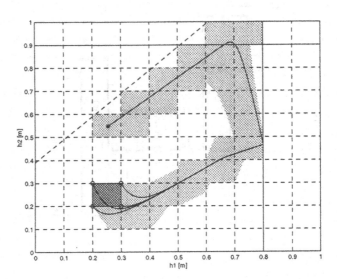

7   Related Work

The presented approach is similar in spirit to the work by Dang and Maler [2] in which reachability analysis of continuous systems is carried out by shifting outwards the boundaries of rectangular parts of the state space depending on the continuous flow on these boundaries. This procedure is called 'face lifting'. A similar concept is the so-called 'bloating' by Greenstreet [4, 5]. Here, the shifting of the boundaries is determined by integrating the original differential equations. Both approaches and the one presented here have in common, that the reachable regions in each iteration step are computed only from the bounds of the predecessor regions and the continuous flow. A major difference is that in our procedure the step size is determined by the partitioning of the state space, whereas in [2] and [4,5] time is discretized. Similarities also exist with the work of Krogh and Chutinan [1] who compute polyhedral approximations of the continuous system's reachable state space (called the 'flow pipe').

8   Conclusions

We have presented and discussed a reachability analysis for switched continuous systems which is based on an on-demand approximation of the continuous flow. It was illustrated by means of a simple, 2-dimensional example, but, in principle, it works for n dimensions. There are two different levels of approximation in our approach. First, the continuous flow is approximated to compute control modes of a rectangular automaton, then the possible values of the continuous variables within a control mode are approximated by the reachability algorithm. In both cases the approximations are conservative in the sense that they overestimate the exact range of values. Thus, the result of our reachability analysis is a clear overapproximation of a system's reachable state space. While the approximation of the reachability algorithm is only due to roundoff and hence very tight, the level of accuracy of the flow approximation depends on the system's structure and the gridsize chosen.

There are some modifications one can think of to improve the algorithm. The main idea which we want to pursue in our future work is to adapt the size of the rectangles depending on the size of the current face and the flow. The approximation routine should be able to return an appropriately sized invariant such that the narrowness or wideness of the ingoing face is taken into account and that the variation of the flow inside the rectangle is minimized. However, in this case care has to be taken such that the algorithm still terminates because the number of the boundaries of the rectangles can become infinite.

Acknowledgments

The work was partially supported by the German Research Council (DFG) in the special program *Analysis and Synthesis of Technical Systems with Continuous-Discrete Dynamics (KONDISK)* and by the temporary graduate school ("Graduiertenkolleg") *Modelling and Model-Based Design of Complex Technical Systems*.

References

1. A. Chutinan and B. H. Krogh. Computing polyhedral approximations of dynamic flow pipes. 1998. Submitted to *37th IEEE Conf. on Decision and Control*.
2. T. Dang and O. Maler. Reachability analysis via face lifting. In T.A. Henzinger and S. Sastry, editors, *HSCC 98: Hybrid Systems—Computation and Control*, Lecture Notes in Computer Science 1386, pages 96–109. Springer-Verlag, 1998.
3. R. Fletcher. *Practical Methods of Optimization*. J. Wiley and Sons, 1987.
4. M.R. Greenstreet. Verifying safety properties of differential equations. In R. Alur and T.A. Henzinger, editors, *CAV 96: Computer Aided Verification*, Lecture Notes in Computer Science 1102, pages 277–287. Springer-Verlag, 1996.
5. M.R. Greenstreet and I. Mitchell. Integrating projections. In T.A. Henzinger and S. Sastry, editors, *HSCC 98: Hybrid Systems—Computation and Control*, Lecture Notes in Computer Science 1386, pages 159–174. Springer-Verlag, 1998.
6. T.A. Henzinger, P.-H. Ho, and H. Wong-Toi. Algorithmic analysis of nonlinear hybrid systems. *IEEE Transactions on Automatic Control*, 43(4):540–554, 1998.
7. T.A. Henzinger, P.W. Kopke, A. Puri, and P. Varaiya. What's decidable about hybrid automata? In *Proceedings of the 27th Annual Symposium on Theory of Computing*, pages 373–382. ACM Press, 1995.

8. S. Kowalewski, O. Stursberg, M. Fritz, H. Graf, I. Hoffmann, J. Preußig, M. Remelhe, S. Simon, and H. Treseler. A case study in tool-aided analysis of discretely controlled continuous systems: The two-tanks problem. In *Hybrid Systems V*, Lecture Notes in Computer Science. Springer-Verlag, 1998.

9. J. Preußig, S. Kowalewski, H. Wong-Toi, and T.A. Henzinger. An algorithm for the approximative analysis of rectangular automata. In *FTRTFT98: Formal Techniques for Real-time and Fault-tolerant Systems*, LNCS. Springer-Verlag, 1998.

10. O. Stursberg, S. Kowalewski, I. Hoffmann, and J. Preußig. Comparing timed and hybrid automata as approximations of continuous systems. In P. Antsaklis, W. Kohn, A. Nerode, and S. Sastry, editors, *Hybrid Systems IV*, Lecture Notes in Computer Science 1273, pages 361–377. Springer-Verlag, 1996.

Refinement and Continuous Behaviour

Mauno Rönkkö[1] and Kaisa Sere[2]

[1] Turku Centre for Computer Science, Department of Computer Science
Åbo Akademi University, Lemminkäisenkatu 14A,FIN-20520 Turku, Finland
mronkko@abo.fi
[2] Department of Computer Science
Åbo Akademi University, Lemminkäisenkatu 14A, FIN-20520 Turku, Finland
Kaisa.Sere@abo.fi

Abstract. Refinement Calculus is a formal framework for the development of provably correct software. It is used by Action Systems, a predicate transformer based framework for constructing distributed and reactive systems. Recently, Action Systems were extended with a new action called the differential action. It allows the modelling of continuous behaviour, such that Action Systems may model hybrid systems. In this paper we investigate how the differential action fits into the refinement framework. As the main result we develop simple laws for proving a refinement step involving continuous behaviour within the Refinement Calculus.

1 Introduction

Action Systems, originally proposed by Back and Kurki-Suonio [2], are predicate transformer based systems for modelling discrete computations. They have been extensively used in the development of reactive and concurrent systems, see Sere et al. [3, 7]. Action Systems support the stepwise refinement paradigm, which is formalised in the *Refinement Calculus* [5]. The refinement in Action Systems preserves total correctness.

The *differential action* introduced recently by Rönkkö and Ravn [14] allows us to use differential equations in capturing continuous phenomena. The differential action has predicate transformer semantics like the other actions, and therefore, it fits seamlessly into the Action Systems framework. Hence, Action Systems with differential actions model hybrid systems where a discrete controller interacts with some continuously evolving environment.

Laws for the refinement of actions involving differential actions have not been investigated, yet. The existence of such laws would not only justify the introduction of the differential action, but also open the refinement technique for the development of hybrid systems with Action Systems. For instance, a hybrid system specification could be developed from a discrete skeleton, details of continuous behaviour could be added in a stepwise manner to an existing specification, and some complex continuous behaviour could be replaced with a combination of simpler discrete and continuous behaviours. The contribution of

F.W. Vaandrager and J.H. van Schuppen (Eds.): HSCC'99, LNCS 1569, pp. 223–237, 1999.
© Springer-Verlag Berlin Heidelberg 1999

this paper is twofold. Firstly, we present total correctness preserving refinement laws for differential actions. Secondly, we show that the refinement of continuous behaviour has a strong connection to path equivalence or *deformation* in topology: two paths are equivalent if there exists a continuous monotonic mapping such that the paths traverse the same values in the same order under the mapping [13]. We limit the investigation to systems in isolation, i.e., systems, which have observable variables, but do not react with other systems. The refinement of reactive components in a hybrid system is a topic of future research.

Overview. We start by defining actions in Section 2. In Section 3 we investigate the refinement between differential actions. Refinement between continuous and discrete behaviour is investigated in Section 4. In Section 5 we present Action Systems along with the refinement laws for systems in isolation. In Section 6 we use a travelling train example to illustrate how the presented refinement laws are used. Section 7 concludes and discusses directions for further work as well as provides pointers to related work.

2   Actions

An action is any statement in Dijkstra's guarded command language [8]. Also, a pure guarded command can be used as an action. Actions operate on a fix set of program variables. Hence, predicates over these variables specify states in a system.

When we consider a given action A, we may speak of a *postcondition* q. It is a predicate specifying desirable states after the execution of the action. The meaning of an action is defined with the *weakest precondition* predicate transformer $\mathrm{wp}(A, q)$. It is a predicate describing the largest set of states from which the action A terminates and reaches the postcondition q. The *termination*, i.e., reaching an arbitrary state, is given by $\mathrm{t}(A) \cong \mathrm{wp}(A, true)$. We may also speak of enabledness of an action, or *guard condition*, i.e., $\mathrm{g}(A) \cong \neg\mathrm{wp}(A, false)$. An action A is said to be enabled in all the states where $\mathrm{g}(A)$ holds.

Let $q[E/X]$ denote the textual substitution of free variables X with expressions E in a predicate q. For predicate p and actions A and B we define:

$\mathrm{wp}(\{p\}, q)$	$\cong p \wedge q$	*(assert)*
$\mathrm{wp}(\mathbf{abort}, q)$	$\cong false$	*(abort)*
$\mathrm{wp}(\mathbf{skip}, q)$	$\cong q$	*(skip)*
$\mathrm{wp}(X := E, q)$	$\cong q[E/X]$	*(assignment)*
$\mathrm{wp}(A; B, q)$	$\cong \mathrm{wp}(A, \mathrm{wp}(B, q))$	*(sequential composition)*
$\mathrm{wp}(A \,[\!]\, B, q)$	$\cong \mathrm{wp}(A, q) \wedge \mathrm{wp}(B, q)$	*(non−determin. choice)*
$\mathrm{wp}(p \rightarrow A, q)$	$\cong p \Rightarrow \mathrm{wp}(A, q)$	*(guarded action)*
$\mathrm{wp}(\mathbf{do}\,A\,\mathbf{od}, q)$	$\cong \exists n : n = \min\{i \mid i \geq 0 \wedge \neg\mathrm{g}(A^{i+1})\}$	*(iteration)*
	$\quad . \mathrm{wp}(A^n, q)$	

where $A^0 \cong \mathbf{skip}$ and $A^{n+1} \cong A; A^n$.

The differential action is used for modelling and reasoning about continuous behaviour. Let e be a predicate that may speak of variables X, and $\dot{X} = F(X)$

be a system of differential equations where $\dot{X}$ is a componentwise first derivative of X and $F(X)$ is a vector of functions with an equal number of components as in X. A differential action, $e : \dot{X} = F(X)$, evolves continuously the values of X provided that the evolution guard e holds. The evolution, which is given by a system of differential equations, stops when the values of X have reached the border of e, see Figure 1.

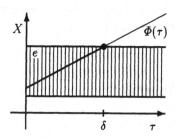

Fig. 1. The differential action $e : \dot{X} = F(X)$ evolves the values of X inside e up to the border values $\varPhi(\delta)$ where $\varPhi$ is the evolution curve and δ is the time when the border of e is reached.

Let τ and δ be real variables and not appearing in X, e or q. The differential action is defined as follows:

$$\text{wp}(e : \dot{X} = F(X), q)$$
$$\widehat{=}\ \exists!\, \varPhi : (\varPhi(0) = X \wedge \dot{\varPhi} = F(\varPhi))$$
$$.\ \exists \delta : \delta = \inf\{\tau \,|\, \tau \geq 0 \wedge \neg e[\varPhi(\tau)/X]\}\ .\ q[\varPhi(\delta)/X]$$

Here, the first existential quantification requires the existence of a unique solution tion $\varPhi$ for $\dot{X} = F(X)$, which is an *autonomous* system of differential equations, i.e., it may not refer to time explicitly. Time is implicit and it is modelled by the variable τ. The second existential quantification requires that the evolution eventually reaches the border of the evolution guard e and that the reached values satisfy the postcondition q. The variable δ is used for denoting the time, when the border of e is reached. Note, that if e does not hold at the beginning of an evolution, δ becomes zero forcing the evolution not to start at all. Hence, the semantics resembles the semantics for an iteration given earlier.

The differential action fulfils the four healthiness conditions for program statements, strictness, monotonicity, conjunctivity, and continuity, that are given by Dijkstra. The proofs are shown elsewhere [16].

Because the differential action has a weakest precondition semantics, we may also speak of the termination, $t(DA)$, and the enabledness, $g(DA)$, of a differential action DA. Clearly, the differential action is deterministic in its ending time. However, the general Action Systems have also a *non-deterministic assignment* [4], which, when composed sequentially with a differential action, is used for modelling continuous behaviour with a non-deterministic ending time.

Depending on the evolution guard, a differential action is open-ended or closed-ended. Let e describe a semi-open interval in a differential action $e : \dot{X} = F(X)$. The differential action is said to be *open-ended* if its evolution terminates always at a state where e does not hold, i.e., $\text{wp}(e : \dot{X} = F(X), \neg e)$ holds. Let $\bar{e}$ be a (semi-)closure of e, such that an evolution of $\bar{e} : \dot{X} = F(X)$ terminates always at the border of the evolution guard, i.e., $\text{wp}(\bar{e} : \dot{X} = F(X), \bar{e})$ holds. Then, $\bar{e} : \dot{X} = F(X)$ is said to be *closed-ended*. Due to the semantics of the differential action, an open-ended differential action is equivalent to a corresponding closed-ended differential action with respect to the weakest precondition. For instance, the predicate $\text{wp}(0 \leq x < 1 : \dot{x} = 1, q) \Leftrightarrow \text{wp}(0 \leq x \leq 1 : \dot{x} = 1, q)$ holds for any predicate q. Hence, without loss of generality, we assume in the rest of the paper that the differential actions are closed-ended.

3   Continuous Refinement

Refinement of actions, as formalised in the Refinement Calculus, is based on the following definition [3]. An action A is *refined* by an action B, denoted by $A \sqsubseteq B$, if the action B can reach all the same post-states at least from the same pre-states as the action A, i.e., $\forall q.\text{wp}(A, q) \Rightarrow \text{wp}(B, q)$. This definition applies to any action, including the differential action, because it fulfils all the required healthiness conditions. However, the complex weakest precondition semantics for the differential action makes the use of it quite cumbersome. Therefore, we would like to have an alternative, but equally powerful, characterisation for the refinement, which is more useful.

It might seem that the definition for the refinement, which speaks only of the pre-states and post-states of an evolution, does not cover the intermediate states. However, the semantics for the differential action imposes a strict ordering on the pre-states with respect to time: the closer the evolution is to the termination the smaller is the set of states that may still lead to a given post-state. Because of this, the refinement of differential actions depends also on the intermediate states indirectly. Therefore, one possibility is to look at the refinement of differential actions at the level of evolutions.

Consider two differential actions $0 \leq x \leq 1 : \dot{x} = 1$ and $0 \leq x \leq 1 : \dot{x} = 2$. If we think of the execution of these actions, all we can observe, is that both of the evolutions traverse the same values for x in the same order. We cannot say anything about the consumed time, since there is no variable measuring time. For instance, we can only observe that from an initial state $x = 0$ both of these actions traverse all the values from 0 to 1 for x. Since we cannot distinguish the evolutions from each other, it seems plausible that one of the actions is a refinement of the other.

In the following we show that proving the evolutions of two differential actions to be observably equal, i.e. path equivalent, is the same as proving the refinement using the definition above. Moreover, since it is easier to show path equivalence than to use the above definition for refinement, we also provide a convenient refinement law for differential actions.

3.1   Refinement

Consider two closed-ended differential actions $DA \hat{=} \bar{e} : \dot{X} = F(X)$ and $DB \hat{=} \bar{i} :$ $\dot{X} = G(X)$. For simplicity, we assume that all the evolutions are finite. In other words, all the evolutions terminate, that is, $t(DA) \wedge t(DB)$ holds. With these assumptions, a necessary requirement for the refinement $DA \sqsubseteq DB$ to hold is that the evolution guards are equal.

Lemma 1. *Let DA and DB be as above. Then, $DA \sqsubseteq DB$ cannot hold, if $\bar{e} \Leftrightarrow \bar{i}$ does not hold.*

Proof. Suppose $\bar{e} \Rightarrow \bar{i}$ does not hold. Then, there exists a predicate p such that $p \Rightarrow \bar{e}$ holds, but $p \Rightarrow \bar{i}$ does not hold. Since all the evolutions are assumed to be finite and both DA and DB are closed-ended, there exists a postcondition q and an evolution of DA whose initial or some intermediate state satisfies $p \wedge \neg q$ and final state satisfies q. On the other hand, because $p \Rightarrow \bar{i}$ does not hold, any state satisfying p in an evolution of DB must be a final state. Therefore, there cannot exist an evolution of DB whose initial or some intermediate state satisfies $p \wedge \neg q$ and final state satisfies q. But then, $wp(DA, q) \Rightarrow wp(DB, q)$ does not hold, and hence, $DA \sqsubseteq DB$ cannot hold. The case $\bar{i} \Rightarrow \bar{e}$ is proven similarly. $\square$

Also, there exists a necessary requirement concerning the evolution curves.

Lemma 2. *Let DA and DB be as above, with $\bar{e} \Leftrightarrow \bar{i}$. The respective finite solution curves for these actions are denoted by Φ and Ψ. Let α be the time of termination for DA, that is, $\alpha = \inf\{\tau \mid \tau \geq 0 \wedge \neg\bar{e}[\Phi(\tau)/X]\}$, and β be the time of termination for DB. Then, $\Phi(\alpha) = \Psi(\beta)$, if the refinement $DA \sqsubseteq DB$ holds.*

Proof. By unfolding $DA \sqsubseteq DB$ we get $\forall q. q[\Phi(\alpha)/X] \Rightarrow q[\Psi(\beta)/X]$. Clearly, this holds only if $\Phi(\alpha) = \Psi(\beta)$. $\square$

Lemma 2 shows an obvious consequence that follows, when the refinement holds between two differential actions. However, we may exploit even further the structure that is added to the state space by the differential action. By doing so, we gain more information concerning the intermediate states in the evolutions as shown by the following lemma.

Lemma 3. *Let DA, DB, Φ, and Ψ be as in Lemma 2. Then, the evolutions Φ and Ψ are continuous, loop-free, and traverse the same states in the same order, if the refinement $DA \sqsubseteq DB$ holds.*

Proof. First we conclude that Φ and Ψ traverse the same states. Because $DA \sqsubseteq DB$ holds, we know that DB must reach all the same post-states at least from the same pre-states as DA. Therefore, Ψ traverses at least the same states as Φ. Because the semantics of a differential action requires uniqueness for evolution curves, we know that for any one state in the state space there is only one evolution curve that passes it. Furthermore, because all the evolutions are finite, we know that all the evolutions are loop-free, and all the states are traversed by some evolution curve. Since the evolution guards are equal, $\bar{e} \Leftrightarrow \bar{i}$, DB must reach all the same post-states exactly from the same pre-states as DA. Otherwise, there would be more than one evolution curve passing one state in the state space. Therefore, Φ and Ψ traverse the same states.

Next we conclude that Φ and Ψ traverse not only the same states but also in the same order. By definition, all the evolutions are continuous. Now, consider a post-state p that is reachable by DA. Due to uniqueness and finiteness of evolutions all the states $\text{wp}(DA, p)$ are connected by one and only one continuous finite loop-free evolution. Let that evolution be Φ_p. Similarly, all the states $\text{wp}(DB, p)$ are connected by a continuous finite loop-free evolution Ψ_p. As concluded above, both Φ_p and Ψ_p traverse exactly the same states. But then, Φ_p and Ψ_p must traverse the same states in the same order, because otherwise Ψ_p either contains loops or is discontinuous.

Therefore, the evolutions Φ and Ψ are continuous, loop-free, and traverse the same states related by R in the same order. $\square$

With the help of Lemma 3 it is easy to show our main theorem, which states that the refinement between differential actions holds if and only if all the corresponding evolution curves are path equivalent.

Theorem 1. *Let DA, DB, Φ, and Ψ be as in Lemma 2. Then, $DA \sqsubseteq DB$ holds, iff $\exists \mu . \Phi \equiv \Psi \circ \mu$ where μ is a continuous monotonic increasing function and $\circ$ is the usual function composition.*

Proof. First the proof to direction " $\Rightarrow$ ". By assumption and Lemma 3 we know that the evolutions Φ and Ψ are continuous, finite, loop-free, and traverse the same states in the same order. Therefore, there exists a mapping μ which is also a continuous monotonic increasing function such that $\Phi \equiv \Psi \circ \mu$. Hence, the claim holds to direction " $\Rightarrow$ ".

Next the proof to direction " $\Leftarrow$ ". As noted earlier, $DA \sqsubseteq DB$ unfolds to $\forall q . q[\Phi(\alpha)/X] \Rightarrow q[\Psi(\beta)/X]$. This holds, if $\Phi(\alpha) = \Psi(\beta)$. By assumption there exists a continuous mapping μ such that $\Phi \equiv \Psi \circ \mu$, and therefore, $\Phi(\alpha) = \Psi(\mu(\alpha))$ holds. But then, β is $\mu(\alpha)$, and thus, $\Phi(\alpha) = \Psi(\beta)$ also holds. Hence, the claim holds also to direction " $\Leftarrow$ ", which concludes the proof. $\square$

This path equivalence property corresponds to the intuition of observability that was discussed earlier. It is also more convenient to use than the original definition for the refinement. For instance, we can now derive a result that proves the earlier example.

Corollary 1. *The refinement $\bar{e} : \dot{x} = c \sqsubseteq \bar{e} : \dot{x} = k$ with constants c and k holds, if $ck > 0$.*

Proof. According to Theorem 1 the refinement $\bar{e} : \dot{x} = c \sqsubseteq \bar{e} : \dot{x} = k$ holds, if there exists a μ such that $\Phi \equiv \Psi \circ \mu$ holds. Here, the corresponding unique solution curves starting from an initial state x are $\Phi(\tau) = x + c\tau$ and $\Psi(\tau) = x + k\tau$. Hence, the mapping μ is $\mu(\tau) = \frac{c}{k}\tau$, and it exists, because $ck > 0$. $\square$

3.2 Data refinement

The results above can also be applied to *data refinement* that speaks of the refinement between actions operating on different state spaces. Let an action A operate on variables X, an action B operate on different variables Y, and a predicate R be a refinement relation that tells how the values of X are related to the values of Y. The action A is said to be data refined by the action B under R, denoted by $A \sqsubseteq_R B$, if B reaches the same post-states as A at least from the same pre-states as A with respect to R [1]. Let q be a postcondition on the variables X. The data refinement is captured by $\forall q.\, R \wedge \mathrm{wp}(A, q) \Rightarrow \mathrm{wp}(B, \exists X.\, R \wedge q)$ where $\exists X.\, R \wedge q$ is a predicate on variables Y. Again, due to the complex differential action semantics it is hard to use this definition as such with differential actions. Therefore, we derive an alternative, but equally powerful, characterisation for the data refinement of differential actions.

Consider two closed-ended differential actions $DA \mathrel{\hat{=}} \bar{e} : \dot{X} = F(X)$ and $DB \mathrel{\hat{=}} \bar{\imath} : \dot{Y} = G(Y)$ with finite evolutions, and a refinement relation R defined everywhere in $\bar{e}$ and $\bar{\imath}$ for the disjoint variables X and Y. A necessary requirement for the refinement $DA \sqsubseteq_R DB$ to hold is that the evolution guards are equal with respect to R.

Lemma 4. *Let DA, DB, and R be as above. Then, $DA \sqsubseteq_R DB$ cannot hold, if $R \wedge \bar{e} \Leftrightarrow R \wedge \bar{\imath}$ does not hold.*

Proof. Suppose $R \wedge \bar{e} \Rightarrow R \wedge \bar{\imath}$ does not hold. Then, there exists a predicate p on the variables X, such that $p \Rightarrow \bar{e}$ holds, but $R \wedge p \Rightarrow \bar{\imath}$ does not hold. Since all the evolutions are assumed to be finite and both DA and DB are closed-ended, there exists a postcondition q on the variables X, and an evolution of DA whose initial or some intermediate state satisfies $p \wedge \neg q$ and final state satisfies q. On the other hand, because $R \wedge p \Rightarrow \bar{\imath}$ does not hold, any state satisfying $\exists X.\, R \wedge p$ in an evolution of DB must be a final state. Therefore, there cannot exist an evolution of DB whose initial or some intermediate state satisfies $\exists X.\, R \wedge p \wedge \neg q$ and final state satisfies $\exists X.\, R \wedge q$. But then, $R \wedge \mathrm{wp}(DA, q) \Rightarrow \mathrm{wp}(DB, \exists X.\, R \wedge q)$ does not hold, and hence, $DA \sqsubseteq_R DB$ cannot hold. The case $R \wedge \bar{\imath} \Rightarrow R \wedge \bar{e}$ is proven similarly. $\square$

There exists also a necessary requirement concerning the evolution curves.

Lemma 5. *Let R, DA and DB with $R \wedge \bar{e} \Leftrightarrow R \wedge \bar{\imath}$ be as in Lemma 4. We denote the evolution curves of DA and DB by Φ and Ψ respectively. Let α be the time of termination for DA, that is, $\alpha = \inf\{\tau \mid \tau \geq 0 \wedge \neg \bar{e}[\Phi(\tau)/X]\}$, and β be the time of termination for DB. Then, $R \Rightarrow R[\Phi(\alpha)/X, \Psi(\beta)/Y]$ holds, if $DA \sqsubseteq_R DB$ holds.*

Proof. $DA \sqsubseteq_R DB$ unfolds to $\forall q.\, R \wedge q[\Phi(\alpha)/X] \Rightarrow R[\Phi(\alpha)/X, \Psi(\beta)/Y]$, which holds only if $R \Rightarrow R[\Phi(\alpha)/X, \Psi(\beta)/Y]$. $\square$

By assuming some properties on the data refinement relation R we may exploit further the structure that is added to the state spaces by the differential actions as shown by the following lemma.

Lemma 6. *Let R, DA, DB, Φ, and Ψ be as in Lemma 5. Furthermore, let the data refinement relation R describe a continuous and bijective relation between $\bar{e}$ and $\bar{i}$. Then, the evolutions Φ and Ψ are continuous, loop-free, and traverse the same states related by R in the same order, if the refinement $DA \sqsubseteq_R DB$ holds.*

Proof. First we conclude that Φ and Ψ traverse the same states related by R. Because $DA \sqsubseteq_R DB$ holds, we know that DB must reach all the same post-states related by R at least from the same pre-states related by R as DA. Therefore, Ψ traverses at least the same states related by R as Φ. Because the semantics of a differential action requires uniqueness for evolution curves and R is continuous and bijective, we know that for any one state in the state space there is only one evolution curve that passes it. Furthermore, because all the evolutions are finite, we know that all the evolutions are loop-free, and all the states are traversed by some evolution curve. Since $R \wedge \bar{e} \Leftrightarrow R \wedge \bar{i}$ holds and R is continuous and bijective, DB must reach all the same post-states related by R exactly from the same pre-states related by R as DA. Otherwise, there would be more than one evolution curve passing one state in either state space. Therefore, Φ and Ψ traverse the same states related by R.

Next we conclude that Φ and Ψ traverse not only the same states related by R but also in the same order. By definition, all the evolutions are continuous. Now, consider a post-state p that is reachable by DA. Due to uniqueness and finiteness of evolutions all the states $wp(DA, p)$ are connected by one and only one continuous finite loop-free evolution. Let that evolution be Φ_p. Similarly, all the states $wp(DB, \exists X. R \wedge p)$ are connected by a continuous finite loop-free evolution Ψ_p. As concluded above, both Φ_p and Ψ_p traverse exactly the same states related by R. But then, Φ_p and Ψ_p must traverse the same states related by R in the same order, because otherwise Ψ_p either contains loops or is discontinuous, or R is not continuous and bijective.

Therefore, the evolutions Φ and Ψ are continuous, loop-free, and traverse the same states related by R in the same order. $\square$

With the help of Lemma 6 it is easy to show the theorem, which states that the refinement between differential actions in different state spaces reduces to a path equivalence problem.

Theorem 2. *Let DA, DB, Φ, Ψ, α, and β be as in Lemma 5, and R be as in Lemma 6. Then, refinement $DA \sqsubseteq_R DB$ holds, iff $\exists \mu. \forall \tau : 0 \leq \tau \leq \alpha. R \Rightarrow R[\Phi(\tau)/X, (\Psi \circ \mu)(\tau)/Y]$ where μ is a continuous monotonic increasing function and $\circ$ is the usual function composition.*

Proof. First the proof to direction " $\Rightarrow$ ". By assumption and Lemma 6 we know that the evolutions Φ and Ψ are continuous, finite, loop-free, and traverse the same states related by R in the same order. Moreover, we know that R describes a continuous mapping between the evolutions. Therefore, there exists a mapping μ which is also a continuous monotonic increasing function such that

$\forall \tau : 0 \leq \tau \leq \alpha . R \Rightarrow R[\Phi(\tau)/X, (\Psi \circ \mu)(\tau)/Y]$. Therefore the claim holds to direction " $\Rightarrow$ ".

Next the proof to direction " $\Leftarrow$ ". As noted earlier, $DA \sqsubseteq DB$ unfolds to $\forall q . R \wedge q[\Phi(\alpha)/X] \Rightarrow R[\Phi(\alpha)/X, \Psi(\beta)/Y]$. and that the refinement holds, if $R \Rightarrow R[\Phi(\alpha)/X, \Psi(\beta)/Y]$ holds. By assumption there exists a continuous mapping μ such that $\forall \tau : 0 \leq \tau \leq \alpha . R \Rightarrow R[\Phi(\tau)/X, (\Psi \circ \mu)(\tau)/Y]$, and therefore, $R \Rightarrow R[\Phi(\alpha)/X, (\Psi \circ \mu)(\alpha)/Y]$ holds. But then, β is $\mu(\alpha)$, and thus, $R \Rightarrow R[\Phi(\alpha)/X, \Psi(\beta)/Y]$ also holds. Hence, the claim holds also to direction " $\Leftarrow$ ", which concludes the proof. $\square$

4 Mixed Refinement

Although the differential action is the only action modelling continuous behaviour, it is just an action among the other actions defined with the weakest precondition predicate transformer. This permits us to reason about *mixed refinement* where continuous behaviour is refined by some discrete behaviour or vice versa. Hence, we may develop hybrid systems from discrete skeletons or from fully continuous mathematical constructions by means of refinement steps. Moreover, we may use refinement to simplify existing hybrid specifications to gain easier analysis for specific purposes.

In this section we focus on the refinement between an assignment statement and a differential action. Without loss of generality, we limit the investigation to that of data refinement.

4.1 Discretisation

In *discretisation* a differential action is refined by some non-differential action. Thus, the intermediate continuous behaviour is removed by refinement. This type of refinement is required when some analog component, say an ordinary electrical motor, is replaced with a discrete component, like for instance a step motor. Discretisation may also be used for simplifying a specification to gain simpler analysis of some overall behaviour.

Theorem 3. *Let a differential action* $e : \dot{X} = F(X)$ *with finite evolutions* Φ *operate on variables* X. *As before, the time of termination for this differential action is denoted by* α. *Let an assignment statement* $Y := E$ *operate on variables* Y, *which are disjoint from* X, *and let* R *be the refinement relation. Then, the data refinement* $e : \dot{X} = F(X) \sqsubseteq_R Y := E$ *holds iff* $R \Rightarrow R[\Phi(\alpha)/X, E/Y]$.

Proof. First to the direction " $\Rightarrow$ ". By unfolding $e : \dot{X} = F(X) \sqsubseteq_R Y := E$ we get an expression $\forall q . R \wedge q[\Phi(\alpha)/X] \Rightarrow R[\Phi(\alpha)/X, E/Y]$, which cannot hold, unless $R \Rightarrow R[\Phi(\alpha)/X, E/Y]$. The claim holds trivially to the direction " $\Leftarrow$ " if the condition $R \Rightarrow R[\Phi(\alpha)/X, E/Y]$ holds. $\square$

4.2   Continualisation

The reverse refinement step for discretisation is called *continualisation*. In continualisation a non-differential action is refined by a differential action causing the introduction of intermediate continuous behaviour to a discrete state change. By using continualisation one can develop a hybrid system specification starting from an entirely discrete specification. The following theorem shows that continualisation is applicable under the same conditions as discretisation.

Theorem 4. *Let a differential action $e : \dot{X} = F(X)$ with finite evolutions Φ operate on variables X. Again, the time of termination for this differential action is denoted by α. Let an assignment statement $Y := E$ operate on variables Y, which are disjoint from X. Furthermore, let R be the refinement relation. Then, the data refinement $Y := E \sqsubseteq_R e : \dot{X} = F(X)$ holds iff $R \Rightarrow R[E/Y, \Phi(\alpha)/X]$.*

The proof of this theorem is analogous to the proof of Theorem 3, and thus, is omitted here.

5   Action Systems

So far we have been discussing the refinement of individual actions. However, a hybrid system can very seldom be modelled by one action. For this purpose we use Action Systems. An *action system* $\mathcal{A}$ is an initialised block of the form

$$\mathcal{A} \;\hat{=}\; |[\,\mathbf{var}\; X : T \;\bullet\; X := E;\; \mathbf{do}\, A\, \mathbf{od}\,]|$$

The expression **var** $X : T$ declares a list of variables X with types T. These variables form the *state space* of $\mathcal{A}$. Some of the variables X may be *observable* from outside. These variables are decorated in the declaration with a superscript asterisk, e.g. a^*. The action $X := E$ initialises all the variables X by expressions E.

Typically, the action A is a non-deterministic choice of a number of actions. After initialisation, an enabled action of A is selected non-deterministically for an execution. There are no fairness assumptions about the selection of actions. The execution of an enabled action, including the differential action, is always atomic. This means that when a differential action is selected and executed, an evolution continues without interruption up to the border of evolution guard. Only after that the other enabled actions have a chance for execution. If two enabled actions refer to disjoint variables, their execution can be in any order or in parallel. Hence, this models parallelism by interleaving.

The execution of $\mathcal{A}$ terminates when none of the actions is enabled, i.e. $\neg\,g(A)$. Similarly, the execution of an action system aborts when an executed action aborts.

5.1   Refinement

The action system formalism allows also the refinement of entire action systems. A refining action system preserves total correctness with respect to the observable variables [4]. Consider two action systems

$A1 \ \widehat{=} \, \| [\, \mathbf{var} \ X^* : T, \ Y1 : V1 \ \bullet \ X, Y1 := E, I1; \ \ \mathbf{do} \, A1 \, \mathbf{od} \,] \|$

$A2 \ \widehat{=} \, \| [\, \mathbf{var} \ X^* : T, \ Y2 : V2 \ \bullet \ X, Y2 := E, I2; \ \ \mathbf{do} \, A2 \, [\!] \, B \, \mathbf{od} \,] \|$,

and a refinement relation R. The refinement $A1 \ \sqsubseteq_R \ A2$ holds if [1]

(1) the initialisations do not contradict with the refinement relation,
(2) all the actions $A2$ have the same effect to the variables as the actions $A1$,
(3) the additional actions B do not have any effect on the state space of $A1$,
(4) the additional actions terminate, and
(5) $A2$ does not terminate, unless $A1$ terminates as well.

These proof obligations are stated formally as

(1) $R[E/X, I1/Y1, I2/Y2]$,
(2) $A1 \ \sqsubseteq_R \ A2$,
(3) $\mathbf{skip} \ \sqsubseteq_R \ B$,
(4) $R \ \Rightarrow \ t(\mathbf{do} \, B \, \mathbf{od})$, and
(5) $R \ \wedge \ g(A1) \ \Rightarrow \ g(A2) \ \vee \ g(B)$.

Let $g1$ and $g2$ be predicates and $S1$ and $S2$ be actions. For proving the refinement of guarded actions $g1 \rightarrow S1 \ \sqsubseteq_R \ g2 \rightarrow S2$ it is adequate to show that the guard is not weakened, $R \wedge g2 \Rightarrow g1$, and that the body of the action is also refined in the context of the new guard, $\{g2\}; S1 \ \sqsubseteq_R \ S2$ [1].

Since the semantics for the differential action is given with the weakest precondition, any of the actions in $A1$, $A2$, and B may contain differential actions. Moreover, we have shown in this paper how to prove a refinement of actions involving a differential action. Therefore, the proof obligations for action systems above cover also the cases where the action systems model some hybrid systems.

6   Example: Travelling Train

We use a simple train example to illustrate how the given theorems are used in practise. In this example a train travels a distance of 2560 units. It starts by accelerating to travelling velocity, 20 units per one time unit, which is then kept for the most of the journey. At the end, the train decelerates to a full stop. Figure 2 depicts how the velocity is changed during the journey.

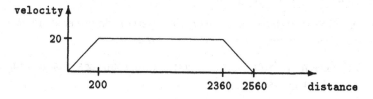

Fig. 2. Velocity during the journey.

In the following we will develop an action system modelling such a travelling train. We start with a discrete action system describing only the state changes, and add the details of continuous behaviour by refinement steps.

6.1 Initial specification

The action system below is the initial specification describing only the discrete state changes during the journey.

$$Train1 \ \widehat{=} \ |[\textbf{var } x1 : \textbf{\textit{R}} \ \bullet \ x1 := 0;$$
$$\textbf{do } x1 = 0 \quad \rightarrow x1 := 200$$
$$[] \quad x1 = 200 \ \rightarrow x1 := 2360$$
$$[] \quad x1 = 2360 \rightarrow x1 := 2560$$
$$\textbf{od}$$
$$]|$$

In this action system the variable $x1$ denotes the travelled distance. The initialisation states that the train is originally located at the very beginning. The first action models the acceleration phase, the second action models the phase where the train travels with a constant velocity, and the last action models the deceleration phase. The action system terminates when the train has travelled the whole distance of 2560 units.

6.2 Introducing continuous movement

In this first refinement step we introduce continuous movement without acceleration and deceleration. Such an action system is the one below, where the differential actions are closed-ended.

$$Train2 \ \widehat{=} \ |[\textbf{var } x2 : \textbf{\textit{R}} \ \bullet \ x2 := 0;$$
$$\textbf{do } x2 = 0 \quad \rightarrow 0 \leq x2 \leq 200 : \dot{x}2 = 1$$
$$[] \quad x2 = 200 \ \rightarrow 200 \leq x2 \leq 2360 : \dot{x}2 = 1$$
$$[] \quad x2 = 2360 \rightarrow 2360 \leq x2 \leq 2560 : \dot{x}2 = 1$$
$$\textbf{od}$$
$$]|$$

In order to prove that this is a refinement of the original specification, we need to define a refinement relation capturing the features that are to be preserved. Because the only requirement for the action system above is to retain the meaning of the variable measuring the distance, the refinement relation is $R1 \ \widehat{=} \ x1 = x2$.

Since both $Train1$ and $Train2$ have an equal number of actions, we can prove the refinement $Train1 \sqsubseteq_{R1} Train2$ in three steps.

i) Initialisations. The initialisations in the action systems do not contradict with the refinement relation, because $R1[0/x1, 0/x2]$ holds.

ii) Actions. Each action in $Train1$ has a corresponding refining action in $Train2$. For instance, consider the case

$$x1 = 0 \rightarrow x1 := 200 \ \sqsubseteq_{R1} \ x2 = 0 \rightarrow 0 \leq x2 \leq 200 : \dot{x}2 = 1 \ .$$

Because $R1 \wedge x2 = 0 \Rightarrow x1 = 0$ holds, the guard is not weakened and we only need to prove $\{x2 = 0\}; x1 := 200 \ \sqsubseteq_{R1} \ 0 \leq x2 \leq 200 : \dot{x}2 = 1$. According to Theorem 4 we have to prove that under the assumption $x2 = 0$ the predicate $R1 \Rightarrow R1[200/x1, \Phi2(\delta2)/x2]$ holds, where $\Phi2$ is the solution curve and $\delta2$ is the termination time for the differential action. Because $x2 = 0$, the solution curve is $\Phi2(\tau) = \tau$. For this, the termination time is $\delta2 = \inf\{\tau \mid \tau \geq 0 \land \neg(0 \leq \tau \leq 200)\}$, that is, $\delta2 = 200$. Thus, the predicate to be proven simplifies to $R1 \Rightarrow R1[200/x1, 200/x2]$, which holds trivially. This concludes the proof for the refinement of the first pair of actions. The refinement of the other two action pairs is proven similarly.

iii) Termination. The termination condition is not strengthened, because the condition $R1 \land g(Train1) \Rightarrow g(Train2)$, that is, $R1 \land (x1 = 0 \lor x1 = 200 \lor x1 = 2360) \Rightarrow (x2 = 0 \lor x2 = 200 \lor x2 = 2360)$, holds.

6.3   Introducing velocity

In this refinement step we add acceleration and deceleration to the specification. Such an action system, where $v3$ denotes the velocity, is the one below.

$$
\begin{aligned}
Train3 \ \widehat{=}\ \ |[\,\mathbf{var}\ &x3, v3 : \boldsymbol{R}, \boldsymbol{R} \ \bullet\ x3, v3 := 0, 0; \\
\mathbf{do}\ &x3 = 0 \quad \rightarrow 0 \leq x3 \leq 200 : \dot{x3}, \dot{v3} = v3, 1 \\
[\!]\ &x3 = 200 \quad \rightarrow 200 \leq x3 \leq 2360 : \dot{x3} = v3 \\
[\!]\ &x3 = 2360 \rightarrow 2360 \leq x3 \leq 2560 : \dot{x3}, \dot{v3} = v3, -1 \\
\mathbf{od}& \\
]&|
\end{aligned}
$$

The refinement relation used for proving that this action system is a refinement of *Train2* must describe what we mean by acceleration and deceleration. Such a refinement relation is the following.

$$
\begin{aligned}
R2 \ \widehat{=}\ (x2 = x3) \land & \\
(0 \leq x3 \leq 200 &\Rightarrow v3 = \sqrt{2 \cdot x3}) \land \\
(200 \leq x3 \leq 2360 &\Rightarrow v3 = 20) \land \\
(2360 \leq x3 \leq 2560 &\Rightarrow v3 = \sqrt{5120 - 2 \cdot x3}).
\end{aligned}
$$

$R2$ describes a continuous and bijective relation between variable $x2$ and variables $x3$ and $v3$ within an interval $0 \leq x2 \leq 2560$. The first conjunct states that the meaning of the variable measuring the distance must remain the same. The second, the third, and the fourth conjuncts describe how the velocity behaves during acceleration, travelling with a constant speed, and deceleration.

The proof of the refinement *Train2* $\sqsubseteq_{R2}$ *Train3* is done in three steps.

i) Initialisations. The initialisations in the action systems do not contradict with the refinement relation, because $R2[0/x2, 0/x3, 0/v3]$ holds.

ii) Actions. Each action in *Train2* has again a corresponding refining action in *Train3* . For instance, consider the case

$$
x2 = 0 \rightarrow 0 \leq x2 \leq 200 : \dot{x2} = 1 \quad \sqsubseteq_{R2} \quad x3 = 0 \rightarrow 0 \leq x3 \leq 200 : \dot{x3}, \dot{v3} = v3, 1 \ .
$$

Because $R1 \land x3 = 0 \Rightarrow x2 = 0$ holds, the guard is not weakened and we only need to prove $\{x3 = 0\}; 0 \leq x2 \leq 200 : \dot{x2} = 1 \sqsubseteq_{R2} 0 \leq x3 \leq 200 : \dot{x3}, \dot{v3} = v3, 1$. According to Theorem 2, it suffices to prove that under the assumption $x3 = 0$

there exists a continuous monotonic increasing function μ, such that $\forall \tau : 0 \leq \tau \leq \alpha$. $R2 \Rightarrow R2[\Phi_{x2}(\tau)/x2, \Psi_{x3}(\mu(\tau))/x3, \Psi_{v3}(\mu(\tau))/v3]$ holds, where Φ_{x2} is the solution curve and α is the termination time for the refined differential action, and Φ_{x3} together with Φ_{v3} are the solution curves for the refining differential action. Because $x3 = 0$, the refinement relation $R2$ states that both $x2 = 0$ and $v3 = 0$, and hence, the solution curves are $\Phi_{x2}(\tau) = \tau$, $\Phi_{x3}(\tau) = \frac{1}{2}\tau^2$, and $\Phi_{v3}(\tau) = \tau$, and the termination time for the refined differential action is $\alpha = 200$. Thus, the predicate to be proven simplifies to $\exists \mu. \forall \tau : 0 \leq \tau \leq 200$. $R2 \Rightarrow R2[\tau/x2, \frac{1}{2}(\mu(\tau))^2/x3, \mu(\tau)/v3]$. After the substitutions and simplifications we get a condition $\exists \mu. \forall \tau : 0 \leq \tau \leq 200$. $\mu(\tau) = \sqrt{2 \cdot \tau}$, which indeed defines a continuous monotonic increasing μ. This concludes the proof for the refinement of the first pair of actions. The refinement of the other two action pairs is proven similarly.

iii) Termination. The termination condition is not strengthened, because $R2 \wedge (x2 = 0 \vee x2 = 200 \vee x2 = 2360) \Rightarrow (x3 = 0 \vee x3 = 200 \vee x3 = 2360)$ holds.

7   Conclusion

In this paper we presented simple, yet powerful, refinement laws for differential actions. Although these laws require the differential equations in the differential actions to be solvable, one can derive specialised laws with no need of solving the differential equations. This was illustrated in Corollary 1. In the data refinement laws, we assumed continuity and bijectivity from the refinement relation. Relaxing on these assumptions yields different refinement laws where path equivalence does not necessarily hold. In this paper we investigated the refinement of one differential action by another. However, it is also possible to think about refinement where one differential action is refined by a sequence of differential actions.

We showed how the refinement laws for differential actions are used in the development of hybrid systems. For this purpose we used a travelling train example. We limited the investigation of using refinement in hybrid setting to systems in isolation. The refinement of reactive components in a hybrid system is a topic of future research.

There are several formalisms that are used in hybrid systems. The connection between some of these formalisms and Action Systems with differential actions has been studied earlier. For instance, the relation to Branicky's unified model [6] was established by Rönkkö and Ravn [15]. Also, the use of HyTech [10] in analysis of Action Systems with differential actions is discussed elsewhere [14]. Refinement in timed and hybrid settings has also been studied in other formalisms [12, 11, 9]. An interesting topic for future research is to see, how the refinement ideas presented here relate to these approaches.

Acknowledgements

We would like to thank Anders P. Ravn, Emil Sekerinski and Magnus Steinby for helpful discussions, as well as the referees, for their comments for the earlier versions of this paper. The work of Kaisa Sere is supported by the Academy of Finland.

References

1. R. J. R. Back. Refinement calculus, part II: Parallel and reactive programs. In J. W. de Bakker, W.-P. de Roever, and G. Rozenberg, editors, *Stepwise Refinement of Distributed Systems: Models, Formalisms, Correctness. Proceedings. 1989*, volume 430 of *Lecture Notes in Computer Science*. Springer–Verlag, 1990.
2. R. J. R. Back and R. Kurki-Suonio. Decentralization of process nets with centralized control. In *Proc. of the 2nd ACM SIGACT-SIGOPS Symp. on Principles of Distributed Computing*, pp. 131–142, 1983.
3. R. J. R. Back and K. Sere. Stepwise refinement of action systems. *Structured Programming*,**12**, pp. 17-30, 1991.
4. R. J. R. Back and J. von Wright. *Trace Refinement of Action Systems*. In J. Parrow, editor, *CONCUR94*, LNCS 666, Springer-Verlag, 1994.
5. R. J. R. Back and J. von Wright. *Refinement Calculus: A Systematic Introduction*. Graduate Texts in Computer Science, Springer-Verlag,1998.
6. M. S. Branicky. *Studies in Hybrid Systems: Modeling, Analysis, and Control*. PhD Thesis. Massachusetts Institute of Technology, EECS Dept, 1995.
7. M. Butler, E. Sekerinski, and K. Sere. An Action System Approach to the Steam Boiler Problem. In J-R. Abrial, E. Borger, H. Langmaack (eds.) *Formal Methods for Industrial Applications: Specifying and Programming the Steam Boiler Control*, LNCS 1165, Springer-Verlag, 1996.
8. E. W. Dijkstra. *A Discipline of Programming*. Prentice–Hall International, 1976.
9. T. A. Henzinger, Z. Manna, and A. Pnueli. *Towards Refining Temporal Specifications into Hybrid Systems*. In R. L. Grossman, A. Nerode, A. P. Ravn, and H. Rischel (eds.) *Hybrid Systems*, LNCS 736, Springer-Verlag, 1993.
10. T. A. Henzinger and P.-H. Ho. HYTECH: The Cornell hybrid technology tool. In P. Antsaklis, W. Kohn, A. Nerode, and S. Sastry, editors, *Hybrid Systems II*, LNCS 999, pp. 265-293, Springer-Verlag, 1995.
11. N.A. Lynch and F.W. Vaandrager. *Forward and backward simulations Part II: Timing-based systems*. Information and Computation, 128(1):1-25, July 1996.
12. Brendan P. Mahony and Ian J. Hayes. *A Case-Study in Timed Refinement: A Mine Pump*. IEEE Transactions on Software Engineering, vol. 18, no. 9, September 1992.
13. M.H.A. Newman. *Elements of the Topology of Plane Sets of Points*. Cambridge University Press, 1961 (First edition 1939).
14. M. Rönkkö and A. P. Ravn. *Action Systems with Continuous Behaviour*. To appear in the *Proceedings of Hybrid Systems V - Fifth International Hybrid Systems Workshop*, LNCS, Springer-Verlag, 1998.
15. M. Rönkkö and A. P. Ravn. *Switches and Jumps in Hybrid Action Systems*. Proceedings of the Estonian Academy of Sciences. Engineering, June 1998, 4, 2, pp. 106-118.
16. M. Rönkkö and K. Sere. *Refinement and Continuous Behaviour*. No. 198 (updated version), Technical Reports, Turku Centre for Computer Science, Åbo Akademi University, Finland, September 1998.

Computing Controllers for Nonlinear Hybrid Systems[*]

Claire Tomlin[1], John Lygeros[2], and Shankar Sastry[2]

[1] Department of Aeronautics and Astronautics
Stanford University, Stanford CA 94305-4035
tomlin@leland.stanford.edu
[2] Department of Electrical Engineering and Computer Sciences
University of California, Berkeley CA 94720-1770
{lygeros,sastry}@eecs.berkeley.edu

Abstract. We discuss a procedure for synthesizing controllers for safety specifications for hybrid systems. The procedure depends on the construction of the set of states of a continuous dynamical system that can be driven to a subset of the state space, avoiding another subset of the state space (the *Reach-Avoid* set). We present a new characterization of the Reach-Avoid set in terms of the solution of a pair of coupled Hamilton-Jacobi partial differential equations. We also discuss a computational algorithm for solving such partial differential equations and demonstrate its effectiveness on numerical examples.

1 Introduction

The synthesis of controllers that meet safety specifications for discrete, continuous and hybrid systems has attracted considerable attention (see [1-4] for an overview). Our work has been based on casting the problem as a two player, zero sum game, between a *controller*, that tries to ensure that the safety specification is satisfied and a *disturbance* (that includes the nondeterminism of the system), that tries to violate the safety specification [5]. In [6] we proposed a procedure for systematically carrying out the controller synthesis for general hybrid systems. The procedure relies on the solution of partial differential equations (PDEs) [7], known as the *Hamilton-Jacobi* equations. Here, we bring the synthesis procedure one step closer to implementation, by proposing a numerical scheme for solving these partial differential equations.

In Section 2, we briefly review the modeling formalism and the controller synthesis problem introduced in [5]. In Section 3 we review the algorithm proposed in [6] for solving the controller synthesis problem. The algorithm requires the computation of the set of states of a continuous dynamical system that can be driven to a given subset of the state space, avoiding another subset of the state

[*] Research supported by NASA under grant NAG 2-1039, by ONR under grant N00014-97-1-0946, by DARPA under contract F33615-98-C-3614 and by a Zonta Postgraduate Fellowship.

space (the *Reach-Avoid* set). We introduce a new procedure for characterizing the Reach-Avoid set, in terms of the solution to a pair of coupled Hamilton-Jacobi PDEs. The advantage of this new characterization (over the single PDE characterization of [7], for example) is that it can deal with situations where the closures of the Reach and Avoid sets overlap, without resorting to approximation. It also remains closer to the Hamilton-Jacobi PDE arising in purely continuous pursuit evasion problems, which makes it easier to carry classical results over to the hybrid domain.

An analytical solution to the Hamilton Jacobi PDEs is likely to be impossible to obtain for most realistic examples. For the class of systems we consider the situation is additionally complicated by the fact that the initial conditions of the PDE may be non-smooth; *shocks*, or discontinuities in the spatial variable as the temporal variable evolves, may develop; the right hand side of the PDE may be non-smooth, due to the optimal (typically *bang-bang*) controls; the right hand side of the PDE may be discontinuous, due to saturation effects introduced to guarantee the monotonicity of the Reach-Avoid set. In Section 4, we present a procedure for numerically computing the Reach-Avoid set, based on the *level set method* of [8]. The advantage of this method is that it can systematically deal with all the technical problems highlighted above, based on the viscosity solution concept for the PDEs. A brief comparison with other techniques proposed in the literature for computing or approximating the reach set of hybrid systems is also given. We demonstrate the application of this approach to a new example from aircraft collision avoidance, developed from [9] (Section 5).

2   Model

For a finite collection V of variables, let $\mathbf{V}$ denote the set of valuations of these variables, i.e. the set of all possible assignments of the variables in V. For example, if x is a state variable taking values in $\mathbb{R}^n$ we write $X = \{x\}$ with $\mathbf{X} = \mathbb{R}^n$. By abuse of notation, we use lower case letters to denote both a variable and its valuation; the interpretation should be clear from the context. We call a variable discrete if its set of valuations is countable and continuous if it is a subset of Euclidean space. We assume the discrete topology for countable sets and the Euclidean metric topology for subsets of Euclidean space. For a topological space X and a set $K \subseteq X$ we denote by K^c the complement, by $\overline{K}$ the closure, by K° the interior, and by $\partial K = \overline{K} \setminus K^\circ$ the boundary of K in the topology of X. Given a set of valuations $W \subseteq \mathbf{V}$ and a subset of the variables $V' \subset V$ we denote by $W|_{V'} \subset \mathbf{V}'$ the restriction of W to V'.

2.1   Hybrid Automata

Definition 1 *A **hybrid automaton**, H, is a collection (X, V, I, f, E, ϕ), with:*

- **State and input variables:** X *and* V *are disjoint collections of state and input variables. We assume that* $X = X_D \cup X_C$ *and* $V = V_D \cup V_C$, *where* X_C

and V_C contain continuous, and X_D and V_D discrete variables. We refer to the valuations $x \in \mathbf{X}$ and $v \in \mathbf{V}$ as the state and the input of the hybrid automaton.

– **Initial states:** *$I \subset \mathbf{X}$ is a set of initial valuations of the state variables.*
– **Continuous evolution:** *$f : \mathbf{X} \times \mathbf{V} \to T\mathbf{X_C}$ is a vector field.*
– **Discrete transitions:** *$E \subset \mathbf{X} \times \mathbf{V} \times \mathbf{X}$ is a set of discrete transitions.*
– **Admissible inputs:** *$\phi : \mathbf{X} \to 2^{\mathbf{V}}$ gives the set of admissible inputs at a given state $x \in \mathbf{X}$.*

To fix notation we let $\mathbf{X_C} \subseteq \mathbb{R}^n$ and $\mathbf{V_C} \subseteq \mathbb{R}^m$. For technical conditions imposed to ensure well posedness see [5].

Definition 2 *A **hybrid time trajectory**, τ, is a finite or infinite sequence of intervals $\tau = \{I_i\}$ of the real line, starting with I_0 and satisfying:*

– *I_i is closed unless τ is a finite sequence and I_i is the last interval, in which case it is left closed but can be right open.*
– *Let $I_i = [\tau_i, \tau_i']$. Then for all i $\tau_i \leq \tau_i'$ and for $i > 0$, $\tau_i = \tau_{i-1}'$.*

We denote by $\mathcal{T}$ the set of all hybrid time trajectories.

Definition 3 *An **execution** of a hybrid automaton H is a collection (τ, x, v) with $\tau \in \mathcal{T}$, $x : \tau \to \mathbf{X}$, and $v : \tau \to \mathbf{V}$ which satisfies:*

– **Initial Condition:** *$x(\tau_0) \in I$.*
– **Discrete Evolution:** *$(x(\tau_{i-1}'), v(\tau_{i-1}'), x(\tau_i)) \in E$, for all i.*
– **Continuous Evolution:** *for all i with $\tau_i < \tau_i'$, x is continuous and v is piecewise continuous in $[\tau_i, \tau_i']$ and for all $t \in [\tau_i, \tau_i')$, $(x(t), v(t), x(t)) \in E$. Moreover, for all $t \in [\tau_i, \tau_i']$ where v is continuous $\frac{d}{dt}(x(t)|_{X_C}) = f(x(t), v(t))$.*
– **Input Constraints:** *for all $t \in \tau$, $v(t) \in \phi(x(t))$.*

We use χ to denote an execution of H and $\mathcal{H}$ to denote the set of all executions of H. We use $x^0 = x(\tau_0)$ to denote the initial state of an execution.

A *property*, P, of a hybrid automaton H is a map:

$$P : \mathcal{H} \to \{\text{True, False}\} \tag{1}$$

We say an execution $\chi \in \mathcal{H}$ satisfies property P if $P(\chi) = \text{True}$; we say a hybrid automaton satisfies a property P if $P(\chi) = \text{True}$ for all $\chi \in \mathcal{H}$. Given a set $F \subset \mathbf{X}$ we define a *safety property*, denoted by $\Box F$, by:

$$\Box F(\chi) = \begin{cases} \text{True if } \forall t \in \tau, x(t) \in F \\ \text{False otherwise} \end{cases}$$

2.2   Controller Synthesis

Assume that we are given a hybrid automaton H, which we refer to as the *plant*, and we are asked to control it using its input variables so that its executions satisfy certain properties. For the purposes of control the input variables of the plant are partitioned into two classes: *controls* and *disturbances*. We write $V = U \cup D$ where U and D are respectively control and disturbance variables. The interpretation is that the controls can be influenced using a *controller*, in an attempt to guide the system, whereas the disturbances are determined by the *environment* and may potentially disrupt the controller's plans.

An instance of the *controller synthesis problem* consists of a pair, (H, P), of a plant hybrid automaton and a property of that automaton. In this paper we restrict our attention to controller synthesis problems where $P = \Box F$. A *static state feedback controller* for H is a map:

$$g : \mathbf{X} \to 2^{\mathbf{U}} \tag{2}$$

Given a plant automaton H and a controller g for H one can define the set of closed loop executions as:

$$\mathcal{H}_g = \{(\tau, x, (u, d)) \in \mathcal{H} | \forall t \in \tau \ u(t) \in g(x(t))\} \tag{3}$$

It is easy to see that this is precisely the set of executions of another hybrid automaton, H_g. We say that controller g solves the synthesis problem $(H, \Box F)$ if H_g satisfies $\Box F$. It can be shown [5] that for controller synthesis problems of the form $(H, \Box F)$, one can restrict attention to memoryless controllers without loss of generality.

A subset $W \subseteq \mathbf{X}$ is *controlled invariant* if the controller synthesis problem $(H, \Box W)$ can be solved when $I = W$. It can be shown [5] that the controller synthesis problem $(H, \Box F)$ can be solved if and only if there exists a unique maximal controlled invariant subset of F. In the next sections we highlight a procedure (introduced in [6]) for computing this subset.

3   Controller Synthesis for Hybrid Systems

3.1   Construction of Controlled Invariant Sets

For the synthesis problem $(H, \Box F)$ we seek to construct the largest set of states for which the control u can guarantee that the property $\Box F$ is satisfied, despite the action of the disturbance d. We first introduce some notation. For any $v = (u, d)$ define the set:

$$Inv(v) = \{x \in \mathbf{X} | v \in \phi(x) \text{ and } (x, v, x) \in E\} \tag{4}$$

For a state $x \in \mathbf{X}$ and input $v = (u, d)$ consider the sets:

$$Next(x, v) = \begin{cases} \{x' \in \mathbf{X} | (x, v, x') \in E\} & \text{if } v \in \phi(x) \\ \emptyset & \text{if } v \notin \phi(x) \end{cases} \tag{5}$$

Inv(v) is the set of states from which continuous evolution is possible under input v, while $Next(x, v)$ is the set of states that can be reached from state x under input v through a discrete transition. Abusing notation slightly, for any set $K \subseteq \mathbf{X}$ and input $v = (u, d)$ we define the *successor of K under v* as the set:

$$Next(K, v) = \bigcup_{x \in K} Next(x, v) \tag{6}$$

For any set $K \subseteq \mathbf{X}$ we define the *controllable predecessor of K*, $Pre_u(K)$, and the *uncontrollable predecessor of K*, $Pre_d(K)$, by:

$$Pre_u(K) = \{x \in \mathbf{X} | \exists u \in \mathbf{U} \; \forall d \in \mathbf{D} \; x \notin Inv(v) \text{ and } Next(K, (u, d)) \subseteq K\} \cap K$$
$$Pre_d(K) = \{x \in \mathbf{X} | \forall u \in \mathbf{U} \; \exists d \in \mathbf{D} \; Next(K, (u, d)) \cap K^c \neq \emptyset\} \cup K^c \tag{7}$$

$Pre_u(K)$ contains all states in K for which u can force a transition back into K. $Pre_d(K)$, on the other hand, contains all states outside K, as well as all states from which a transition outside of K is possible whatever u does. It is easy to show that:

Proposition 1 $Pre_u(K) \cap Pre_d(K) = \emptyset$.

The controllable and uncontrollable predecessors will be used in the discrete part of the algorithm for determining controlled invariant subsets. For the continuous part we introduce the *Reach-Avoid* operator:

Definition 4 (Reach-Avoid) *For two disjoint sets $B \subseteq \mathbf{X}$ and $G \subseteq \mathbf{X}$, define the Reach-Avoid operator as:*

$$Reach(B, G) = \{x^0 \in \mathbf{X} \mid \forall u \in \mathcal{U}_f \; \exists d \in \mathcal{D} \text{ and } t \geq 0 \text{ such that } \\ x(t) \in B \text{ and } x(s) \notin G \text{ for all } s \in [0, t]\} \tag{8}$$

*Here $\mathcal{U}_f$ denotes the set of all **U**-valued feedback strategies, $\mathcal{D}$ denotes the set of piecewise continuous functions from the real line to **D** and $x(\cdot)$ the (unique) continuous state trajectory starting at $x(0) = x^0$ under input (u, d).*

The set $Reach(B, G)$ contains the states from which, for all $u(\cdot)$, there exists a $d(\cdot)$, such that the state trajectory can be driven to B while avoiding an "escape" set G.

Consider the following algorithm.

$$\begin{aligned}
&\text{Let} \quad W^0 = F, W^{-1} = \emptyset, i = 0. \\
&\text{While } W^i \neq W^{i-1} \text{ do} \\
&\qquad W^{i-1} = W^i \setminus Reach(Pre_d(W^i), Pre_u(W^i)) \tag{9} \\
&\qquad i = i - 1 \\
&\text{end}
\end{aligned}$$

In the first step of this algorithm, we remove from F all states for which there is a disturbance $d(\cdot)$ which through continuous evolution can bring the system either outside F, or to states from which a transition outside F is possible, without first touching the set of states from which a transition keeping the system inside F can be forced. Since at each step $W^{i-1} \subseteq W^i$, the set W^i decreases monotonically as i decreases. If the algorithm terminates, we denote the fixed point by W^*. In this case, W^* can be shown to be the largest controlled invariant subset contained in F.

3.2   Characterization of the Reach-Avoid Set

To implement this algorithm, we need to calculate Pre_u, Pre_d, and $Reach$. The calculation of Pre_u and Pre_d is conceptually straight forward and can be done by inverting the transition relation E. In this section we discuss the computation of the $Reach$ set. By definition, the discrete state remains constant along continuous evolution. Therefore, $Reach(Pre_d(W^i), Pre_u(W^i))$ can be computed in "parallel" for each discrete state (see [10] for a similar partition proposed for optimal control problems). In the following analysis, we describe this calculation for *one* discrete state $q \in \mathbf{X_D}$. Abusing notation, we use from now on x to denote the continuous part of the state, with the discrete part frozen at q.

For two disjoint sets $B \subseteq \mathbf{X_C}$ and $G \subseteq \mathbf{X_C}$, assume that there exist differentiable functions $l_B : \mathbf{X_C} \to \mathbb{R}$ and $l_G : \mathbf{X} \to \mathbb{R}$ such that $B \triangleq \{x \in \mathbf{X_C}|l_B(x) \leq 0\}$ and $G \triangleq \{x \in \mathbf{X_C}|l_G(x) \leq 0\}$[1]. Consider the following system of coupled Hamilton-Jacobi equations:

$$-\frac{\partial J_B(x,t)}{\partial t} = \begin{cases} H_B^*(x, \frac{\partial J_B(x,t)}{\partial x}) & \text{for } \{x \in X \mid J_B(x,t) > 0\} \\ \min\{0, H_B^*(x, \frac{\partial J_B(x,t)}{\partial x})\} & \text{for } \{x \in X \mid J_B(x,t) \leq 0\} \end{cases} \tag{10}$$

and

$$-\frac{\partial J_G(x,t)}{\partial t} = \begin{cases} H_G^*(x, \frac{\partial J_G(x,t)}{\partial x}) & \text{for } \{x \in X \mid J_G(x,t) > 0\} \\ \min\{0, H_G^*(x, \frac{\partial J_G(x,t)}{\partial x})\} & \text{for } \{x \in X \mid J_G(x,t) \leq 0\} \end{cases} \tag{11}$$

where $J_B(x,0) = l_B(x)$ and $J_G(x,0) = l_G(x)$, and

$$H_B^*(x, \frac{\partial J_B}{\partial x}) = \begin{cases} 0, & \text{for } \{x \in X \mid J_G(x,t) \leq 0\} \\ \max_{u \in \mathbf{U}} \min_{d \in \mathbf{D}} \frac{\partial J_B}{\partial x} f(x,u,d), & \text{otherwise} \end{cases} \tag{12}$$

$$H_G^*(x, \frac{\partial J_G}{\partial x}) = \begin{cases} 0, & \text{for } \{x \in X \mid J_B(x,t) \leq 0\} \\ \min_{u \in \mathbf{U}} \max_{d \in \mathbf{D}} \frac{\partial J_G}{\partial x} f(x,u,d), & \text{otherwise} \end{cases} \tag{13}$$

Equation (10) describes the evolution of the set B under the Hamiltonian H_B^*. This is the solution to the "$\max_u \min_d$" game for reachability in purely continuous systems (see for example [6]), with the modification that $H_B^* = 0$ in $\{x \in \mathbf{X_C} \mid J_G(x,t) \leq 0\}$. This ensures that the evolution of $J_B(x,t)$ is frozen once this set is reached. Similarly, equation (11) describes the evolution of the set G under the Hamiltonian H_G^*. Here a "$\min_u \max_d$" is used, since it is assumed

---

[1] More generally, B and G may be expressed as the maximum of a set of differentiable functions, as discussed in Section 4.

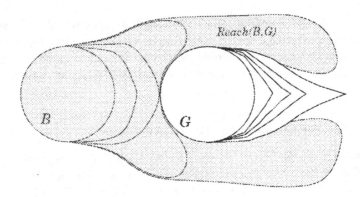

Fig. 1. The computation of $Reach(B, G)$ in a single discrete state q.

that the control tries to push the system into G, to escape from B. $H_G^* = 0$ in $\{x \in \mathbf{X_C} \mid J_B(x, t) \le 0\}$ to ensure that the evolution of $J_G(x, t)$ is frozen once this set is reached. In both cases we perform the calculation in "backwards time": we start the calculation at a fixed final time $t = 0$ and evolve backwards to a free initial time $t < 0$.

Note that in both games, the disturbance is given the advantage by assuming that the control plays first. In the following sequence of Lemmas, we prove that the resulting set $\{x \in \mathbf{X_C} \mid J_B(x, t) < 0\}$ contains neither G nor states for which there is a control which drives the system into G; and the set $\{x \in \mathbf{X_C} \mid J_G(x, t) < 0\}$ contains neither B nor states for which there is a disturbance which drives the system into B. We then prove that $\{x \in \mathbf{X_C} \mid J_B(x, t) < 0\}$ is the set $Reach(B, G)$. Figure 1 illustrates an example.

Assume that differentiable functions J_B and J_G satisfying the above partial differential equations exist. For all $t \le 0$, let

$$B(t) \triangleq \{x \in \mathbf{X_C} \mid J_B(x, t) \le 0\} \quad G(t) \triangleq \{x \in \mathbf{X_C} \mid J_G(x, t) \le 0\} \quad (14)$$

Note that $B = B(0)$ and $G = G(0)$.

Lemma 1 For all $t_2 \le t_1 \le 0$, $B(t_1) \subseteq B(t_2)$ and $G(t_1) \subseteq G(t_2)$.

Proof: Since $\frac{\partial J_B}{\partial t} \ge 0$ when $J_B(x, t) \le 0$ and $\frac{\partial J_G}{\partial t} \ge 0$ when $J_G(x, t) \le 0$, both $J_B(x, t)$ and $J_G(x, t)$ are monotone non-increasing functions of $-t$ when $J_B(x, t) \le 0$ and $J_G(x, t) \le 0$. Thus, as t decreases, the sets $B(t)$ and $G(t)$ do not decrease. ∎

Lemma 2 If $B^\circ(0) \cap G^\circ(0) = \emptyset$ then for all $t \le 0$, $B^\circ(t) \cap G^\circ(t) = \emptyset$.

Proof: Assume, for the sake of contradiction, that $x^0 \in B^\circ(t) \cap G^\circ(t)$ for some $t = t_1 < 0$, i.e. that

$$J_B(x^0, t_1) < 0 \quad \text{and} \quad J_G(x^0, t_1) < 0 \quad (15)$$

We first show that $J_B(x^0, 0) \geq 0$ and $J_G(x^0, 0) \geq 0$ (i.e., x^0 is outside of both B and G at $t = 0$). Suppose this is not true, i.e. suppose for example that $J_B(x^0, 0) < 0$ and $J_G(x^0, 0) \geq 0$. Then for all $t \leq 0$

$$\frac{\partial J_G(x^0, t)}{\partial t} = 0 \tag{16}$$

which implies that $J_G(x^0, t) = J_G(x^0, 0) \geq 0$ for all $t \leq 0$, which contradicts (15). A similar argument holds for the case in which $J_B(x^0, 0) \geq 0$ and $J_G(x^0, 0) < 0$.

Thus $J_B(x^0, 0) \geq 0$ and $J_G(x^0, 0) \geq 0$. Since $J_B(x, t_1) < 0$, there exists $t_2 \in [t_1, 0]$ such that $J_B(x^0, t_2) = 0$, and for all $t \in [t_1, t_2]$, $J_B(x^0, t) \leq 0$. Thus for at least some interval in $[t_1, t_2]$, $J_G(x^0, t) > 0$ (to allow $J_B(x^0, t)$ to decrease in this interval) and

$$\frac{\partial J_G(x^0, t)}{\partial t} = 0 \tag{17}$$

But this contradicts the assumption that $x^0 \in G^\circ(t_1)$. A symmetric argument holds for $J_G(x^0, t)$.   ∎

Lemma 3 For all $t \leq 0$, $B(t) \cap G(t) = \partial B(t) \cap \partial G(t)$. Moreover, for all $t' \leq t$, $B(t) \cap G(t) \subseteq \partial B(t') \cap \partial G(t')$.

Proof:

$$B(t) \cap G(t) = (B^\circ(t) \cap G^\circ(t)) \cup (\partial B(t) \cap \partial G(t)) \cup (B^\circ(t) \cap \partial G(t)) \cup (\partial B(t) \cap G^\circ(t))$$

From Lemma 2, $(B^\circ(t) \cap G^\circ(t)) = \emptyset$.

Assume that for some $t = t_1 < 0$, $x^0 \in B^\circ(t) \cap \partial G(t)$. Therefore, $J_B(x^0, t_1) < 0$ and $J_G(x^0, t_1) = 0$. Therefore, there exists $t_2 \in [t_1, 0]$ such that $J_B(x^0, t_2) = 0$ and for all $t \in [t_1, t_2]$, $J_B(x^0, t) \leq 0$. Thus for some interval of $[t_1, t_2]$, $J_G(x^0, t) > 0$ and

$$\frac{\partial J_G(x^0, t)}{\partial t} = 0 \tag{18}$$

which contradicts the assumption that $x^0 \in \partial G(t_1)$. Thus $B^\circ(t) \cap \partial G(t) = \emptyset$. A symmetric argument holds for $x^0 \in \partial B(t) \cap G^\circ(t)$ for $t = t_1 < 0$, thus $\partial B(t) \cap G^\circ(t) = \emptyset$.

Therefore, $B(t) \cap G(t) = \partial B(t) \cap \partial G(t)$. $B(t) \cap G(t) \subseteq \partial B(t') \cap \partial G(t')$ for $t' \leq t$ follows from Lemma 1.   ∎

Theorem 1 (Characterization of Reach-Avoid) Assume that $J_B(x, t)$ (or $J_G(x, t)$ respectively) satisfies the Hamilton-Jacobi equation (10) (or (11) respectively), and that it converges uniformly in x as $t \to -\infty$ to a function $J_B^*(x)$ (or $J_G^*(x)$ respectively). Then,

$$Reach(B, G) = \{x \in \mathbf{X_C} \mid J_B^*(x) < 0\} \tag{19}$$

Proof: Let $x^0 \in \{x \in \mathbf{X_C} \mid J_B^*(x) < 0\}$. Therefore, by construction, for all $u(\cdot) \in \mathcal{U}$ there exists $d(\cdot) \in \mathcal{D}$ such that the state trajectory $x(\cdot)$, starting at x^0, will eventually enter B. Also, by Lemma 2, $J_G(x^0) > 0$. Thus $\forall u(\cdot) \in \mathcal{U}$, $\exists d(\cdot) \in \mathcal{D}$, such that the state trajectory $x(\cdot)$ starting at x^0 never enters G. Thus, $\{x \in \mathbf{X_C} \mid J_B^*(x) < 0\} \subseteq Reach(B, G)$.

Now let $x^0 \in \{x \in \mathbf{X_C} \mid J_B^*(x) \geq 0\}$. Assume for the sake of contradiction that for all $u(\cdot) \in \mathcal{U}$, there exists a $d(\cdot) \in \mathcal{D}$ such that the trajectory $x(\cdot)$, starting at x^0, enters B. Since for all $x \in B$, $J_B^*(x) < 0$, there exists a time $t_1 > 0$ at which this trajectory crosses $\{x \in \mathbf{X_C} \mid J_B^*(x) = 0\}$. However, for all x such that $J_B^*(x) = 0$, there must exist a $u \in \mathbf{U}$ such that for all $d \in \mathbf{D}$, $f(x, u, d)$ points outside of $\{x \in \mathbf{X_C} \mid J_B^*(x) < 0\}$. This contradicts the assumption of existence of a $d(\cdot)$ which drives the system to B. Thus, $Reach(B, G) \subseteq \{x \in \mathbf{X_C} \mid J_B^*(x) < 0\}$. ∎

Using the function J_B^* obtained once the algorithm has converged, a controller which renders W^* invariant can be constructed:

$$g(x) = \begin{cases} \{u \in \phi(x)|_U \mid \forall d \in \phi(x)|_D \; Next(x, (u, d)) \subseteq W^*\}, \text{ if } x \in (W^*)^o \\ \{u \in \phi(x)|_U \mid \forall d \in \phi(x)|_D \left(\frac{\partial J_B^*(x)}{\partial x} f(x, (u, d)) \geq 0 \wedge x \in Inv(u, d) \right) \\ \quad \vee (Next(x, (u, d)) \subseteq W^* \wedge x \notin Inv(u, d)) \}, \text{ if } x \in \partial W^* \\ \phi(x)|_U, \text{ if } x \in (W^*)^c \end{cases}$$

Here $\wedge$ stands for the logical AND and $\vee$ for the logical OR.

In general, one cannot expect to solve for W^* using a finite computation. The class of hybrid systems for which algorithms like the one presented here are guaranteed to terminate is known to be restricted [4]. Techniques have been proposed to resolve this problem, making use of approximation schemes to obtain estimates of the solution (some are discussed in the next section). In practice, we are helped by the fact that we are usually interested in finite time computations, rather than computing for $t \to -\infty$ or until a fixed point is reached.

Another problem is the requirement that the controller resulting from our algorithm be *non-Zeno* (does not enforce the safety requirement by preventing time from diverging). The algorithm proposed here has no way of preventing such behavior. A practical method of resolving the Zeno problem is adding a requirement that the amount of time the system remains in each discrete state is bounded below by a positive number (representing, for example, the clock period of a digital computer).

4   Computation using Level Set Methods

In practice, the usefulness of the algorithm for hybrid controller synthesis depends on our ability to efficiently compute the optimal control and disturbance trajectories $(u^*(\cdot), d^*(\cdot))$, that arise from the solution of the Hamilton-Jacobi partial differential equations. Numerical solutions are potentially complicated by the fact that the right hand side of proposed PDEs is non-smooth, the initial data F may have non-smooth boundary, $(u^*(\cdot), d^*(\cdot))$ may be discontinuous, and the solution may develop shocks over time. New optimal control tools [11] can make the computation of $(u^*(\cdot), d^*(\cdot))$ feasible, at least numerically. In this section, we discuss a numerical technique developed by Osher and Sethian which computes the correct viscosity solution to the proposed PDEs, ensuring that discontinuities are preserved. We then compare it briefly with other approximation techniques found in the literature.

4.1   A Level Set Method for Boundary Approximation

Consider the Hamilton-Jacobi equation:

$$-\frac{\partial J(x,t)}{\partial t} = \begin{cases} H^*(x, \frac{\partial J(x,t)}{\partial x}) & \text{for } \{x \in X_C \mid J(x,t) > 0\} \\ \min\{0, H^*(x, \frac{\partial J(x,t)}{\partial x})\} & \text{for } \{x \in X_C \mid J(x,t) \le 0\} \end{cases} \tag{20}$$

with boundary condition $J(x,0) = l(x)$. A *viscosity solution* [12, 13] to (20) is the solution as $\epsilon \to 0$ to the partial differential equation:

$$-\frac{\partial J_\epsilon(x,t)}{\partial t} = \begin{cases} H^*(x, \frac{\partial J_\epsilon(x,t)}{\partial x}) + \epsilon \Delta J_\epsilon(x,t) & \text{for } \{x \in X_C \mid J_\epsilon(x,t) > 0\} \\ \min\{0, H^*(x, \frac{\partial J_\epsilon(x,t)}{\partial x})\} + \epsilon \Delta J_\epsilon(x,t) & \text{for } \{x \in X_C \mid J_\epsilon(x,t) \le 0\} \end{cases} \tag{21}$$

with boundary condition $J_\epsilon(x,0) = l_\epsilon(x)$.

The level set methods of Osher and Sethian [8] ([14] provides a comprehensive survey) are a set of computation schemes for propagating interfaces in which the speed of propagation is governed by a partial differential equation. These numerical techniques compute the viscosity solution to the Hamilton-Jacobi partial differential equation, ensuring that shocks are preserved. The methods have proved fruitful in many applications, including shape recovery problems in computer vision [15], and plasma etching problems in micro chip fabrication [16].

The key idea of the level set method is to embed the curve or surface to be evolved, for example the $n-1$-dimensional boundary of the reach set, as the zero level set of a function in n-dimensional space. The advantage of this formulation is that the n-dimensional function always remains a function as long as its speed of propagation is smooth, while the $n-1$-dimensional boundary may develop shocks or change topology under this evolution. The numerical methods of [14] choose the solution of (20) to be the one obtained from (21) as the viscosity coefficient ϵ vanishes. Below we present an outline of the method for a two-dimensional example.

In order for the numerical scheme to closely approximate the gradient $\frac{\partial J^*(x,t)}{\partial x}$, especially at points of discontinuity, an appropriate approximation to the spatial derivative must be used. Consider an example in two dimensions, with X_C discretized into a grid with spacing Δx_1 and Δx_2. As before we use $x = (x_1, x_2)$ to denote an element of X_C. The *forward difference operator* D^{+x_i} is defined (for x_1, similarly for x_2) as:

$$D^{+x_1} J(x,t) = \frac{J((x_1 + \Delta x_1, x_2), t) - J(x,t)}{\Delta x_1} \tag{22}$$

The *backward difference operator* D^{-x_i} is defined (for x_1, similarly for x_2) as

$$D^{-x_1}J(x,t) = \frac{J(x,t) - J((x_1 - \Delta x_1, x_2), t)}{\Delta x_1} \qquad (23)$$

The *central difference operator* D^{0x_i} is defined (for x_1, similarly for x_2) as

$$D^{0x_1}J(x,t) = \frac{J((x_1 + \Delta x_1, x_2), t) - J((x_1 - \Delta x_1, x_2), t)}{2\Delta x_1} \qquad (24)$$

At each grid point in x, the partial derivatives $\frac{\partial}{\partial x_1}$ and $\frac{\partial}{\partial x_2}$ may be approximated to first order using either the forward, backward, or central difference operators. The correct choice of operator depends on the direction of $f(x, u^*, d^*)$ (in our case it depends on $-f(x, u^*, d^*)$ since we compute backwards in time). If $-f(x, u^*, d^*)$ flows from left to right (from smaller to larger values of x_1), then then D^{-x_1} should be used to approximate $\frac{\partial J(x,t)}{\partial x_1}$; and if $-f(x, u^*, d^*)$ flows from bottom to top (from smaller to larger values of x_2), then then D^{-x_2} should be used to approximate $\frac{\partial}{\partial x_2}$ (and vice versa). Such an approximation is called an *upwind* scheme, since it uses information upwind of the direction that information propagates.

The algorithm for the two dimensional example proceeds as follows. Choose a domain of interest in $\mathbf{X_C}$ and discretize the domain with a grid of spacing $\Delta x_1, \Delta x_2$. Let x_{ij} represent the grid point $(i\Delta x_1, j\Delta x_2)$ and let $\tilde{J}(x_{ij}, t)$ represent the numerical approximation of $J(x_{ij}, t)$.

Set $t = 0$ and compute the initial condition $\tilde{J}(x_{ij}, 0) = l(x_{ij})$.

While for some x_{ij}, $\tilde{J}(x_{ij}, t) \neq \tilde{J}(x_{ij}, t - \Delta t)$ perform the following steps:

1. Compute $u^*(x_{ij}, D^{0x_1}\tilde{J}(x_{ij}, t), D^{0x_2}\tilde{J}(x_{ij}, t))$ and $d^*(x_{ij}, D^{0x_1}\tilde{J}(x_{ij}, t), D^{0x_2}\tilde{J}(x_{ij}, t))$.
2. Calculate $f(x_{ij}, u^*, d^*)$
3. If $(-f(x_{ij}, u^*, d^*))$ flows from greater to lesser values of x_1, let $\frac{\partial}{\partial x_1} = D^{+x_1}$, otherwise let $\frac{\partial}{\partial x_1} = D^{-x_1}$.
4. If $(-f(x_{ij}, u^*, d^*))$ flows from greater to lesser values of x_2, let $\frac{\partial}{\partial x_2} = D^{+x_2}$, otherwise let $\frac{\partial}{\partial x_2} = D^{-x_2}$.
5. Compute $\tilde{J}(x_{ij}, t - \Delta t)$:
   For x_{ij} such that $\tilde{J}(x_{ij}, t) > 0$,

$$\tilde{J}(x_{ij}, t - \Delta t) = \tilde{J}(x_{ij}, t) + \Delta t \frac{\partial \tilde{J}(x_{ij}, t)}{\partial x} f(x_{ij}, u^*, d^*) \qquad (25)$$

For x_{ij} such that $\tilde{J}(x_{ij}, t) \leq 0$,

$$\tilde{J}(x_{ij}, t - \Delta t) = \begin{cases} \tilde{J}(x_{ij}, t) + \Delta t \frac{\partial \tilde{J}(x_{ij}, t)}{\partial x} f(x_{ij}, u^*, d^*) \\ \qquad \text{if } \frac{\partial \tilde{J}(x_{ij}, t)}{\partial x} f(x_{ij}, u^*, d^*) < 0 \\ \tilde{J}(x_{ij}, t) \quad \text{otherwise} \end{cases} \qquad (26)$$

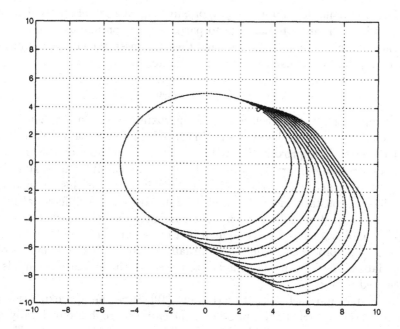

Fig. 2. $\{x \in \mathbf{X_C} \mid J(x,t) \leq 0\}$ shown in the (x_r, y_r)-plane for $[\underline{v}_1, \overline{v}_1] = [2,4]$, $[\underline{v}_2, \overline{v}_2] =$ $[1,5]$ and $\psi_r = 2\pi/3$.

Figure 2 displays the result of applying this algorithm to the two-aircraft example (presented in detail in Section 5) with zero angular velocity and $[\underline{v}_1, \overline{v}_1] =$ $[2,4]$, $[\underline{v}_2, \overline{v}_2] = [1,5]$ and $\psi_r = 2\pi/3$.

The above discussion presents the very basic idea of level set methods; for special forms of the Hamilton-Jacobi equation, extremely efficient variants of this method exist [14]. In particular, the *narrow band* and *fast marching* methods speed up the algorithm by confining the computation to a narrow band around the evolving front.

4.2   Other Methods

A number of other techniques have been proposed for numerically approximating the set of states reachable by a hybrid system, primarily for the purpose of verifying safety properties. We review some of them below.

Polyhedral Approximations Consider a hybrid automaton with no control inputs. For each valuation of the discrete state, the continuous dynamics of such an automaton can be captured by a differential inclusion:

$$\dot{x} \in g(x) = \{f(x,d) \mid d \in \mathbf{D}\} \tag{27}$$

One technique for numerically approximating the set of states that can be reached by such an automaton is to partition the continuous state space into polyhedral regions (e.g. rectangles) and then to conservatively approximate $g(x)$ in each region by a constant inclusion of the form:

$$\dot{x} \in [g_{\min}, g_{\max}] \tag{28}$$

The computation of the reach set for the approximate system can then be carried out exactly using tools developed for linear hybrid automata. In [17] it is shown that the error in approximating the set of states reachable by the true system in a bounded time interval can be made arbitrarily small by approximating the differential inclusion arbitrarily closely. An advantage of this method is that the class of constant inclusions used to approximate the differential inclusion is known to be decidable, thus one can guarantee that the reachable set of the approximate system can be computed in a finite number of steps. However, the amount of preprocessing required to initially approximate the dynamics may be formidable, especially if the time horizon over which the reach set needs to be calculated is large.

A related approach is that of [18] and [19]. Here the approximation of the dynamics is not carried out a-priori over the entire space, but only locally around the boundary of the reachable set as the latter propagates in time. This could potentially lead to substantial computational savings. The class of sets generated by these techniques is again polyhedral, but no a-priori bounds are given on the accuracy of the approximation.

Approximating non-smooth sets with smooth sets We have shown that the reach set at any time $t \in (-\infty, 0]$ may have a non-smooth boundary due to switches in (u^*, d^*), non-smooth initial data, or the formation of shocks. The level set method scheme propagates these discontinuities, yet its implementation may require a very small time step to do this accurately. In [7] we present a method for over-approximating such non-smooth sets with sets for which the boundary is continuously differentiable. Suppose that there exist differentiable functions l_G^i, $i = 1, \ldots, k$ such that for G a closed subset of $\mathbf{X_C}$:

$$G = \{x \in \mathbf{X_C} \mid \forall i \in \{i = 1, \ldots, k\}, \, l_G^i(x) \leq 0\} \tag{29}$$

Following [20, 21] we define two smooth functions:

$$G^\epsilon(x) = \epsilon \ln \left[\sum_{i=1}^{k} e^{l_G^i(x)/\epsilon} \right]$$

$$G_\epsilon(x) = G^\epsilon(x) - \epsilon \ln k$$

Define:

$$G_\epsilon = \{x \in \mathbf{X_C} \mid G_\epsilon(x) \leq 0\}$$
$$G^\epsilon = \{x \in \mathbf{X_C} \mid G^\epsilon(x) \leq 0\}$$

One can show that $G_\epsilon \subseteq G \subseteq G^\epsilon$ and[2] $\lim_{\epsilon \to 0} G_\epsilon = \lim_{\epsilon \to 0} G^\epsilon = G$. By applying Algorithm 9 to smooth inner and outer approximations of the set G, one can calculate smooth inner and outer approximations to the reach set.

A similar idea is to use ellipsoids as inner and outer approximations to the reach set [22,23]. [23] presents efficient algorithms for calculating both the minimum volume ellipsoid containing given points, and the maximum volume ellipsoid in a polyhedron, using matrix determinant maximization subject to linear matrix inequality constraints.

5   Two-Aircraft Conflict Resolution

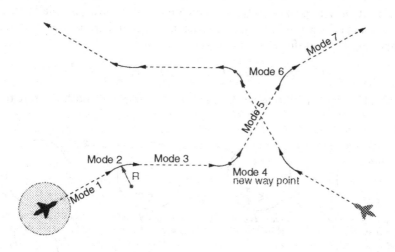

Fig. 3. Two aircraft in seven modes of operation: in modes 1, 3, 5, and 7 the aircraft follow a straight course and in modes 2, 4, and 6 the aircraft follow arcs of circles. The initial relative heading is preserved throughout.

Consider two aircraft flying in collision course on the same horizontal plane (Figure 3). To avoid the collision the aircraft go through a coordinated avoidance maneuver: when they come within a certain distance of each other, they both start to turn to the right, following a trajectory which is a sequence of arcs of circles of fixed radii, and straight lines (*trimmed flight* segments). We assume that aircraft 1 initiates the avoidance maneuver and that the aircraft communicate and switch modes simultaneously. We also assume that the angles of the avoid maneuver are fixed, so that the straight path of mode 3 is at a $-45°$ angle to the straight path of mode 1, and that of mode 5 is at a $45°$ to that of mode 1. Also, the radius of each arc is fixed at a pre-specified value, and the lengths of the segments in modes 3 and 5 are equal to each other, but unspecified. Given

---

[2] Considering appropriate topologies for 2^{X_c}.

some uncertainty in the actions of the aircraft, we would like to generate the relative distance between aircraft at which the aircraft may switch safely from mode 1 to mode 2, and the minimum lengths of the segments in modes 3 and 5, to ensure that a 5 nautical mile separation is maintained.

The system can be modeled by a hybrid automaton with seven discrete states ($\mathbf{X_D} = \{straight1, arc1, straight2, arc2, straight3, arc3, straight4\}$) and four continuous states, the relative position, (x_r, y_r), and heading, ψ_r, of the two aircraft, and a clock variable, z, to keep track of how long the aircraft have stayed in each mode. Overall, $\mathbf{X_C} = \mathbb{R}^2 \times [0, 2\pi] \times \mathbb{R}$. A discrete control input $\sigma \in \mathbf{U_D} = \{0, 1\}$ can be used to initiate the maneuver. There is also a continuous control input, the groundspeed of aircraft 1, $v_1 \in \mathbf{U_C} = [\underline{v_1}, \overline{v_1}]$ and a continuous disturbance input, the groundspeed of aircraft 2, $v_2 \in \mathbf{D} = [\underline{v_2}, \overline{v_2}]$. The speed of aircraft 2 is treated as a disturbance because we assume that aircraft 1 can estimate it only approximately. The dynamics are shown pictorially in Figure 4, in the usual location invariant, transition guard and transition reset relation convention. The unsafe set G is given by:

$$G = \mathbf{X_D} \times \{(x_r, y_r, \psi_r, z) \in \mathbf{X_C} | x_r^2 + y_r^2 \le 5^2\} \qquad (30)$$

To simplify the calculation we assume that the speed of both aircraft remains

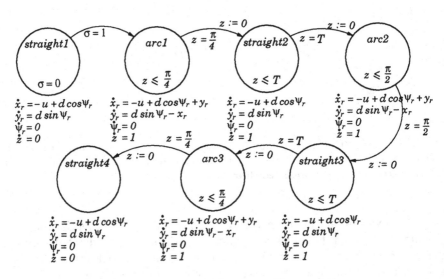

Fig. 4. Hybrid automaton model for the conflict resolution maneuver.

constant during the circular parts of the maneuver, but can take on any allowable value in the straight parts. In other words, $\phi(x) = \mathbf{U_D} \times \{(\hat{v}_1, \hat{v}_2)\}$ if $x|_{X_D} \in \{arc1, arc2, arc3\}$ and $\phi(x) = \mathbf{U_D} \times \mathbf{U_C} \times \mathbf{D}$ otherwise.

Our goal is to compute the relative distance at which the maneuver must start, the length of the straight legs *straight2* and *straight3*, as well as the groundspeed v_1^* along those legs, to ensure safety. The unsafe sets computed

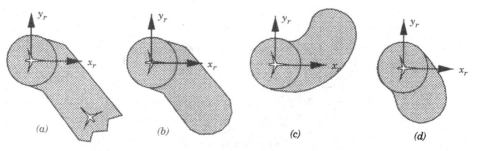

Fig. 5. $J_{G_i}(x) \leq 0$ for (a) Modes 1 and 7 ($i = 1, 7$), $\omega_1 = \omega_2 = 0$ and $[\underline{v}_1, \bar{v}_1] = [2, 4]$, $[\underline{v}_2, \bar{v}_2] = [1, 5]$ (the jagged edge means the set extends infinitely); (b) Modes 3 and 5 ($i = 3, 5$), $\omega_1 = \omega_2 = 0$ and $[\underline{v}_1, \bar{v}_1] = [2, 4]$, $[\underline{v}_2, \bar{v}_2] = [1, 5]$; (c) Mode 4 ($i = 4$), $\omega_1 = \omega_2 = 1$ and $v_1 = v_2 = 5$; and (d) Modes 2 and 6 ($i = 2, 6$), $\omega_1 = \omega_2 = -1$ and $v_1 = v_2 = 5$. In all cases, $\psi_r = 2\pi/3$.

by our algorithm for each one of the discrete modes are shown in Figure 5 for $i = 1, \ldots, 7$. In the straight modes, the sets are calculated using (v_1^*, v_2^*) of Section 4 (and thus show a close resemblance to the set in Figure 2). The corners in the set for the straight modes (1, 3, 5 and 7) indicate discontinuities in the optimal controls v_1^* and v_2^*. Figure 6 displays the fixed point $W^* = W^{-7}$ for the initial mode *straight1*. The controller that switches between the modes is also illustrated in Figure 6. The time spent in the straight legs of the maneuver T, may be chosen to minimize the deviation from the original route, while ensuring that $(W^*)^c$ is small.

6   Concluding Remarks

A common concern for all approximations for the computation of the set of reachable states is that, for safety properties, one typically would like a conservative over approximation. This will ensure that if the approximate set of reachable states satisfies the property the exact set will also satisfy it. This requirement is built into most of the approximation techniques discussed in Section 4. It is not as easy to satisfy with the level set method, however. One needs to keep accurate bounds of the numerical errors and grow the final estimate of the reach set appropriately.

An additional issue one needs to consider in the context of controller synthesis is the controlled invariance of the approximation. This issue has not to our knowledge been addressed by any of the methods proposed in the literature (since they are primarily concerned with verification). One would like to ensure that the numerical estimate of the reach set (for the case of the level set method, this could be some interpolation between the collection of discrete points produced by the algorithm) is controlled invariant. If this is indeed the case, one would also like to obtain a controller that renders the approximation invariant.

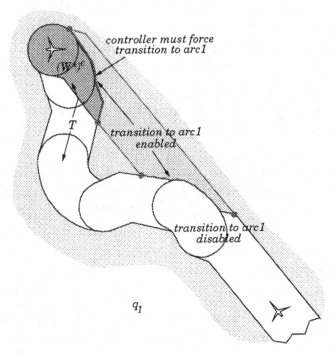

controller must force
transition to arc1

T

transition to arc1
enabled

transition to arc1
disabled

q_1

Fig. 6. $(W^*)^c = (W^{-7})^c$ in *straight1*.

References

1. W. Thomas, "On the synthesis of strategies in infinite games", in *Proceedings of STACS 95, Volume 900 of LNCS*, Ernst W. Mayr and Claude Puech, Eds., pp. 1–13. Springer Verlag, Munich, 1995.
2. T. Başar and G. J. Olsder, *Dynamic Non-cooperative Game Theory*, Academic Press, second edition, 1995.
3. O. Maler, A. Pnueli, and J. Sifakis, "On the synthesis of discrete controllers for timed systems", in *STACS 95: Theoretical Aspects of Computer Science*, Ernst W. Mayr and Claude Puech, Eds., Lecture Notes in Computer Science 900, pp. 229–242. Springer Verlag, Munich, 1995.
4. H. Wong-Toi, "The synthesis of controllers for linear hybrid automata", in *Proceedings of the IEEE Conference on Decision and Control*, San Diego, CA, 1997.
5. J. Lygeros, C. Tomlin, and S. Sastry, "Controllers for reachability specifications for hybrid systems", *Automatica*, vol. 35, no. 3, 1999, To appear.
6. C. Tomlin, J. Lygeros, and S. Sastry, "Synthesizing controllers for nonlinear hybrid systems", in *Hybrid Systems: Computation and Control*, T. Henzinger and S. Sastry, Eds., Lecture Notes in Computer Science 1386, pp. 360–373. Springer Verlag, New York, 1998.
7. J. Lygeros, C. Tomlin, and S. Sastry, "On controller synthesis for nonlinear hybrid systems", in *Proceedings of the IEEE Conference on Decision and Control*, Tampa, FL, 1998, pp. 2101–2106.

8. S. Osher and J. A. Sethian, "Fronts propagating with curvature-dependent speed: Algorithms based on Hamilton-Jacobi formulations", *Journal of Computational Physics*, vol. 79, pp. 12–49, 1988.

9. Claire J. Tomlin, *Hybrid Control of Air Traffic Management Systems*, PhD thesis, Department of Electrical Engineering, University of California, Berkeley, 1998.

10. Michael S. Branicky and Sanjoy K. Mitter, "Algorithms for optimal hybrid control", in *Proceedings of the IEEE Conference on Decision and Control*, 1995, pp. 2661–2666.

11. Adam L. Schwartz, *Theory and Implementation of Numerical Methods Based on Runge-Kutta Integration for Solving Optimal Control Problems*, PhD thesis, Department of Electrical Engineering, University of California, Berkeley, 1996.

12. M. G. Crandall and P.-L. Lions, "Viscosity solutions of Hamilton-Jacobi equations", *Transactions of the American Mathematical Society*, vol. 277, no. 1, pp. 1–42, 1983.

13. M. G. Crandall, L. C. Evans, and P.-L. Lions, "Some properties of viscosity solutions of Hamilton-Jacobi equations", *Transactions of the American Mathematical Society*, vol. 282, no. 2, pp. 487–502, 1984.

14. J. A. Sethian, *Level Set Methods: Evolving Interfaces in Geometry, Fluid Mechanics, Computer Vision, and Materials Science*, Cambridge University Press, New York, 1996.

15. R. Malladi, J. A. Sethian, and B. C. Vemuri, "Shape modeling with front propagation: A level set approach", *IEEE Transactions on Pattern Analysis and Machine Intelligence*, vol. 17, no. 2, pp. 158–175, 1995.

16. J. M. Berg, A. Yezzi, and A. R. Tannenbaum, "Phase transitions, curve evolution, and the control of semiconductor manufacturing processes", in *Proceedings of the IEEE Conference on Decision and Control*, Kobe, 1996, pp. 3376–3381.

17. A. Puri, P. Varaiya, and V. Borkar, "ε-approximation of differential inclusions", in *Proceedings of the IEEE Conference on Decision and Control*, New Orleans, LA, 1995, pp. 2892–2897.

18. T. Dang and O. Maler, "Reachability analysis via face lifting", in *Hybrid Systems: Computation and Control*, S. Sastry and T.A. Henzinger, Eds., number 1386 in LNCS, pp. 96–109. Springer Verlag, 1998.

19. M.R. Greenstreet and I. Mitchell, "Integrating projections", in *Hybrid Systems: Computation and Control*, S. Sastry and T.A. Henzinger, Eds., number 1386 in LNCS, pp. 159–174. Springer Verlag, 1998.

20. D. P. Bertsekas, *Constraint Optimization and Lagrange Multiplier Methods*, Academic Press, New York, 1982.

21. E. Polak, *Optimization: Algorithms and Consistent Approximations*, Springer Verlag, New York, 1997.

22. A. B. Kurzhanski and I. Valyi, *Ellipsoidal calculus for estimation and control*, Birkhauser, Boston, 1997.

23. L. Vandenberghe, S. Boyd, and S.-P. Wu, "Determinant maximization with linear matrix inequality constraints", *SIAM Journal on Matrix Analysis and Applications*, vol. 19, no. 2, pp. 499–533, 1998.

Stabilization of Orthogonal Piecewise Linear Systems: Robustness Analysis and Design

C.A. Yfoulis*[1], A. Muir[1], P.E. Wellstead[1], and N.B.O.L. Pettit[2]

[1] Control Systems Centre, UMIST, P.O.Box 88, Manchester, M60 1QD, U.K.
yfoulis@csc.umist.ac.uk
[2] CF-T Danfoss A/S, DK-6430, Nordborg, Denmark
pettit@danfoss.dk

Abstract. In previous work the stabilization of Orthogonal Piecewise Linear (OPL) systems was considered and a simple design technique was outlined. In this paper the problems of robustness analysis and design for OPL systems are investigated. It is shown that, due to simplicity in the algebra involved, piecewise-linear Lyapunov functions offer considerable ease in addressing robustness. Assuming real structured parametric uncertainties in general affine linear state-space models, time-varying or state-dependent uncertainties as well as modeling errors can be taken into account, while retaining the same spirit in the design procedure. Bounds for the uncertain parameters can be easily found using linear programming and the computational complexity is kept low. These issues complete the OPL framework and confirm that it constitutes a simple design technique for addressing stability, performance and robustness while taking into account control limitations.

1 Introduction

The research described here was stimulated by practical experience with Piecewise Linear (PL) systems as they arise naturally in industry. Such systems combine PL control rules with linear dynamics, are hard to analyze and can lead to unpredictable behavior. The OPL approach was conceived specifically to produce a simple design method for such PL systems.

The OPL structure, described in [10, 11] is a special PL structure obtained when the state-space is partitioned into a number of regions by means of hyperplanes parallel to the state coordinate axes. All regions formed are hyperrectangles and independent local conditions for stability can be imposed by using local PL Lyapunov functions, called in [10] Local Polyhedral (LP) Lyapunov functions . If these conditions are satisfied by applying independent local controls, then global stability is proved by constructing a global PL Lyapunov function, which is the result of "gluing" together the local functions. The design

---

* This research was supported by the EPSRC under grant GR/K 36300 and studentship Ref. No. 97700206 for the first author. The support of the UMIST graduate research fund is also acknowledged.

F.W. Vaandrager and J.H. van Schuppen (Eds.): HSCC'99, LNCS 1569, pp. 256–270, 1999.
© Springer-Verlag Berlin Heidelberg 1999

gives rise to a variable-structure and discontinuous control law. Although not explicitly considered during the design, sliding modes are likely to occur and their stability is always guaranteed. Simple linear programming is used to generate controls and the convergence rate can be maximized, leading to exponential stability with optimized performance.

PL Lyapunov functions were used in [1,6] for robust stability analysis or design. Similar ideas are further extended and studied herein. It is shown that independent local robustness analysis and design is a straightforward extension of the ideas outlined in [10,11]. After introducing a general state-space model which can accommodate uncertainties in all entries of all state-space matrices involved, it is shown how the effect of all uncertainties can be aggregated and related to the stability margin in the stability conditions.

The stability analysis problem is treated first. Bounds on allowable perturbations are found such that stability of a nominally stable system is maintained. Next the stability design problem is studied. Assuming uncertainties with known bounds, a design procedure is described which generates controls such that the stability conditions are satisfied for all values of the uncertain parameters, while control limitations are respected. Both problems can be easily solved in a computationally tractable manner, provided that the number of uncertain parameters is kept relatively small. Moreover, the non-conservative nature of the results is noted.

The paper is organized as follows: In section 2 the OPL framework introduced in [7] is briefly described. In section 3 the nice properties of PL Lyapunov functions are discussed and the robustness analysis and design problems are subsequently outlined. Finally, two illustrative examples are given in section 4 and the paper is concluded with some remarks on further developments.

2 Stabilization of OPL systems

2.1 The OPL framework

The OPL framework is described in [7,10] . Some basic concepts are repeated here for completeness.

Definition 1. *An* **orthogonal complex** *in a finite dimensional real Euclidean space $\mathbb{R}^n$ is defined to be the set $P = \{P_i \subset \mathbb{R}^n : i = 1, \ldots, (2d-1)^n\}$ of $(2d-1)^n$ bounded convex polytopes P_i (referred to as* regions *throughout) created when $2d$ $(d \in \mathbb{N})$ hyperplanes orthogonal to each state are introduced in the state-space. The hyperplanes perpendicular to the x_i coordinate take the form $x_i = c_{i,j}$ and $x_i = -c_{i,-j}$, $i = 1, 2, \ldots, n$, $j = 1, \ldots, d$, $c_{i,\pm j} > 0$. There are a total of $2(d \times n)$ hyperplanes. P_0 is the central region which contains the origin. All regions are assumed to be closed.*

Compact coverings $X_k = \prod_{i=1}^{n} \left[-c_{i,-(k+1)}, \ c_{i,(k+1)} \right]$ comprising a series of nested polytopes surrounding P_0 and corresponding *rings* $R_k = X_k \setminus X_{k-1}$ are introduced. Every region is assigned its own affine linear dynamics

$$\dot{x} = A_i\, x + B_i\, u + c_i \ , \quad x \in P_i \tag{1}$$

where $x \in \mathbb{R}^n$, $u \in \mathbb{R}^m$, $c_i \in \mathbb{R}^n$, $A_i \in \mathbb{R}^{n \times n}$, $B_i \in \mathbb{R}^{n \times m}$, $P_i \in P$. The control input u takes the form

$$u = F_i x + g_i \quad , \quad x \in P_i \tag{2}$$

where $F_i \in \mathbb{R}^{m \times n}$, $g_i \in \mathbb{R}^m$ and the closed-loop system becomes

$$\dot{x} = f(x) = (A_i + B_i F_i) x + (B_i g_i + c_i) \quad , \quad x \in P_i \tag{3}$$

It is assumed that $c_0 = 0$, $g_0 = 0$ so that $f(0) = 0$ and that the origin is the only equilibrium point for the linear system in P_0. In general, every local region P_i contains its own local control law u. The aim of the switching controller in the regions surrounding P_0 is then to act as a set of safety rules which ensure boundedness of the state inside P_0.

In terms of a hybrid model, the OPL system can be described as *autonomous switching* between a set of piecewise linear affine models [2].

2.2   Stabilization using PL Lyapunov-like functions

The problem of stabilization of OPL systems has been studied in terms of the rings present in the system. The control objective was to design each ring R_k to be *asymptotically stable*, i.e. all trajectories emanating from any point in R_k are eventually attracted to the lower ring R_{k-1}. This design requirement was systematically addressed using appropriately designed Lyapunov-like functions, called Local polyhedral (LP) and Ring polyhedral (RP)Lyapunov functions. These further specified the form of the stability conditions for every local region in the OPL system. The RP and LP polyhedral functions can be thought of as *ring-induced* and *region-induced* polyhedral functions respectively, since they are especially designed to ensure stable behaviour for those domains of the state space.

The multiple Lyapunov-like functions used in this work are similar in spirit to the ones used in [5] and certainly a special form of the general functions proposed in [2] for stability analysis of hybrid systems. However, they are used herein for stability design of a special class of hybrid systems, for which it is shown that simple design methodologies exist, due to PL dynamics and the PL form of the proposed Lyapunov functions.

The form of the RP-Lyapunov functions is quite simple and the number of facets involved is kept quite small ($2 \times n$). This is in contrast to classical approaches for control synthesis using PL-Lyapunov functions ([1] and references therein) , in which it has been proved that polyhedral Lyapunov functions are a universal class for linear uncertain systems control design, but practically this implies that the number of different linear segments must be sufficiently large and, although techniques for determination of controls exist, they give rise to complicated procedures which suffer from computational explosion w.r.t dimension.

In this work, a more practical and simple policy has been adopted. By selecting simple forms of Lyapunov functions, induced by the PL dynamics partition, we avoid computational explosion and simplify the design procedure. Moreover, in addition to stability, the form of the Lyapunov function used (PL with local validity) allows independent local design while offering ease in addressing robustness and performance in addition to stability.

However, the existence of LP and RP Lyapunov functions is not guaranteed. The method is applicable only if controllable directions can be found for all linear systems. Moreover, if a number of conditions have to be simultaneously satisfied for a single region, control solutions might not exist. Aggregation of regions is important for reducing the exponential increase of the number of regions and control rules, since we do not expect the number of partitioning hyperplanes to increase significantly with dimension in general. In [10] it is shown that due to the special OPL structure and the Lyapunov function selection, computation demands are polynomial in the state dimension n, provided that the total number of regions is not too high. Simple linear programming can be used for generating controls.

Two **sufficient** conditions, regarding the asymptotic stabilization of a ring and the whole OPL system have been stated in [10].

Proposition 1. *A ring R_k is asymptotically stable if there exists an RP-Lyapunov function $V_k(x)$ for that ring.*

Proposition 2. *An OPL system is asymptotically stable if every ring R_k of the system is asymptotically stable.*

Proof. When all rings are asymptotically stable, transition from the upper rings to the lower ones occurs on the common boundary between them and it is obvious that the global behaviour is absolutely predictable. This implies that phenomena such as limit cycles, chaotic motion, unstable sliding motion, parasitic equilibria etc. are not possible. Thus, we conclude that the system is asymptotically stable, i.e. all trajectories finally enter the central region in finite time and are eventually settled at the origin.

3   Robustness issues

3.1   Robust stability and PL Lyapunov functions

Robust stability is often addressed in the *robust control Lyapunov function* (*RCLF*) framework. Assume the system dynamics are given in the form ([4])

$$\dot{x} = f(x) + g_w(x)w + g_u(x)u \qquad (4)$$

Given a locally Lipschitz, proper and positive definite scalar function $V : \mathbb{R}^n \to \mathbb{R}$, its *level set* $V^{-1}[c_1, c_2] \doteq \{x \in \mathbb{R}^n : c_1 \leq V(x) \leq c_2\}$, uncertainties w in a polytope $\mathcal{W} \subset \mathbb{R}^l$ and control limitations $u(t) \in \mathcal{U} \subset \mathbb{R}^m$ we repeat the following definition ([4]) :

Definition 2. *Consider a positive definite function* $W(x)$. *The function* $V(x)$ *is a* robust control Lyapunov function (RCLF) *with stability margin* $W(x)$ *with controls in* $\mathcal{U}$ *over* $V^{-1}[c_1, c_2]$ *for the system (4) if there exists a control law* $\mu : \mathbb{R}^n \to \mathcal{U}$ *such that*

$$\sup_{x \in V^{-1}[c_1, c_2]} \max_{w \in \mathcal{W}} \{ L_f V(x) + L_{g_w} V(x) w + L_{g_u} V(x) \mu(x) + W(x) \} \leq 0 \quad (5)$$

where $L_f V(x)$ the *Lie derivative* of $V(x)$ along $f(x)$.

For general nonlinear systems, stability analysis based on evaluation of (5) suffers from high computational complexity, especially when the form of the Lyapunov function is not exactly specified and needs to be designed as well (usually using an iterative procedure).

In the following, affine piecewise linear uncertain dynamics and polytopic Lyapunov functions are considered. Polytopic functions of the form

$$V(x) = \max_{1 \leq i \leq s} \{ \gamma_i^T \cdot x \} \quad (6)$$

divide the state-space into a number of cones in which $V(x) = \gamma_i^T \cdot x$ is linear. The stability conditions imposed locally in every such cone $\mathcal{C}_i$ are

$$\gamma_i^T \cdot \dot{x} < 0 \quad \forall x \in \mathcal{C}_i \quad (7)$$

A useful state-space model for structured uncertainty is to consider a number of uncertain parameters affecting the entries of the matrices involved. Let the linear dynamics in some $\mathcal{C}_i$ be

$$\dot{x} = A \cdot x + B \cdot u + c \quad \forall x \in \mathcal{C}_i \quad (8)$$

with

$$A = A^{(n)} + \sum_{i=1}^{n_A} \kappa_i E_i^{(A)} , \quad B = B^{(n)} + \sum_{j=1}^{n_B} \lambda_j E_j^{(B)} ,$$

$$c = c^{(n)} + \sum_{l=1}^{n_c} \mu_l e_l^{(c)} \quad (9)$$

where $A^{(n)}$, $B^{(n)}$, $c^{(n)}$ are the nominal values and κ_i, λ_j, μ_l are uncertain, state-dependent or time-varying parameters affecting some or all of the entries of the nominal matrices, as specified in the $E_i^{(A)}$, $E_j^{(B)}$, $e_l^{(c)}$ matrices. This framework is a generalization of the ideas in [3,12]. All parameters can affect more than one entry simultaneously. The uncertain parameters are considered independent of each other and take values in bounded intervals

$$\kappa_i \in [\kappa_i^-, \kappa_i^+] \quad i = 1, \ldots, n_A$$

$$\lambda_j \in [\lambda_j^-, \lambda_j^+] \quad j = 1, \ldots, n_B \quad (10)$$

$$\mu_l \in [\mu_l^-, \mu_l^+] \quad l = 1, \ldots, n_c$$

Under the assumption that the perturbed matrices depend linearly on the parameters this framework is quite general, i.e. it can cover a large number of possibilities. In addition, all uncertain parameters can represent external physical variables of interest.

Let us consider a convex, closed and bounded region $\mathcal{R} \subseteq \mathcal{C}_i \subset \mathbb{R}^n$. Then the polytopic Lyapunov function is linear $\forall x \in \mathcal{R}$, say $V(x) = \gamma^T \cdot x$ and the stability condition (7) becomes

$$(\gamma^T \cdot A) \cdot x + (\gamma^T \cdot B) \cdot u + (\gamma^T \cdot c) < 0 \quad \forall x \in \mathcal{R} \qquad (11)$$

If a local controller $u = F \cdot x + g$ has been designed for $\mathcal{R}$ then (11) yields

$$\left[\gamma^T \cdot (A + B \cdot F)\right] \cdot x + \left[\gamma^T \cdot (c + B \cdot g)\right] < 0 \quad \forall x \in \mathcal{R} \qquad (12)$$

From (9),(12)

$$p^T \cdot x + q + \sum_i \kappa_i \cdot \left(e_i^{(A)} \cdot x\right)$$
$$+ \sum_j \lambda_j \cdot \left(e_j^{(B)} \cdot x + r_j\right) + \sum_l s_l \cdot \mu_l < 0 \quad \forall x \in \mathcal{R} \quad (13)$$

where

$$p^T = \gamma^T \cdot (A^{(n)} + B^{(n)} F), \; q = \gamma^T \cdot (c^{(n)} + B^{(n)} g)$$
$$e_i^{(A)} = \gamma^T \cdot E_i^{(A)}, \; e_j^{(B)} = \gamma^T \cdot E_j^{(B)} \cdot F \qquad (14)$$
$$r_j = \gamma^T \cdot E_j^{(B)} \cdot g, \; s_l = \gamma^T \cdot e_l^{(c)}$$

Assuming a *polytopic* $\mathcal{R}$ the following result is obtained

Proposition 3. *The local stability condition (13) is satisfied $\forall x \in \mathcal{R}$ if and only if it is satisfied for all vertices of $\mathcal{R}$.*

Proof. Necessity : Obvious, since the vertices belong to the region.

Sufficiency : Any point $x \in \mathcal{R}$ can be written as

$$x = \sum_{\nu=1}^{n_v} \rho_\nu \cdot x_\nu , \; \sum_\nu \rho_\nu = 1, \; \rho_\nu \geq 0 \; \forall \nu$$

where n_v is the number of vertices of $\mathcal{R}$. If (13) is satisfied for all vertices x_ν, i.e. $\mathcal{A}(x_\nu) < 0$ (where $\mathcal{A}(x)$ the left-hand side expression in (13)) then

$$\mathcal{A}(x) = \sum_{\nu=1}^{n_v} \rho_\nu \cdot \mathcal{A}(x_\nu) < 0 \quad \forall x \in \mathcal{R} \qquad (15)$$

as can be easily verified from the linear form of $\mathcal{A}(x)$.

Let us denote by w the vector of all uncertain parameters

$$w = [\kappa_1 \ldots \kappa_{n_A} \lambda_1 \ldots \lambda_{n_B} \mu_1 \ldots \mu_{n_c}]^T \qquad (16)$$

Following the result of Proposition 3 a set of linear inequalities in w is obtained from (13) (one for each vertex) which further specifies a convex polyhedral region in the parameter's space for which the stability condition (12) is satisfied. This is due to the simple algebraic conditions resulting from the use of polyhedral Lyapunov functions . Moreover, the resulting polyhedral region specifies the solution in an exact and non-conservative (necessary and sufficient) manner as follows easily from Proposition 3. This region contains all possible combinations of bounding interval solutions (of the form (10)) for robust stability and is further exploited in the next sections for the OPL analysis and design problems.

3.2   Robust OPL analysis

We consider the following problem :

Problem 1. Consider an OPL system and a control law u that satisfies the stability properties in a local region $P_i \in P$ for the nominal system. The **robust OPL analysis problem** consists of determining allowable bounds for the uncertain parameters of the perturbed system (9) such that local stability is preserved.

In an OPL setting, analysis or design is carried out locally and independently in each region. Let S_p, $p = 1, \ldots, n_p$ denote the sectors in which P_i can be divided [10]. Every sector has its own local condition in the form

$$\gamma_p^T \cdot \dot{x} < 0 , \quad \forall x \in S_p \qquad (17)$$

Let the linear system dynamics and feedback law in P_i be

$$\dot{x} = A x + B u + c , u = F x + g , \quad x \in S_p \qquad (18)$$

so that the closed-loop system becomes

$$\dot{x} = A_c x + b_c , \quad x \in S_p \qquad (19)$$

It follows easily from the discussion in the previous section that a set of linear inequalities in the uncertain parameters can be obtained for every sector. If these are subsequently combined, i.e. the intersection of the feasible regions for each sector is taken, a feasible polyhedral region in the parameter's space is defined. Bounded intervals for the uncertain parameters can be easily specified by fitting hyper-rectangles into the feasible region. An infinite number of possible solutions of course exists and many techniques for specifying them including linear programming can be used. The idea is demonstrated in the following simple example :

Example 1 : Consider the region shown in Figure 1 with the following controlled dynamics

$$\dot{x} = A_c \cdot x , \quad A_c = A_c^{(n)} + \kappa_1 \cdot E_1 + \kappa_2 \cdot E_2 \qquad (20)$$

where

$$A_c^{(n)} = \begin{pmatrix} -3 & -2 \\ 1 & 0 \end{pmatrix}, \ E_1 = \begin{pmatrix} -1 & 1 \\ 0 & 0 \end{pmatrix}, \ E_2 = \begin{pmatrix} 1 & 0 \\ 0 & 1 \end{pmatrix} \qquad (21)$$

Two uncertain or time-varying parameters κ_1, κ_2 are considered. Both affect the element $A_c(1,1)$. The stability condition in sector S_1 is $\dot{x}_1 < 0$, $\forall x \in S_1$ and it is satisfied for the nominal system matrix $A_c^{(n)}$. When the perturbed matrix A_c is considered the stability conditions for all the vertices determine the feasible region shown in Figure 1. Two interval solutions (which can be readily found using linear programming) are also shown.

The dotted line shown corresponds to the new bounds of the feasible region if the values -3 and -2 in the first row of $A_c^{(n)}$ are replaced by -2 and -1. The stability condition is then satisfied for the nominal system and the feasible region is reduced in size. It is interesting to observe that the orientation of the bounding lines is maintained and only their position is modified, i.e. they are shifted. Moreover, the rectangles determining the parameter intervals are not bounded from below.

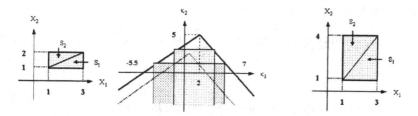

Fig. 1. OPL region of example 1 (left) and the fea- **Fig. 2.** OPL region of example 2 sible region in the parameter's space (right)

3.3 Robust OPL design

For control design, we consider the following :

Problem 2. Consider a local region $P_i \in P$ in an uncontrolled OPL system (18) in which parametric uncertainty is specified as in (9) with known intervals (10) for the uncertain parameters. The **robust OPL design problem** consists of finding a feedback control law u such that

- The local stability condition in P_i is satisfied for all values of the uncertain parameters;
- Control limitations $|u| < U$ (i.e. $|u_i| < U_i$ for all its components) are respected;
- The rate of convergence (which is a measure of the system's local performance) is maximized.

Let us recall (13) in which $e_j^{(B)}(F)$, $r_j(g)$ $\forall j$ need to be designed according to the requirements of Problem 2. The vector of uncertain parameters w satisfies $w \in W$, where W contains all values for the uncertain parameters in the bounded intervals. As described in Proposition 3, (13) has to be satisfied at all vertices of the region of interest. Thus, for every sector S_p of P_i we require

$$\max_{w \in W} \left\{ p^T \cdot x_\nu + q + \sum_i \kappa_i \cdot \left(e_i^{(A)} \cdot x_\nu \right) \right.$$
$$\left. + \sum_j \lambda_j \cdot \left(e_j^{(B)} \cdot x_\nu + r_j \right) + \sum_l s_l \cdot \mu_l \right\} < 0 \quad \forall x_\nu \in S_p \quad (22)$$

where x_ν, $\nu = 1, \ldots, n_\nu$ denote the vertices of S_p. If $w = [w_A \ w_B \ w_c]$ is divided in three parts containing separately the parameters affecting A, B, c, (22) becomes

$$\max_{w_A \in W_A} \{K\} + \max_{w_B \in W_B} \{\Lambda\} + \max_{w_c \in W_c} \{M\} < 0 \quad \forall x_\nu \in S_p \quad (23)$$

where W_A, W_B, W_c contain the corresponding sets of intervals. The maximization in (23) can be performed separately since the uncertain parameters are independent. It is

$$\max_{w_A \in W_A} \{K\} = p_1^T \cdot x_\nu + \sum_i \max\{\zeta_i \kappa_i^-, \zeta_i \kappa_i^+\} \quad (24)$$

and

$$\max_{w_c \in W_c} \{M\} = q_1 + \sum_l \max\{s_l \mu_l^-, s_l \mu_l^+\} \quad (25)$$

where $p_1^T = \gamma_p^T A^{(n)}$, $q_1 = \gamma_p^T c^{(n)}$, $\zeta_i = e_i^{(A)} \cdot x_\nu$. Both quantities can be calculated easily prior to control law selection. The second quantity contains the control elements F, g to be found and is written as

$$\max_{w_B \in W_B} \{\Lambda\} = p_2^T \cdot x_\nu + q_2 + \sum_j [\lambda_j^+ \Theta_j^+ - \lambda_j^- \Theta_j^-] \quad (26)$$

where $p_2^T = \gamma_p^T B^{(n)} F$, $q_2 = \gamma_p^T B^{(n)} g$ and
$V_j = e_j^{(B)} \cdot x_\nu + r_j = \Theta_j^+ - \Theta_j^-$, $\Theta_j^+, \Theta_j^- \geq 0$. The new auxiliary variables Θ_j^+, Θ_j^- are added so that the partial maxima in the sum can be calculated with respect to the unknown controls F, g. It is a compact technique (see also [10]) that makes the calculation of (23) straightforward and it is suitable for linear programming implementation. Note that (23)– (26) correspond to a *necessary and sufficient* condition for a solution to the *robust* OPL design problem. Control limitations can be easily added as extra conditions in the linear program.

Furthermore, instead of $\dot{V}(x) < 0$ the condition $\dot{V}(x) < -\epsilon V(x)$ is implemented in the linear program and the maximum ϵ is searched for. This corresponds to maximization of the decay rate, i.e. to optimal performance. Thus, robustness and performance can be addressed in this simple manner. In addition, the implementation is computationally tractable, since only the vertices of the region of interest need to be checked.

The stability margin mentioned in (5) can be directly related to robustness and performance. (13) becomes

$$\mathcal{A}^{(n)}(x) + W(x) < 0 \text{ with } W(x) = W_1(x) + W_2(x) \qquad (27)$$

where $\mathcal{A}^{(n)}(x) = p^T x + q$ corresponds to nominal values , the maximum bound due to the uncertainties is given by $W_1(x) = \max_{w \in W}\{\mathcal{A}(x)\} - \mathcal{A}^{(n)}(x)$ and $W_2(x) = \epsilon \cdot V(x)$ relates to the maximum decay rate. These quantities provide the means for quantifying robustness and performance and determine certain trade-offs between them. In addition, the total stability margin $W(x)$ can be related to the available control energy. Thus, all important qualities and limitations can be linked in this simple manner and utilized during the design procedure for simpler and more effective design.

In the following example the previously mentioned concepts are applied in a simple case :

Example 2 : Consider the region shown in Figure 2 with the following uncontrolled dynamics

$$\dot{x} = A^{(n)} \cdot x + b^{(n)} \cdot u + c^{(n)} \qquad (28)$$

where

$$A^{(n)} = \begin{pmatrix} 2 & 1 \\ 0 & 3 \end{pmatrix}, \, b^{(n)} = \begin{pmatrix} 1 \\ 1 \end{pmatrix}, \, c^{(n)} = \begin{pmatrix} 0.5 \\ -0.3 \end{pmatrix} \qquad (29)$$

Let us assume independent uncertainties for all entries of $A^{(n)}$, $b^{(n)}$, $c^{(n)}$, e.g. 10% symmetric uncertainty. Then the following uncertain parameters can be assigned

$$\kappa_1 \in [-0.2\,, 0.2]\,, \ \kappa_2 \in [-0.1\,, 0.1]\,, \ \kappa_3 \in [-0.1\,, 0.1]\,,$$
$$\kappa_4 \in [-0.3\,, 0.3]\,, \ \lambda_1 \in [-0.1\,, 0.1]\,, \ \lambda_2 \in [-0.1\,, 0.1]\,, \qquad (30)$$
$$\mu_1 \in [-0.05\,, 0.05]\,, \ \mu_2 \in [-0.03\,, 0.03]$$

so that the following perturbed matrices are obtained

$$A = A^{(n)} + \sum_{i=1}^{4} \kappa_i E_i^{(A)} \,, \ B = B^{(n)} + \sum_{j=1}^{2} \lambda_j E_j^{(B)} \,,$$
$$c = c^{(n)} + \sum_{l=1}^{2} \mu_l e_l^{(c)} \qquad (31)$$

where the uncertainty structure is given by

$$E_1^{(A)} = \begin{pmatrix} 1 & 0 \\ 0 & 0 \end{pmatrix}, \, E_2^{(A)} = \begin{pmatrix} 0 & 1 \\ 0 & 0 \end{pmatrix}, \, E_3^{(A)} = \begin{pmatrix} 0 & 0 \\ 1 & 0 \end{pmatrix},$$
$$E_4^{(A)} = \begin{pmatrix} 0 & 0 \\ 0 & 1 \end{pmatrix}, \, e_1^{(b)} = e_1^{(c)} = \begin{pmatrix} 1 \\ 0 \end{pmatrix}, \, e_2^{(b)} = e_2^{(c)} = \begin{pmatrix} 0 \\ 1 \end{pmatrix} \qquad (32)$$

The stability conditions are $\dot{x}_1 < 0 \; \forall \boldsymbol{x} \in S_1$ and $\dot{x}_2 < 0 \; \forall \boldsymbol{x} \in S_2$ and they are not satisfied with the nominal values. Using a linear program we found that $U \geq 11$ for satisfying the condition in S_1 assuming nominal values and $U \geq 13$ with 10% uncertainty. Similarly for S_2 $U \geq 12$ and $U \geq 15$ respectively. Typical controllers and decay rates found (assuming $|u| \leq 15$) are $u_1 = -3.17 \cdot x_1 - 1.22 \cdot x_2 - 0.611$, $u_2 = -0.11 \cdot x_1 - 3.74 \cdot x_2 + 0.3$, $\epsilon_1 = 0.65$, $\epsilon_2 = 0.0675$. These numbers quantify performance locally. This simple example shows the simplicity in addressing stability, robustness and performance issues using the proposed technique.

4   Examples

Two illustrative examples are next given. The first is a linear system containing two PL elements ("hard" nonlinearities) while the second is a smooth nonlinear system approximated in a piecewise-linear form.

4.1   Mechatronic actuator

A simplified model of a mechatronic actuator [6] is assumed, in which the aim is high positioning precision with no overshoot and robustness with respect to the unknown sliding friction F_R. F_L is a constant external force. The system is shown in Figure 3. The dynamics in state space form are

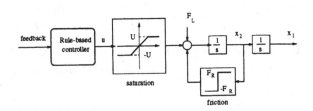

Fig. 3. The simplified mechatronic actuator model

$$\dot{x}_1 = x_2$$
$$\dot{x}_2 = F_L - F_R \cdot sign(x_2) + sat(u)$$

where $x_1 = x, x_2 = \dot{x}$ are position and velocity respectively. By taking into account the control limitations and applying a similarity transformation $y_1 = x_1 + x_2$, $y_2 = x_2$ in order to make both coordinate axes controllable (see [10]) the system's state-space equation becomes

$$\dot{\boldsymbol{y}} = \begin{pmatrix} 0 & 1 \\ 0 & 0 \end{pmatrix} \cdot \boldsymbol{y} + \begin{pmatrix} 1 \\ 1 \end{pmatrix} \cdot u + \begin{pmatrix} 1 \\ 1 \end{pmatrix} \cdot c \tag{33}$$

and now both directions are controllable. Without loss of generality, we assume saturation bounds $|u| \leq 10$ and a region of interest given by $-5 \leq y_i \leq 5$, $i =$

1, 2. By splitting the region of interest into 9 regions we create a ring and form an OPL system with its associated RP-Lyapunov function. The initial and transformed working domains together with the level surfaces of the corresponding Lyapunov functions are shown in Figure 4. Two different design techniques have

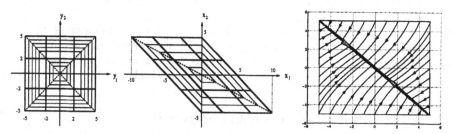

Fig. 4. The transformed and actual working domains and corresponding RP-Lyapunov functions

Fig. 5. Simulated trajectories of the controlled mechatronic actuator using constant controls

been used in [10], namely constant controls and state feedback. The robustness ideas discussed in previous sections were applied in order to maximize the stability margin under the given control saturation limitations. Thus, optimal performance and robustness to uncertain parameters of interest have been achieved. For more details the reader can refer to [10].

The simulated trajectories of the controlled system using constant controls are shown in Figure 5. It is essentially a sliding mode controller along the line $y_1 + y_2 = 0$ or $\dot{x} + 0.5x = 0$, which is a stable sliding surface. Thus, in the attempt to generate the most robust controller based on the stability conditions imposed by the use of a PL Lyapunov function and under some limitations we ended up with a sliding mode controller, having very good robustness properties, as can be verified using simulations with large disturbances. This control law is much simpler compared to the one used in [6] and other highly complicated controllers. The system can be also controlled using state-feedback including the central region. The design is not going to be as robust as with constant controls, as can be easily verified by reapplying the same conditions. However, the linear programs will optimize the stability margin, thus resulting in maximum robustness margins as well.

The simulated trajectories of the controlled system are shown in Figure 6. Note that sliding modes are present. Sliding modes can be avoided by modifying the control law in two regions only. The new simulated trajectories are shown in Figure 7.

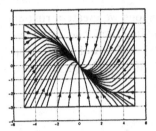

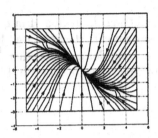

Fig. 6. Simulated trajectories of the controlled mechatronic actuator using state feedback

Fig. 7. Simulated trajectories of the controlled mechatronic actuator using state feedback and extra conditions for sliding mode avoidance

4.2   A Mass-spring-damper nonlinear system

The nonlinear system considered in [9], p.8 is studied here. It is a mass-spring-damper nonlinear system governed by the following equation of motion

$$\ddot{x} = -0.1 \cdot \dot{x}^3 - 0.02 \cdot x - 0.67 \cdot x^3 + u \tag{34}$$

in which the stiffness coefficient of the spring and the damping coefficient of the damper are nonlinear. The aim of this example is to show how the OPL ideas can be applied to smooth nonlinear systems and generate simple control laws.

The system equations in state-space form are

$$\dot{x}_1 = x_2$$
$$\dot{x}_2 = -0.1 \cdot x_2^3 - 0.02 \cdot x_1 - 0.67 \cdot x_1^3 + u$$

where $x_1 = x$ and $x_2 = \dot{x}$ denote again position and velocity respectively. A similarity transformation $y_1 = 2 \cdot x_1 + x_2$, $y_2 = x_2$ is applied for obtaining two controllable directions y_1, y_2.

The initial region of interest $-1.5 \le x_i \le 1.5$, $i = 1,2$ [9] becomes $-4.5 \le y_1 \le 4.5$, $-1.5 \le y_2 \le 1.5$ (see Figure 8). The new working domain is divided into three rings for subsequent OPL design. It is shown in Figure 9. The transformed dynamics are

$$\dot{y}_1 = -0.01 \cdot y_1 + 2.01 \cdot y_2 + u + g(\boldsymbol{y})$$
$$\dot{y}_2 = -0.01 \cdot y_1 + 0.01 \cdot y_2 + u + g(\boldsymbol{y})$$

where $g(\boldsymbol{y}) = -0.67 \cdot [\frac{1}{2}(y_1 - y_2)^3] - 0.1 \cdot y_2^3$. The nonlinear dynamics can be approximated as linear in every OPL region by using least squares. The error of the approximation is taken into account by considering an uncertain parameter with appropriate bounds in the constant term of the linear equation. E.g. if $g(\boldsymbol{y})$ is approximated as $g(\boldsymbol{y}) = r_1 \cdot y_1 + r_2 \cdot y_2 + r_3$ in some OPL region with error bounds $e \in [e^-, e^+]$ then the equations become

$$\dot{\boldsymbol{y}} = \begin{pmatrix} r_1 - 0.01 & r_2 + 2.01 \\ r_1 - 0.01 & r_2 + 0.01 \end{pmatrix} \cdot \boldsymbol{y} + \begin{pmatrix} 1 \\ 1 \end{pmatrix} \cdot u + \begin{pmatrix} 1 \\ 1 \end{pmatrix} \cdot r_3$$

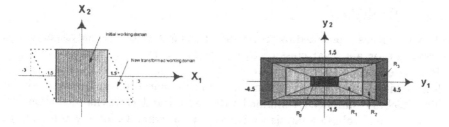

Fig. 8. Initial and transformed working domains   **Fig. 9.** OPL partition in three rings

and an uncertain parameter $\mu_1 \in [e^-, e^+]$ affecting r_3 is introduced. Thus, the ideas of robust OPL design can be comfortably applied and the approximation errors are treated as uncertainties.

Although a large number of OPL regions results when considering 3 rings (Figure 9), an attempt to reduce the number of control laws by aggregating many of them has been made. It was found that simple constant controls can robustly stabilize the system. A discontinuous constant control law

$$u = \begin{cases} +u_i & \text{if } y_1 + 3y_2 \leq 0, \\ -u_i & \text{if } y_1 + 3y_2 > 0 \end{cases} \tag{35}$$

has been chosen, where u_i is the constant value used in ring R_i. In this example $u_0 = 5$, $u_1 = 8$, $u_2 = 12$, $u_3 = 18$. Discontinuous switching along the line $y_1 + 3y_2 = 0$ leads to a stable sliding mode. By smoothing the switching using a hyperbolic tangent function [8] smooth dynamical behaviour is obtained, while retaining the robustness properties of the sliding mode strategy. The trajectories of the controlled system using many initial conditions on the outer boundary of the uppermost ring are shown in Figure 10.

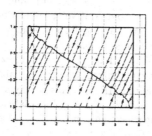

Fig. 10. Simulated trajectories of the controlled system

The main advantages of this control law over the control technique proposed in [9] are its simplicity, better robustness and performance properties and the ability to quantify robustness in a more straightforward and non-conservative manner.

5 Conclusions

This paper has attempted to provide solutions to the robust analysis and design problems for a special class of Piecewise Linear systems, namely Orthogonal Piecewise Linear systems. The use of PL Lyapunov functions has been shown to lead to simple stability conditions, linear in the uncertain parameters involved. The algebraic simplicity combined with the linear local dynamics allow independent local robust analysis and design in a computationally tractable and non-conservative manner. The stability margin can be related to the robustness and performance properties in a flexible manner, rendering the control technique proposed as simple and effective. Although the OPL framework has been inspired by practical systems containing "hard" nonlinearities, the technique can be also applied to smooth nonlinear systems and approximation errors are taken easily into account.

Further research will be devoted to the generalization of the ideas to larger classes of PL and Hybrid systems.

References

1. Blanchini, F. : Nonquadratic Lyapunov functions for Robust Control. Automatica **31(3)** (1995) 451-461
2. Branicky, M. : Multiple Lyapunov functions and Other Analysis Tools for Switched and Hybrid Systems. IEEE Trans. Automatic Control **43(4)** (1998) 475-482
3. Gardiner, J. : Computation of Stability Robustness Bounds for State-Space Models with Structured Uncertainty. IEEE Trans. Automatic Control **42(2)** (1997) 253-256
4. McConley, M.W., Dahleh, M.A., Feron E. : Polytopic Control Lyapunov Functions for Robust Stabilization of a Class of Nonlinear Systems. In Proc. ACC' 97 416-419
5. Johansson, M., Rantzer A. : Computation of Piecewise Quadratic Lyapunov functions for Hybrid Systems. IEEE Trans. Automatic Control **43(4)** (1998) 555-559
6. Kiendl, H., Ruger J.J. : Stability analysis of fuzzy control systems using facet functions. Fuzzy sets and systems **70** (1995) 275-285
7. Pettit, N.B.O.L., Muir A. : Simple control of nonlinear systems. In Proc. ECC' 97
8. Song, G., Mukherjee R. : A comparative study of Conventional Nonsmooth Time-Invariant and Smooth Time-Varying Robust compensators. IEEE Trans. Control Systems Technology **6:4** (1998) 571-576
9. Tanaka, K., Ikeda T., Wang H.O. : Robust stabilization of a class of uncertain nonlinear systems via fuzzy control: Quadratic stabilizability, H_∞ control theory and linear matrix inequalities. IEEE Trans. Fuzzy systems **4:1** (1996) 1-13
10. Yfoulis, C.A., Muir, A., Pettit, N.B.O.L., Wellstead, P.E. : Stabilization of Orthogonal Piecewise Linear systems using Piecewise-Linear Lyapunov-like functions. Control Systems Centre Internal report 875 (1998). WWW home page: http://www.csc.umist.ac.uk
11. Yfoulis, C.A., Muir, A., Pettit, N.B.O.L., Wellstead, P.E. : Stabilization of Orthogonal Piecewise Linear systems using Piecewise-Linear Lyapunov-like functions. In Proc. CDC 98 (1998)
12. Zhou, K., Khargonekar, P.P. : Stability Robustness Bounds for Linear State-Space Models with Structured Uncertainty. IEEE Trans. Automatic Control **32(7)** (1987) 621-623

Author Index

Springer
and the
environment

At Springer we firmly believe that an
international science publisher has a
special obligation to the environment,
and our corporate policies consistently
reflect this conviction.

We also expect our business partners –
paper mills, printers, packaging
manufacturers, etc. – to commit
themselves to using materials and
production processes that do not harm
the environment. The paper in this
book is made from low- or no-chlorine
pulp and is acid free, in conformance
with international standards for paper
permanency.

Springer

Lecture Notes in Computer Science

For information about Vols. 1–1488
please contact your bookseller or Springer-Verlag

Vol. 1489: J. Dix, L. Fariñas del Cerro, U. Furbach (Eds.), Logics in Artificial Intelligence. Proceedings, 1998. X, 391 pages. 1998. (Subseries LNAI).

Vol. 1490: C. Palamidessi, H. Glaser, K. Meinke (Eds.), Principles of Declarative Programming. Proceedings, 1998. XI, 497 pages. 1998.

Vol. 1491: W. Reisig, G. Rozenberg (Eds.), Lectures on Petri Nets I: Basic Models. XII, 683 pages. 1998.

Vol. 1492: W. Reisig, G. Rozenberg (Eds.), Lectures on Petri Nets II: Applications. XII, 479 pages. 1998.

Vol. 1493: J.P. Bowen, A. Fett, M.G. Hinchey (Eds.), ZUM '98: The Z Formal Specification Notation. Proceedings, 1998. XV, 417 pages. 1998.

Vol. 1494: G. Rozenberg, F. Vaandrager (Eds.), Lectures on Embedded Systems. Proceedings, 1996. VIII, 423 pages. 1998.

Vol. 1495: T. Andreasen, H. Christiansen, H.L. Larsen (Eds.), Flexible Query Answering Systems. IX, 393 pages. 1998. (Subseries LNAI).

Vol. 1496: W.M. Wells, A. Colchester, S. Delp (Eds.), Medical Image Computing and Computer-Assisted Intervention – MICCAI'98. Proceedings, 1998. XXII, 1256 pages. 1998.

Vol. 1497: V. Alexandrov, J. Dongarra (Eds.), Recent Advances in Parallel Virtual Machine and Message Passing Interface. Proceedings, 1998. XII, 412 pages. 1998.

Vol. 1498: A.E. Eiben, T. Bäck, M. Schoenauer, H.-P. Schwefel (Eds.), Parallel Problem Solving from Nature – PPSN V. Proceedings, 1998. XXIII, 1041 pages. 1998.

Vol. 1499: S. Kutten (Ed.), Distributed Computing. Proceedings, 1998. XII, 419 pages. 1998.

Vol. 1500: J.-C. Derniame, B.A. Kaba, D. Wastell (Eds.), Software Process: Principles, Methodology, and Technology. XIII, 307 pages. 1999.

Vol. 1501: M.M. Richter, C.H. Smith, R. Wiehagen, T. Zeugmann (Eds.), Algorithmic Learning Theory. Proceedings, 1998. XI, 439 pages. 1998. (Subseries LNAI).

Vol. 1502: G. Antoniou, J. Slaney (Eds.), Advanced Topics in Artificial Intelligence. Proceedings, 1998. XI, 333 pages. 1998. (Subseries LNAI).

Vol. 1503: G. Levi (Ed.), Static Analysis. Proceedings, 1998. IX, 383 pages. 1998.

Vol. 1504: O. Herzog, A. Günter (Eds.), KI-98: Advances in Artificial Intelligence. Proceedings, 1998. XI, 355 pages. 1998. (Subseries LNAI).

Vol. 1505: D. Caromel, R.R. Oldehoeft, M. Tholburn (Eds.), Computing in Object-Oriented Parallel Environments. Proceedings, 1998. XI, 243 pages. 1998.

Vol. 1506: R. Koch, L. Van Gool (Eds.), 3D Structure from Multiple Images of Large-Scale Environments. Proceedings, 1998. VIII, 347 pages. 1998.

Vol. 1507: T.W. Ling, S. Ram, M.L. Lee (Eds.), Conceptual Modeling – ER '98. Proceedings, 1998. XVI, 482 pages. 1998.

Vol. 1508: S. Jajodia, M.T. Özsu, A. Dogac (Eds.), Advances in Multimedia Information Systems. Proceedings, 1998. VIII, 207 pages. 1998.

Vol. 1510: J.M. Zytkow, M. Quafafou (Eds.), Principles of Data Mining and Knowledge Discovery. Proceedings, 1998. XI, 482 pages. 1998. (Subseries LNAI).

Vol. 1511: D. O'Hallaron (Ed.), Languages, Compilers, and Run-Time Systems for Scalable Computers. Proceedings, 1998. IX, 412 pages. 1998.

Vol. 1512: E. Giménez, C. Paulin-Mohring (Eds.), Types for Proofs and Programs. Proceedings, 1996. VIII, 373 pages. 1998.

Vol. 1513: C. Nikolaou, C. Stephanidis (Eds.), Research and Advanced Technology for Digital Libraries. Proceedings, 1998. XV, 912 pages. 1998.

Vol. 1514: K. Ohta, D. Pei (Eds.), Advances in Cryptology – ASIACRYPT'98. Proceedings, 1998. XII, 436 pages. 1998.

Vol. 1515: F. Moreira de Oliveira (Ed.), Advances in Artificial Intelligence. Proceedings, 1998. X, 259 pages. 1998. (Subseries LNAI).

Vol. 1516: W. Ehrenberger (Ed.), Computer Safety, Reliability and Security. Proceedings, 1998. XVI, 392 pages. 1998.

Vol. 1517: J. Hromkovič, O. Sýkora (Eds.), Graph-Theoretic Concepts in Computer Science. Proceedings, 1998. X, 385 pages. 1998.

Vol. 1518: M. Luby, J. Rolim, M. Serna (Eds.), Randomization and Approximation Techniques in Computer Science. Proceedings, 1998. IX, 385 pages. 1998.

1519: T. Ishida (Ed.), Community Computing and Support Systems. VIII, 393 pages. 1998.

Vol. 1520: M. Maher, J.-F. Puget (Eds.), Principles and Practice of Constraint Programming - CP98. Proceedings, 1998. XI, 482 pages. 1998.

Vol. 1521: B. Rovan (Ed.), SOFSEM'98: Theory and Practice of Informatics. Proceedings, 1998. XI, 453 pages. 1998.

Vol. 1522: G. Gopalakrishnan, P. Windley (Eds.), Formal Methods in Computer-Aided Design. Proceedings, 1998. IX, 529 pages. 1998.

Vol. 1524: G.B. Orr, K.-R. Müller (Eds.), Neural Networks: Tricks of the Trade. VI, 432 pages. 1998.

Vol. 1525: D. Aucsmith (Ed.), Information Hiding. Proceedings, 1998. IX, 369 pages. 1998.

Vol. 1526: M. Broy, B. Rumpe (Eds.), Requirements Targeting Software and Systems Engineering. Proceedings, 1997. VIII, 357 pages. 1998.